P9-CDC-443

VIOLENT
EARTH

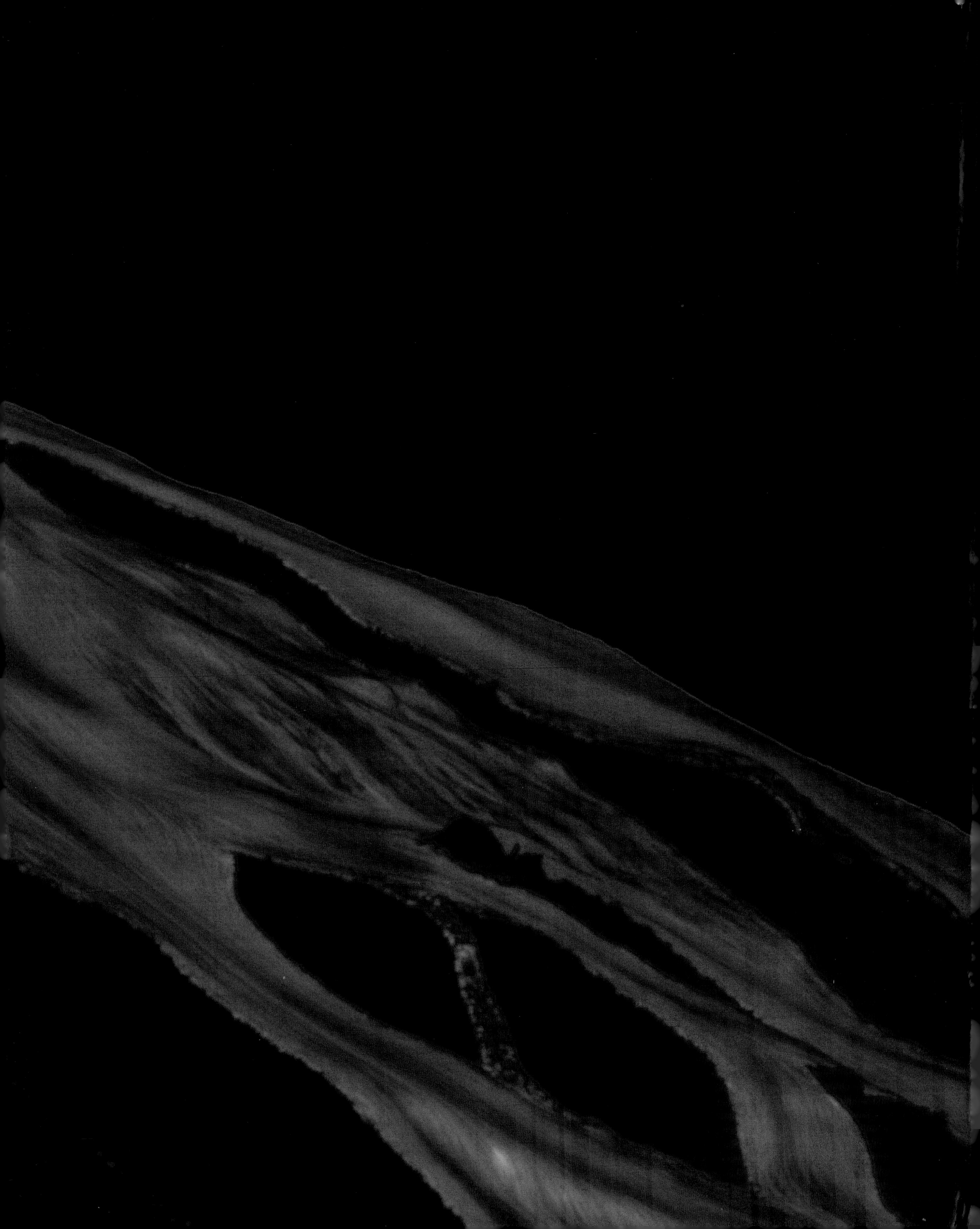

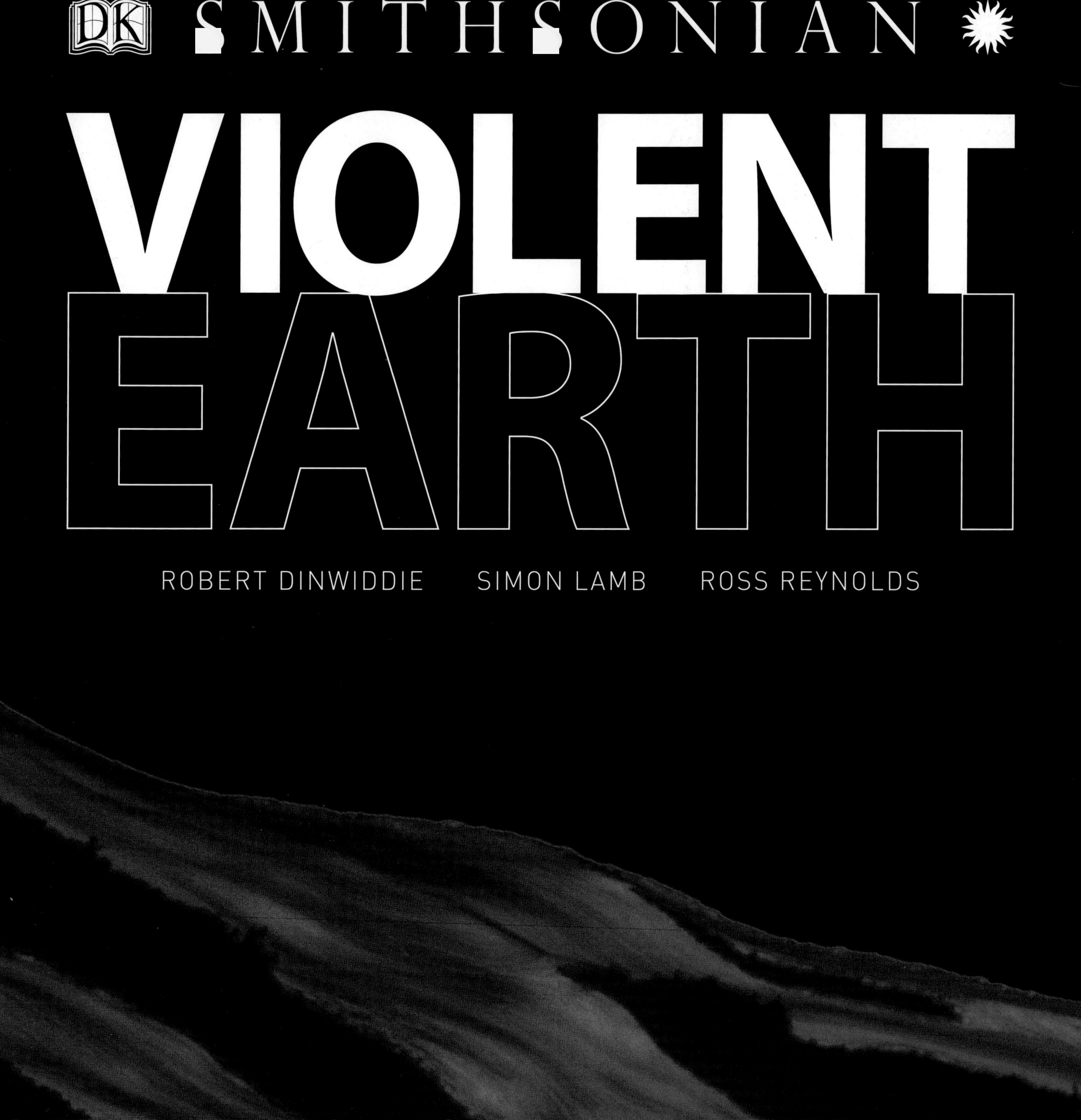
DK SMITHSONIAN
VIOLENT EARTH
ROBERT DINWIDDIE SIMON LAMB ROSS REYNOLDS

LONDON, NEW YORK, MELBOURNE, MUNICH AND DELHI

Smithsonian consultants Paul Kimberly, Dr. Don E. Wilson, Dr. M. G. (Jerry) Harasewych

Editorial consultant David Rothery

Project editors Nicola Hodgson, Scarlett O'Hara
Project art editors Mandy Earey, Richard Horsford, Clare Marshall
Editors Shaila Brown, Jenny Finch, Wendy Horobin, Ashwin Khurana
Designers Daniela Boraschi, Vicky Short
US editors Shannon Beatty, Rebecca Warren

Picture researchers Myriam Megharbi, Roland and Sarah Smithies
Illustrators Peter Bull Art Studio, Barry Croucher/The Art Agency, Mike Garland, Mick Posen/The Art Agency
Cartography Encompass Graphics Ltd, Simon Mumford

Production editor John Goldsmid
Production controller Erika Pepe

Managing editor Julie Ferris
Managing art editor Owen Peyton Jones
Associate publishing director Liz Wheeler
Art director Phil Ormerod
Publishing director Jonathan Metcalf

DK INDIA
Editorial manager Rohan Sinha
Senior editors Kingshuk Ghoshal, Garima Sharma
Editors Megha Gupta, Shatarupa Chaudhuri, Samira Sood
Design consultant Shefali Upadhyay
Design manager Arunesh Talapatra
Senior designer Sudakshina Basu
Designers Arijit Ganguly, Amit Malhotra, Nidhi Mehra, Kavita Dutta, Zaurin Thoidingjam
Production manager Pankaj Sharma
DTP manager Balwant Singh
DTP designers Nand Kishor Acharya, Bimlesh Tiwary

First American Edition, 2011

Published in the United States by DK Publishing, 375 Hudson Street, New York, New York 10014

11 12 13 14 15 10 9 8 7 6 5 4 3 2

002—178093—Oct/2011

Published in Great Britain by Dorling Kindersley Limited

A catalog record for this book is available from the Library of Congress

ISBN 978-0-7566-8685-7

DK books are available at special discounts when purchased in bulk for sales promotions, premiums, fund-raising, or educational use. For details, contact: DK Publishing Special Markets, 375 Hudson Street, New York, New York 10014 or SpecialSales@dk.com

Printed and bound in China by LEO

Discover more at
www.dk.com

THE SMITHSONIAN INSTITUTION
Established in 1846, the Smithsonian Institution—the world's largest museum and research complex—includes 19 museums and galleries and the National Zoological Park. The total number of objects, works of art, and specimens in the Smithsonian's collections is estimated at 137 million, the bulk of which is contained in the National Museum of Natural History, which holds more than 126 million specimens and objects. The Smithsonian is a renowned research center, dedicated to public education, national service, and scholarship in the arts, sciences, and history.

CONTENTS

4 EARTHQUAKES

5 RESTLESS OCEANS

6 EXTREME WEATHER

7 REFERENCE

1

DYNAMIC PLANET

<< Earth from space
The Sun casts an orange glow on waves of cloud
along the east flanks of the Andes mountain rang

EARTH'S ORIGIN

About 4.6 billion years ago, within the spinning disk of dust and gas that surrounded the newly forming Sun, small chunks of material collided and amalgamated, culminating in the collisions of larger bodies called protoplanets. These collisions, in turn, led to the release of tremendous amounts of heat energy, and as a result Earth was born in a hot, molten state.

BIRTH OF OUR PLANET

The origins of our planet are closely tied to the formation of the whole Solar System—the Sun, the eight planets orbiting it, and many other bodies, such as comets and asteroids. Astronomers now agree that the Solar System started forming about 4.6 billion years ago out of an immense, slowly spinning cloud of gas and dust within the Milky Way galaxy. Gradually, gravity caused the cloud to contract and spin faster, and as its central region became denser, it also became hotter. This region eventually became the Sun. Surrounding the central region was a spinning disk of gas, dust, and ice. Within the disk, grains of ice and dust stuck together to form solid particles of ever-increasing size—pebbles, rocks, boulders, and eventually bodies called planetesimals, which can be anywhere in size from several feet to hundreds of miles wide. These came together through collisions to form protoplanets, which were roughly the size of our present-day Moon. Protoplanets underwent a series of violent collisions to form the four inner planets (Mercury, Venus, Earth, and Mars) and the cores of the outer giant planets, such as Jupiter.

It is not known how many protoplanets came together to form Earth, but it may have been a dozen or so. With each collision, a tremendous amount of heat was generated as the kinetic energy of the colliding bodies converted to heat energy. In addition, as the number of protoplanets diminished, and their size grew, each one contracted under the influence of its own gravity, a process that also generated heat. Eventually, a final collision is thought to have occurred between an object almost the size of Earth—a precursor of our planet that is sometimes called proto-Earth or young Earth —and a protoplanet about the size of Mars, known as Theia. The result of this final collision was the Earth-Moon system, comprising Earth itself and its orbiting moon.

THE FORMATION OF THE SOLAR SYSTEM

The most widely accepted theory for how the Solar System originated is called the nebular hypothesis, shown here. It explains all the most obvious features of the system, such as why it is flat, and why all the planets orbit the Sun in the same direction. According to this model, objects such as Earth formed from the gradual accretion (sticking together) of small rock and ice particles to form larger ones called planetesimals, then yet larger ones called protoplanets, and finally planets.

1 **Solar nebula forms** The Solar System originated from a spinning cloud of gas and dust

2 **The Sun takes shape** As the cloud contracted, it spun faster. Its central region heated up to form the Sun

3 **Planetesimals form** Dust and ice accreted within a surrounding disk to make planetesimals

4 **Rocky planets** Colliding planetesimals and protoplanets formed four rocky planets

5 **Giant planets** In the outer region of the disk, gas accumulated around rock and ice cores to form giant planets

6 **Unused debris** Some leftover planetesimals formed a cloud of comets

PROTOPLANET COLLISION
Not long after proto-Earth had formed, a protoplanet about the size of Mars collided with it to form the Earth and Moon.

CORE FORMATION

It is likely that a process called differentiation occurred in each of the larger protoplanets that came together to form Earth. Differentiation, which is only possible in molten bodies, involves heavier materials, such as iron and nickel, sinking to the middle to form a core. With each major protoplanet collision, enough heat would have been generated to keep the combined body molten, and the two cores would have quickly merged. This means that Earth is likely to have had a core soon after it came into existence.

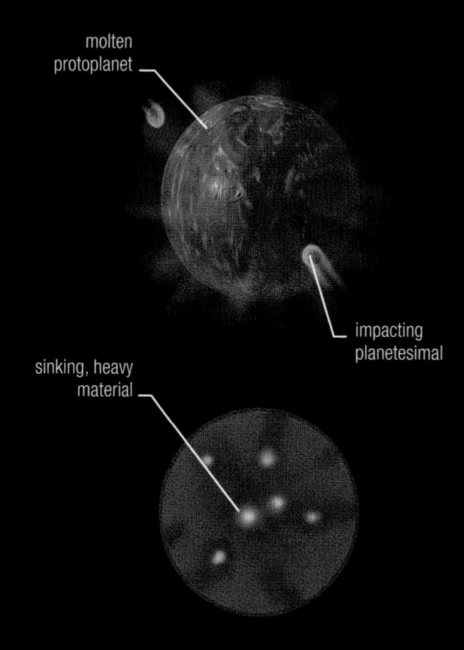

ACCUMULATION OF MATERIAL
Planetesimals collided to form protoplanets. Further impacts from them generated heat that kept the protoplanets molten. As they grew, gravity caused the protoplanets to contract, producing more heat.

HEAVY MATERIAL SINKS
Heavy materials, such as iron and nickel, present in molten protoplanets, or introduced by impacting bodies, sank toward the center. This process, sometimes called an iron catastrophe, generated more heat.

CORE FULLY FORMED
Each large protoplanet formed an iron and nickel core, with lighter materials forming an outer layer called the mantle. As protoplanets collided to form the proto-Earth, and finally Earth, their cores merged.

FORMATION OF THE MOON

About 30 to 80 million years after it formed, the proto-Earth is thought to have suffered a final impact by a Mars-sized protoplanet—Theia. The ejected material formed a ring around what was now Earth, and came together to form a single body, the Moon. Earth remained molten for some time, having acquired from the collision a larger iron core than before. The Moon was much closer to Earth than it is today—perhaps 10 times closer. This produced extremely strong tides at Earth's surface, causing additional heating, and perhaps the development of tectonic plates.

FIRST LAND AND OCEANS

During the first 800 million years of Earth's history, much of Earth's surface was a molten mass of liquid rock. Gradually, it began to cool, forming islands of land surrounded by hot oceans and a poisonous atmosphere.

THE FIRST LANDMASSES

No rocks seem to have survived from the first 300 million years of Earth's existence, since any solid crust that formed was soon broken up by impacts from comets and asteroids. Earth's interior was much hotter than it is today, causing high volcanic activity that continually reworked the surface. But as the bombardment diminished and the surface cooled, the crust began to stabilize. Pieces of continental crust, the basis of landmasses, had formed by 3.8 billion years ago, but some oceanic crust is likely to have formed earlier (see Earth's Oldest Rocks below). The development of oceanic crust coincided with the start-up of slow heat convection processes in Earth's mantle that marked the beginnings of plate tectonics (see pp.20–21). As new oceanic crust formed where the earliest plates moved apart, as much again was drawn back into the mantle by convectional movements. The descent of water-laden oceanic crust into the mantle caused melting there, and the rise of molten mantle rock to the surface led to the formation of continental crust. Initially, this probably existed in the form of arcs of volcanic islands. The continuing movements of the early plates caused these landmasses to converge, gradually forming the ancient centers, called cratons, of today's continents.

EARTH'S OLDEST ROCKS

In 2008, some rocks found near Inukjuak, Hudson Bay, Canada, were dated at 4.28 billion years old, making them the oldest known in the world. A slice through one of the rock formations is shown above. Their age gives an idea of when Earth's surface started to become more stable. Geologists believe the rocks were originally formed as volcanic lavas at the bottom of an ocean. Since then they have been heavily metamorphosed (altered by heat and pressure).

DEVELOPMENT OF THE ATMOSPHERE

Earth's first atmosphere consisted mainly of hydrogen and helium. However this was blown away by the solar wind—a stream of particles from the Sun—and replaced by a second atmosphere, made of gases from volcanoes. These included nitrogen, carbon dioxide, and water vapor, as well as some hydrogen and helium—much of which escaped into space—and smaller amounts of other gases. Much of the water condensed and precipitated, forming the first oceans. Initially, little or no free oxygen was present—the amount increased over billions of years once the first microorganisms evolved and started converting carbon dioxide into oxygen by photosynthesis.

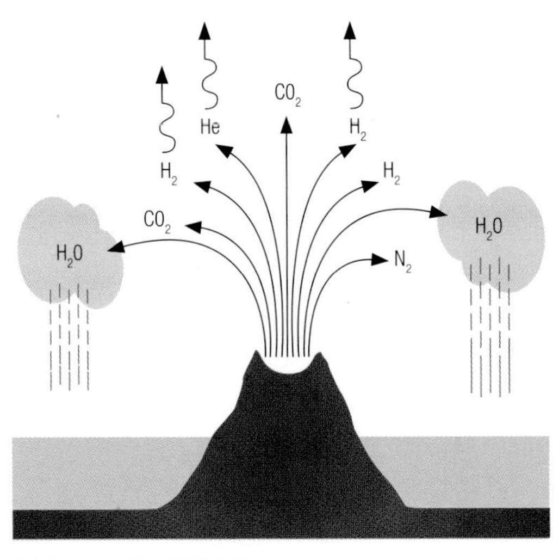

SECOND ATMOSPHERE
Gases that erupted from volcanoes and contributed to Earth's second atmosphere include nitrogen (N_2), water vapor (H_2O), carbon dioxide (CO_2), and small amounts of various other gases.

FORMATION OF THE OCEANS

Between 4.3 and 4 billion years ago, Earth had cooled enough for water molecules in the atmosphere to condense, fall onto Earth's surface, and persist as free-standing bodies of water. Most of this was probably released by volcanoes. Grains of the mineral zircon laid down in a watery environment dating back more than 4 billion years, show that some surface water was present then. Pillow lavas from Greenland, many up to 3.8 billion years old, could only have been formed by volcanic eruptions under water.

WATERY PLANET
Earth is thought to have had substantial oceans as long as 4 billion years ago.

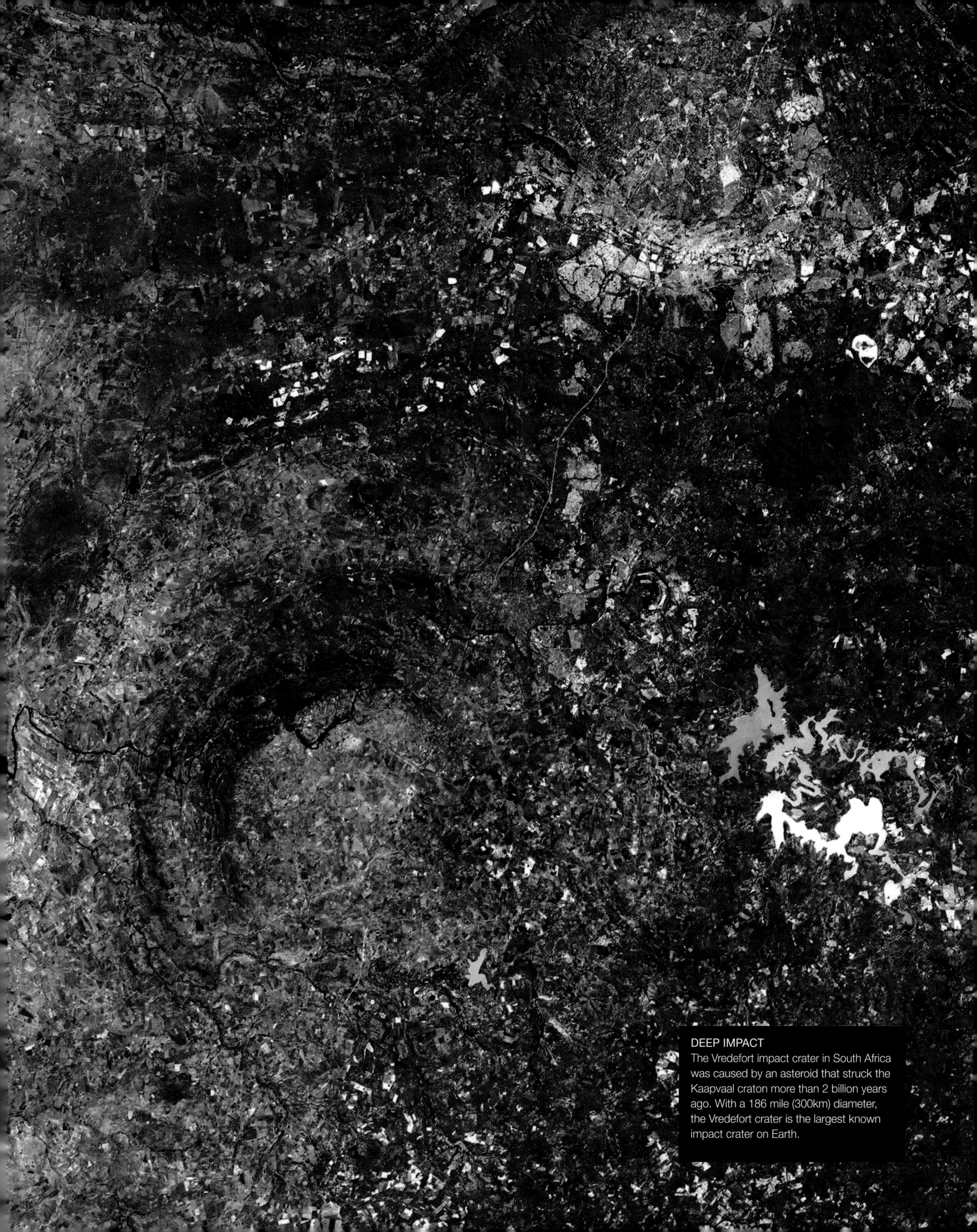

DEEP IMPACT

The Vredefort impact crater in South Africa was caused by an asteroid that struck the Kaapvaal craton more than 2 billion years ago. With a 186 mile (300km) diameter, the Vredefort crater is the largest known impact crater on Earth.

EARTH'S STRUCTURE

Earth was born in a molten state, setting the stage for a layered internal structure. Based on chemical composition and density, it has three main layers, called the crust, mantle, and core. Of these, the core and mantle have always been present, while the crust has gradually developed throughout the planet's history.

THREE-LAYERED STRUCTURE

In terms of proportions, the contributions that the crust, mantle, and core make to Earth's overall bulk are similar to those made by the shell, white, and yolk of an egg. Like an eggshell, the crust is a thin layer at Earth's surface, contributing 0.2–1.1 percent of the total depth. It is made of many different types of rock, most of which are relatively light and rich in silicon, with an average density of 1.5–1.7oz per cu in (2.7–3g per cu cm). There are two main types of crust—continental and oceanic. Of these, the thick continental crust is composed predominantly of low-density rocks, such as granite. The thinner oceanic crust consists mainly of relatively high-density rocks, such as basalt.

The mantle lies below Earth's crust. It extends from the base of the crust to the core-mantle boundary, at a depth of about 1,860 miles (2,990km). The mantle is considered to have two main layers: upper and lower. Rock samples from the upper mantle, occasionally brought to the surface by volcanic activity, show that it is made of silicate-based rocks rich in magnesium. Finally, the core extends from the core-mantle boundary down to Earth's center, approximately 3,950 miles (6,360km) beneath the surface. The density of the core varies from 5.8–7.5oz per cu in (10–13g per cu cm), and it is known to consist of an inner solid part and an outer liquid part. No samples from the core are available, but it is thought to be composed of nickel-iron alloys.

Lower mantle
Composed of semisolid silicate rocks, with a temperature between 3,600 and 6,300°F (2,000 and 3,500°C)

Outer core
Made up of liquid iron and nickel, its temperature ranges from 6,300–7,200°F (3,500–4,000°C)

Inner core
Consists of solid iron with a little nickel, ranging in temperature from 7,200–8,500°F (4,000–4,700°C)

Upper mantle
Composed of solid to semisolid silicate rocks, with a temperature range of 750 to 3,500°F (400 to 2000°C)

INSIDE THE PLANET
Earth's principal layers are the core and the mantle—both consisting of an inner and outer section—and the crust, which falls into continental and oceanic varieties. Each part differs in composition and temperature.

CHEMICAL COMPOSITION OF EARTH'S LAYERS

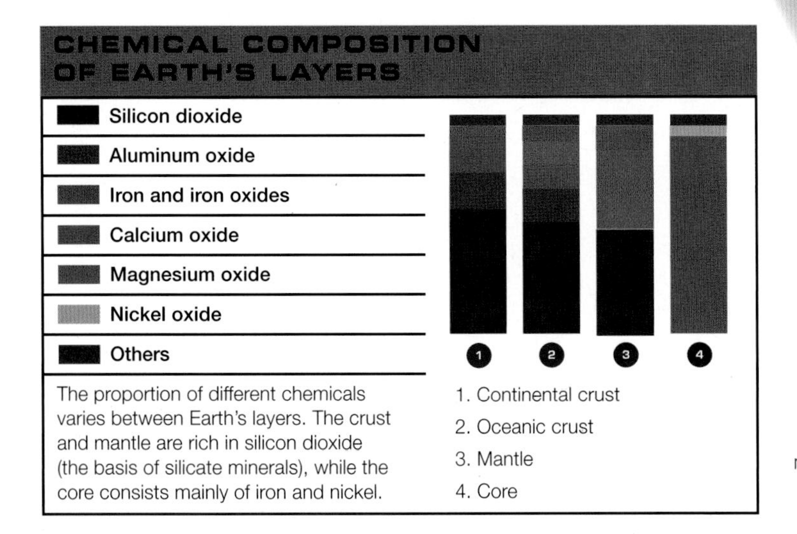

The proportion of different chemicals varies between Earth's layers. The crust and mantle are rich in silicon dioxide (the basis of silicate minerals), while the core consists mainly of iron and nickel.

Oceanic crust
Made up of solid rock, such as basalt, its temperature ranges from 32 to 800ºF (0 to 400ºC)

SHAPE AND FORM

Earth's overall shape is determined by the effects of gravity and rotation. The force of gravity pulls Earth into an almost perfect sphere, while the planet's rotation, once every 24 hours, reduces the effect of gravity around the equator. This makes the equatorial region bulge outward by several miles. As a result, Earth's diameter at the equator is 7,926 miles (12,756km), compared to 7,900 miles (12,713km) pole-to-pole.

Other processes at Earth's surface also produce height and depth variations that currently amount to about a 12 mile (20 km) difference between the highest peaks and the deepest oceanic trenches. Larger height and depth variations than this can never last long, because gravity, coupled with the effects of erosion, continually acts to even Earth's surface out. Earth's equatorial bulge complicates matters geographically as well. For example, Mount Everest is the world's highest mountain at 29,035ft (8,850m) when measured above sea level. But when measured from the center of Earth, Mount Chimborazo in Ecuador is the highest, since it sits in a region where Earth bulges the most. Chimborazo's summit is 3,967 miles (6,384km) from Earth's center, while Everest's is 3,966 miles (6,382km).

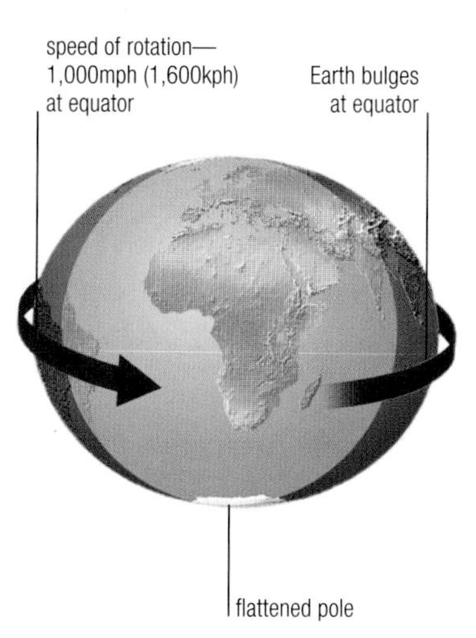

CHANGING SHAPE
Earth's rotation produces inertial (centrifugal) forces at the surface that modify its shape, so that it bulges slightly at the equator and is flattened at the poles. The planet's shape differs from a perfectly spherical shape by about 0.3 percent.

ALTERNATIVE IDEAS ABOUT EARTH'S INTERIOR

Various strange suggestions have been made in the past about Earth's internal structure. In the 17th century, eminent astronomer Edmond Halley speculated that Earth consists of a series of thin, nested, spherical shells, with gas occupying the intervening spaces. Theologian Thomas Burnet proposed that Earth contains huge chasms full of water. But in 1798, Henry Cavendish, a physicist, showed that Earth's average density is more than five times that of water—effectively quashing any ideas that it consisted mainly of gas or watery liquids.

Internal fire
The German Jesuit scholar Athanasius Kircher produced drawings suggesting that Earth's interior contained several interlinked fiery chambers.

ATHANASIUS KIRCHER

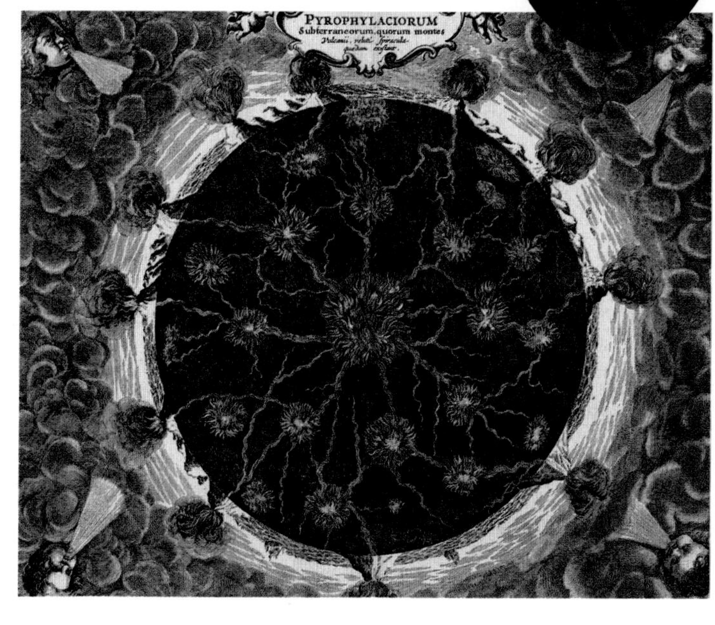

Continental crust
Formed from solid rock, such as granite, its temperature ranges between -130 and 1700ºF (-90–900ºC)

THE CORE AND MANTLE

In 1866, French geologist Gabriel Daubrée proposed that beneath Earth's surface are layers of different rocks that become progressively denser with depth, culminating in a central region made of iron and nickel. His ideas have proved correct and much more has been discovered about the mantle and core since then.

INVESTIGATING THE CORE

Since Earth's core cannot be observed directly, most of what is known about it has come from studying seismic waves produced by earthquakes. As well as seismic waves that travel along the surface of the Earth, called surface waves, there are two types that travel through Earth's interior, called primary (P) and secondary (S) waves. Early investigations showed that when P waves travel straight through Earth, they slow down as they pass through the center: this was the first proof of the existence of a dense core. Also noted were "shadow zones" on Earth's surface, where P and S waves from a particular earthquake could not be detected. Analysis of these zones helped estimate the core's diameter—now known to be about 4,300 miles (6,940km)—and indicated that the core's outer region, at least, must be liquid. Later studies showed that the core has a solid inner part made mainly of iron, with a diameter of about 1,500 miles (2,440km). This inner core is growing as the liquid outer core solidifies around it. Its estimated temperature is 7,200–8,500°F (4,000–4,700°C), while the outer core's is 6,300-7,200°F (3,500–4,000°C).

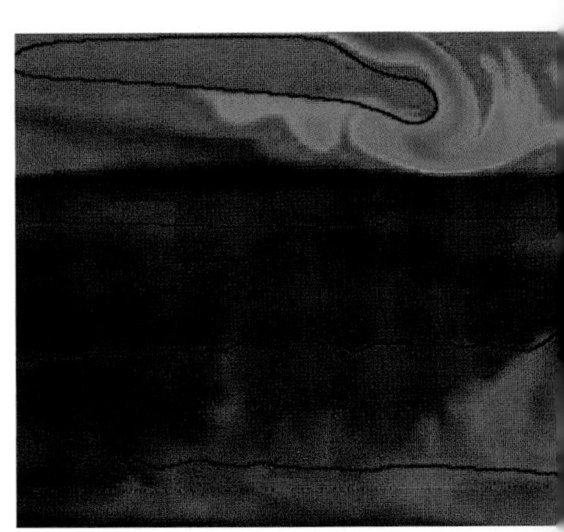

SEISMIC TOMOGRAM
Seismic tomography is a technique that provides images of slices through Earth's interior by analyzing the motion of seismic waves. It can be used to trace processes such as heat convection in the mantle.

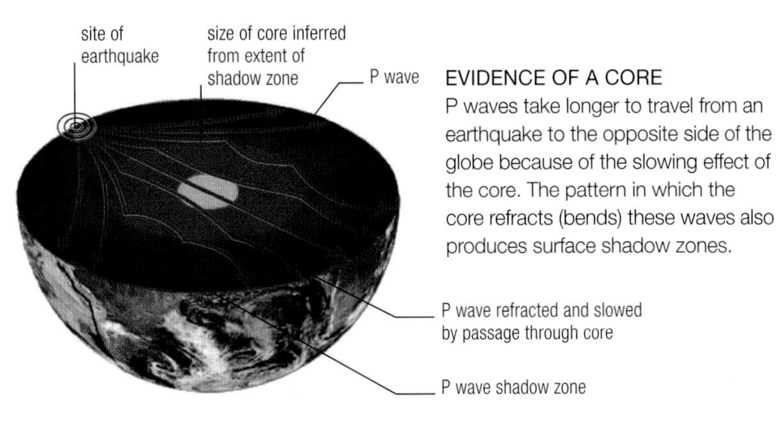

EVIDENCE OF A CORE
P waves take longer to travel from an earthquake to the opposite side of the globe because of the slowing effect of the core. The pattern in which the core refracts (bends) these waves also produces surface shadow zones.

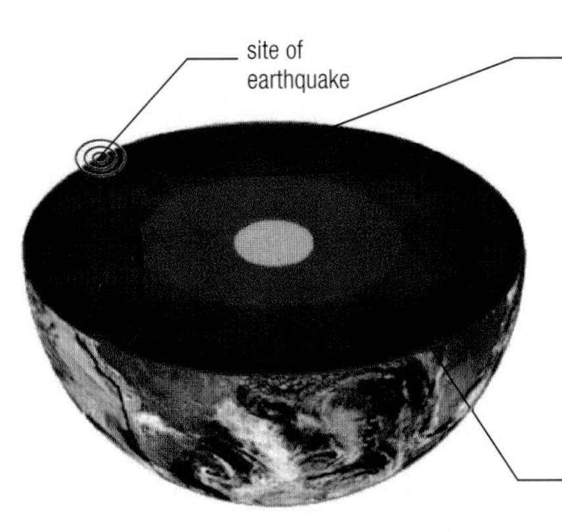

EVIDENCE OF LIQUID IN CORE
S-type seismic waves cannot pass through liquids and are not detected on the opposite side of the globe. This indicates that at least the outer region of the core is liquid.

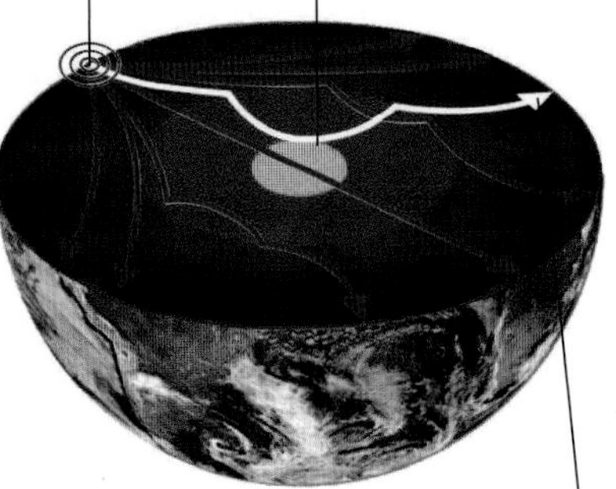

EVIDENCE OF INNER CORE
A faint P wave can be detected in the shadow zone following an earthquake. When this was first discovered, it was deduced that the detected P wave must have reflected off the surface of an inner part of the core.

EARTH'S MAGNETIC FIELD

Earth's magnetic field is thought to be caused by powerful electric currents that run through the liquid metal in its outer core. These currents are the result of swirling motions within the liquid as it convects heat outward from the inner core. Because these swirling motions occur somewhat independently of the planet's rotational movement, the positions of Earth's north and south magnetic poles gradually shift over time. Occasionally, the magnetic field flips—possibly as a result of turbulence in the liquid metal—so that the magnetic north ends up at the geographical south pole.

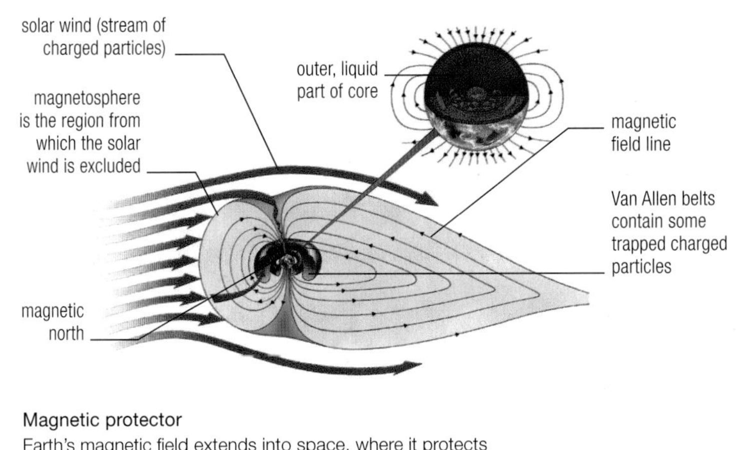

Magnetic protector
Earth's magnetic field extends into space, where it protects the planet from the solar wind—a stream of potentially harmful, energetic, charged particles emanating from the Sun.

INSIDE THE INNER CORE

In 2008, scientists at the University of Illinois announced that a study of seismic waves passing through Earth's inner core suggests that it is not a uniform sphere of solid iron and nickel, but has two distinct parts, inner and outer. These are thought to have slightly different crystalline structures (arrangements of metal atoms). The inner region of the inner core has a diameter of about 733 miles (1,180km) and constitutes 0.08 percent of the volume of Earth.

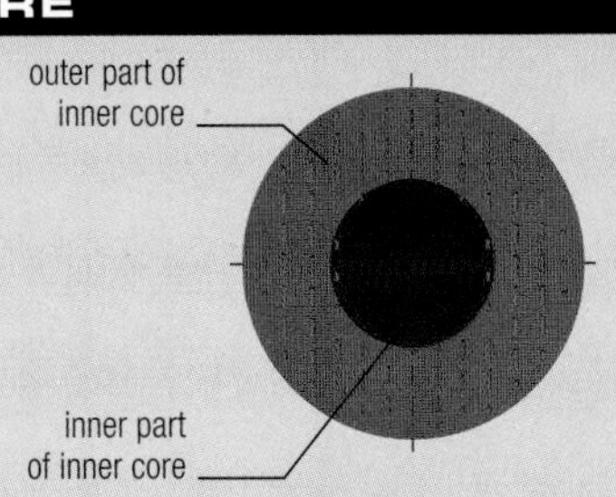

EARTH'S MANTLE

The mantle has two main parts—upper and lower—with the boundary between them at a depth of about 400 miles (660km). The upper mantle consists mainly of a rock called peridotite, while the lower mantle is thought to be dominated by rocks with a more compact structure. The uppermost layer of the upper mantle is firmly attached to the crust above it to form a rigid, brittle, combined unit called the lithosphere (see pp.16–17). Below it is a less rigid, warmer region of the upper mantle called the asthenosphere. Temperatures within the mantle vary from less than 1,800°F (1,000°C) at the boundary with the crust, to about 6,300°F (3,500°C) at the core–mantle boundary.

MANTLE ROCK
Stray mantle rocks, such as this piece of green-tinted peridotite, which are brought to the surface by upheavals connected to volcanism, are called mantle xenoliths.

CORE AND MANTLE
Heat energy flowing out of Earth's core is thought to drive slow, circular, convectional movements of material within the mantle (red arrows). These, in turn, are believed to drive the movement of tectonic plates at Earth's surface.

EARTH'S OUTER SHELL

Earth has a rigid outer shell, which includes not just its surface layer, the crust, but also a thick layer of solid rock from the top of the mantle that is fused to the crust. Together, the crust and upper mantle form a strong structural entity called the lithosphere.

TYPES OF CRUST

Earth has two types of crust—continental crust, which forms the dry land and continental shelves, and thinner oceanic crust, which forms the ocean floor. Continental crust varies in thickness between 16 and 43 miles (25 and 70km) and is composed of igneous, metamorphic, and sedimentary rock types, although igneous rocks such as granite and diorite predominate. Oceanic crust is denser, varies in thickness from 4 to 7 miles (6 to 11km), and consists of just a few types of igneous rock. Earth recycles its oceanic crust much more frequently than its continental crust. This accounts for the relative paucity of rock types in oceanic crust and also for the fact that no oceanic crust is more than 200 million years old, whereas some continental rocks are more than 4 billion years old.

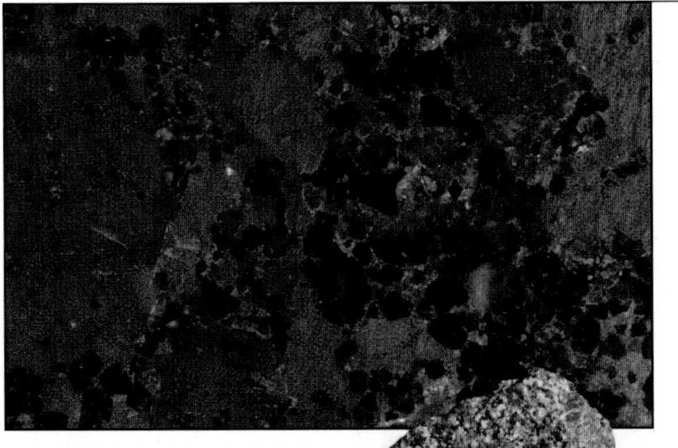

GRANITE
A common plutonic rock, granite is found within the continental crust.

DIORITE
This dark gray rock commonly forms where two tectonic plates collide.

CONTINENTAL LITHOSPHERE
This type of lithosphere has a top layer of continental crust, which ranges from 16 to 43 miles (25 to 70km) in thickness and is underlain by about 50 miles (80km) of the uppermost mantle layer.

EARTH

Continental crust Consists of a wide variety of rock types

Mohorovičić discontinuity

Uppermost mantle layer Consists mainly of the coarse-grained igneous rock peridotite

Asthenosphere Warm, deformable layer of the upper mantle just below the lithosphere

CONTINENTAL LITHOSPHERE

THE MOHOROVIČIĆ DISCONTINUITY

The boundary between crust and mantle is called the Mohorovičić Discontinuity, or Moho for short. Croatian geophysicist Andrija Mohorovič discovered it in 1909. By studying records of seismic waves generated by an earthquake, Mohorovičić noticed that some waves arrived at detecting stations earlier than others. He reasoned that these waves must have traveled down into a denser region of Earth's interior (the mantle), where they could travel faster, before coming back to the surface. Seismic waves of the P-wave variety, can travel through the mantle at an average of about 5 miles per second (8km per second) compared with less than about 3.7 miles per second (6km per second) through the crust.

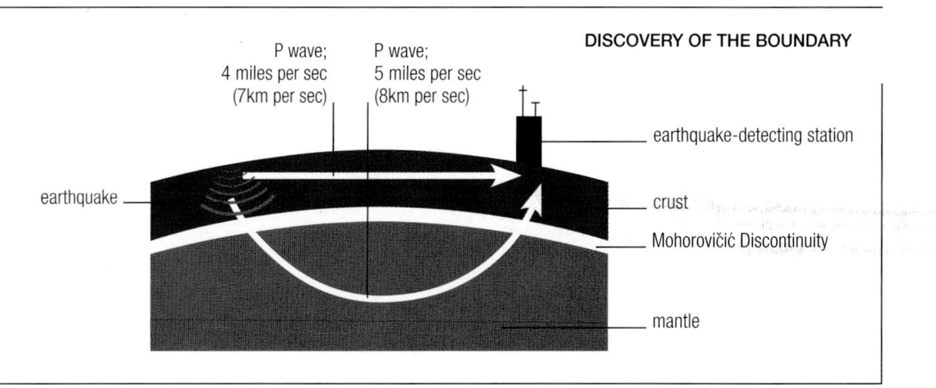

OCEANIC LITHOSPHERE
This type of lithosphere has a top layer of oceanic crust, which ranges from 4 to 7 miles (6 to 11km) in thickness and is underlain by about 25 to 60 miles (40 to 100km) of the uppermost mantle layer.

Sediments

Oceanic crust
Consists mainly of gabbro and basalt

Mohorovičić discontinuity

OCEANIC LITHOSPHERE

Uppermost mantle layer

Asthenosphere

THE LITHOSPHERE

In chemical composition, Earth's crust and mantle are two distinct and important layers, but in terms of mechanics and the processes occurring at Earth's surface, the lithosphere is a more fundamental unit than crust or mantle. This is because tectonic plates (see pp.20–21) are made of lithosphere. The two types of lithosphere—continental and oceanic—correspond to the two types of crust that form their top layers. The lithosphere floats on top of a less rigid layer of the mantle called the asthenosphere.

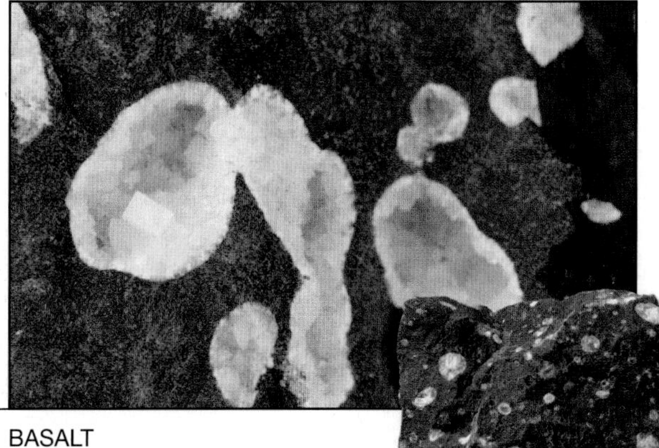

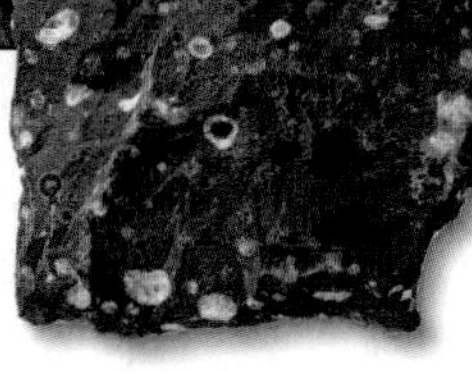

BASALT
A dark, fine-grained rock, basalt (seen here with white mineral inclusions) makes up much of the upper part of oceanic crust.

GABBRO
Chemically similar to basalt, gabbro constitutes much of the lower two-thirds of oceanic crust.

CAVE OF CRYSTALS
Earth's continental crust is riddled with natural wonders. Here, massive beams of selenite (gypsum) dwarf explorers in a cavern, some 1,000ft (300m) underground, known as the Cave of Crystals. Discovered in 2000, adjacent to the Naica silver mine in northern Mexico, the chamber contains the largest natural crystals ever found.

TECTONIC PLATES

Earth's outer shell is not a single entity. It is broken up into irregularly shaped pieces, called tectonic plates that fit together like a jigsaw.

SIGNIFICANCE OF PLATES

Earth's surface has looked like a cracked eggshell since it began to cool 4 billion years ago. It is currently divided into eight or nine major plates and several dozen smaller ones (see pp.26–27). The plates are in constant motion, drifting by a few inches every year. These movements are driven by heat flows within Earth's mantle that pull some plates apart and push others together. When this occurs there is a huge release of energy, resulting in earthquakes and volcanic eruptions as the plates grind past, collide with, or dip beneath each other. Many plates carry continental crust, so the face of Earth as we see it today is very different from how it looked millions of years ago due to the constant destruction and reformation of the plates. The theory of plate tectonics, which explains earthquakes, volcanism, mountain building, deep-sea trenches, and many other geologic phenomena, is a development of the earlier theory of continental drift (see below).

BOUNDARY BETWEEN PLATES
This rift through Iceland is part of the dividing line between the North American and Eurasian plates. Most of the boundary runs along the Atlantic Ocean floor.

DEVELOPING THE CONTINENTAL DRIFT THEORY

270 MILLION YEARS AGO — 200 MILLION YEARS AGO — TODAY

Between the 16th and 19th centuries, geographers noted that Africa's coastline seems to "fit" with that of South America, as though the two continents had once been close together. In 1912, German scientist Alfred Wegener published his theory of continental drift, with evidence that South America had once been joined to Africa, and Europe to North America. However, he could not explain how the continents later moved apart. His theory was not taken seriously until, in the late 1920s, it was realized that convection in Earth's mantle might provide a mechanism for the movement. In the 1960s, it was shown that new oceanic lithosphere is continuously made at seafloor ridges and then pushed away from the ridges. This also pushes the continents attached to the oceanic lithosphere.

PATTERNS OF PLATE MOVEMENT

The pattern of plate movements can be broadly divided into three main categories, divergent, convergent, and transform. At some boundaries between plates, called divergent boundaries, two plates gradually move apart (see pp.28–29). These boundaries are mainly found on the ocean floor and are known as mid-ocean spreading ridges. New oceanic lithosphere is created and added to the edges of plates as they move away from each other. Any continents attached to the plates move with them. This has caused all past and present movements of landmasses.

Elsewhere, at convergent boundaries (see pp.30–31), the edges of two plates move toward each other. They either crumple up to form mountains when two continents collide, or one plate is subducted beneath the edge of a neighboring plate when the more dense ocean crust collides with the less dense continental crust. The plate boundaries where subduction occurs are always on the sea floor and marked by deep trenches. The subduction process causes major earthquakes and volcanic activity. At a third type of plate boundary, called a transform boundary, the edges of two plates grind past each other in opposite directions. At these boundaries too, there is a high incidence of earthquakes.

VOLCANOES

Much volcanic activity on land and some underwater is caused by the edge of one plate pushing beneath another. This triggers a series of subterranean processes that encourage formation of volcano-forming magma (molten rock).

RIFT VALLEYS

When a plate starts splitting apart in the middle of a continent, the crust becomes stretched, fissures appear, and crustal blocks collapse downward. The result is a rift valley, such as the East African Rift Valley.

MOUNTAINS

Many large mountain ranges, such as the Himalayas and the Alps, are the result of collisions between continents carried on different plates as those plates converged on each other.

UNDERWATER VENTS

Sometimes plate movements cause hotspots of magma (melted rock) under the seafloor. The result can be plumes of hot, mineral-laden water, or masses of gas-filled bubbles emanating from rocks underwater.

DEVELOPMENT OF MODERN LANDMASSES

Around 3.8 billion years ago, the process that led to the development of modern landmasses had begun. Earth had a few small areas of land, called cratons, at its surface and a vast expanse of ocean. Tectonic plates forming Earth's outer shell were probably thinner than they are today. The cratons survive as some of the most ancient regions of modern continents.

A PROCESSION OF SUPERCONTINENTS

The process started when mantle convection—the slow movement of material within the Earth—began to push the early plates around at the surface. As this happened, the cratons were shifted around. Sometimes they joined up, sometimes they split, to produce ever-changing arrangements of land and oceans. Every so often the movements aggregated them into "supercontinents," subsequently breaking these up again. Past supercontinents include Vaalbara (which existed about 3 billion years ago), Kenorland (2.6 billion years ago), and Columbia (1.7 billion years ago). Not much is known about their exact shapes, location, or configuration. The first supercontinent about which a little more information is available is Rodinia, which formed about 1.1 billion years ago.

Laurentia The core craton for what is now North America

Baltica Contained crust that now makes up part of northern and eastern Europe and Russia

Gondwana Contained the cores of what became modern southern continents

2 CAMBRIAN LANDMASSES (515 MYA)
The next supercontinent was Pannotia, most of which lay in the southern hemisphere. Around 540 million years ago, it split up, with three pieces—Laurentia, Baltica, and Siberia—detaching from it. By around 515 million years ago, these became islands surrounded by the Panthalassic and Iapetus oceans. A large southern landmass called Gondwana was left behind.

Northern Rodinia Moved to north and west

New spreading ridge Helped push two parts of Rodinia apart

Southern Rodinia Swung toward the south pole

1 RODINIA BREAKS UP (700 MYA)
Rodinia lasted for about 350 million years before breaking up. It lay predominantly in the southern hemisphere, but the detailed arrangement of its components is unclear. Their estimated locations about 50 million years after the break-up are shown here. The outlines of some modern landmasses are also shown, to indicate their relationship more ancient formations.

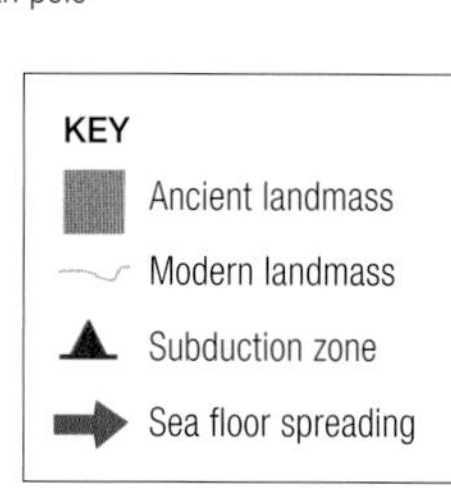

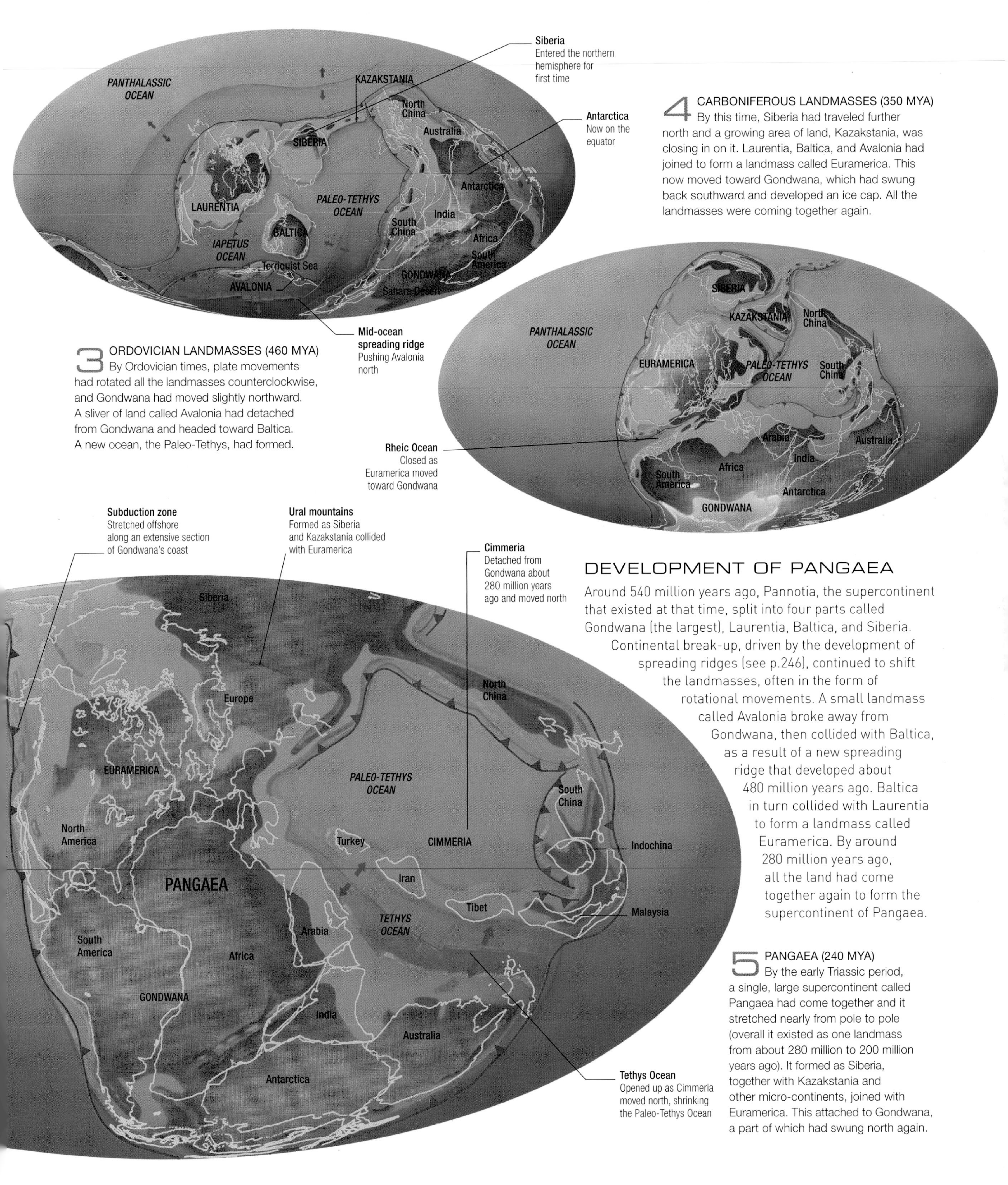

3 ORDOVICIAN LANDMASSES (460 MYA)
By Ordovician times, plate movements had rotated all the landmasses counterclockwise, and Gondwana had moved slightly northward. A sliver of land called Avalonia had detached from Gondwana and headed toward Baltica. A new ocean, the Paleo-Tethys, had formed.

4 CARBONIFEROUS LANDMASSES (350 MYA)
By this time, Siberia had traveled further north and a growing area of land, Kazakstania, was closing in on it. Laurentia, Baltica, and Avalonia had joined to form a landmass called Euramerica. This now moved toward Gondwana, which had swung back southward and developed an ice cap. All the landmasses were coming together again.

DEVELOPMENT OF PANGAEA

Around 540 million years ago, Pannotia, the supercontinent that existed at that time, split into four parts called Gondwana (the largest), Laurentia, Baltica, and Siberia. Continental break-up, driven by the development of spreading ridges (see p.246), continued to shift the landmasses, often in the form of rotational movements. A small landmass called Avalonia broke away from Gondwana, then collided with Baltica, as a result of a new spreading ridge that developed about 480 million years ago. Baltica in turn collided with Laurentia to form a landmass called Euramerica. By around 280 million years ago, all the land had come together again to form the supercontinent of Pangaea.

5 PANGAEA (240 MYA)
By the early Triassic period, a single, large supercontinent called Pangaea had come together and it stretched nearly from pole to pole (overall it existed as one landmass from about 280 million to 200 million years ago). It formed as Siberia, together with Kazakstania and other micro-continents, joined with Euramerica. This attached to Gondwana, a part of which had swung north again.

PANGAEA RIFTS APART

Around 200 million years ago, signs of stress began to appear in parts of the supercontinent Pangaea, signaling its imminent break-up. Rifts and weaknesses started to develop in a region corresponding to what is now the northeastern seaboard of the US. Magma flowing from Earth's interior began pouring into weaknesses between the rock layers. These eventually solidified to form thick igneous rock deposits. But it was not until about 180 million years ago that rifts opened a significant body of water between what is now the east coast of North America on one side, and northwest Africa on the other. This widening seaway eventually became the central part of the Atlantic Ocean.

Around 130 million years later, the Southern Atlantic also started to open up—this time as part of the break-up of Gondwana. Further north, parts of what were to become North America and northwestern Europe remained joined until about 65 million years ago, when an additional arm of the Atlantic began to open in the region of what is now the Norwegian Sea. This rifting process separated Greenland and eastern Canada from western Europe and Scandinavia.

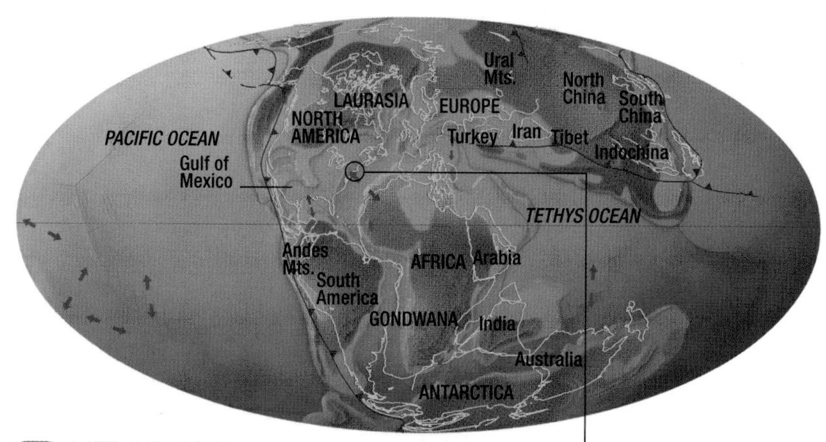

6 LATE JURASSIC LANDMASSES (150 MYA)
By late Jurassic times, the first phase in the opening of the Atlantic had split Pangaea, with North America moving away from what was to become west Africa. In the southern hemisphere, Gondwana had also begun to split apart.

PALISADES SILL, NEW JERSEY
This volcanic rock formation was created during the splitting of North America from what is now northwest Africa about 200 million years ago. Its overall thickness is about 980ft (300m).

ARCTIC OCEAN
Asian-Alaskan land bridge
Rocky Mountains
Gulf of Mexico
PACIFIC OCEAN
Proto-Caribbean Sea
NORTH AMERICA
Greenland
EURASIA
North China
South China
Indochina
NORTH ATLANTIC
Arabia
AFRICA
TETHYS OCEAN
SOUTH AMERICA
Madagascar
India
Australia
SOUTH ATLANTIC
ANTARCTICA

7 CRETACEOUS LANDMASSES (95 MYA)
At this time, Gondwana was in the final stages of its break-up. The South Atlantic formed as South America moved away from Africa; the North Atlantic widened and India was now an island, moving north toward Asia. Greenland was still close to Europe.

GONDWANA BREAKS UP

The ancient continent of Gondwana, which contained all of today's southern continents (Africa, South America, Antarctica, and Australia), together with Madagascar and India, started disintegrating about 180 million years ago. In the first stage of the split, the closely conjoined continents of Africa and South America began to rift apart from the rest of Gondwana, which formed a separate landmass, East Gondwana. From about 130 million years ago, South America started to split from Africa and within about 20 million years, there was an open and ever-widening ocean—the South Atlantic—between the two. Around this time, East Gondwana also split, with India detaching itself from Antarctica and Australia, and moving rapidly north. The final stage in the break-up started about 90 million years ago when Australia and Antarctica began to separate, with Australia heading north. In many parts of the former Gondwana, formations of prominent volcanic rocks can be seen that lay testament to the various stages in the separation process, such as the Paraná trap basalts of Brazil.

PARANÁ TRAPS OF SOUTHERN BRAZIL
About 130 million years ago, the rifting process that parted South America from Africa formed these thick deposits of basaltic rocks. Vast amounts of fluid lava flowed from ground fissures to create them.

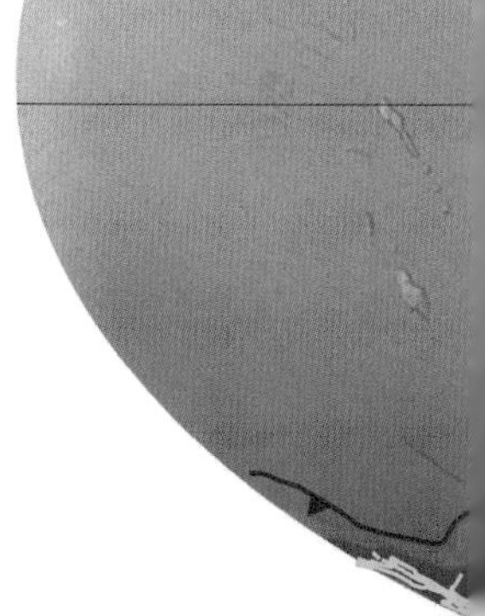

TODAY'S CONTINENTS TAKE SHAPE

Numerous movements have occurred over the past 65 million years to bring about the present-day configuration of Earth's landmasses. Some of the major changes have included further widening of the Atlantic, the collision of India with Asia to form the Himalayas, the convergence of Africa with Europe, the joining of North and South America at the Panama Isthmus, and the parting of Africa and Arabia followed by Arabia's collision with southwest Asia. Meanwhile in east and southeast Asia, a combination of volcanicism and a complex series of plate movements has led to the present-day arrangement of landmasses in that part of the world.

ROCK MEASUREMENTS

Piecing together past continental movements depends on scientists gathering vast amounts of data about rocks across the world. Relevant data can include a rock's age, its orientation in the ground, and its magnetic properties. As continents shift, the rocks in those continents turn as well, and where magnetic minerals in the rocks may once have pointed north, they may now point elsewhere. By plotting such shifts, scientists can gradually reconstruct the past.

PYRENEES, SPAIN
Around 65 million years ago, a section of continent called Iberia—which had separated from other landmasses some 80 million years earlier—collided with southern France, pushing up the Pyrenees mountain chain.

GIANT'S CAUSEWAY, NORTHERN IRELAND
These basaltic columns formed from lava erupted from volcanic fissures during the rifting process that opened up the most northerly parts of the Atlantic some 60 million years ago.

8 PALEOGENE LANDMASSES (50 MYA)
Greenland had by now moved some distance from Europe, while India continued its rapid journey north. Africa was converging on southern Europe. Australia also moved north.

EARTH'S PLATES TODAY

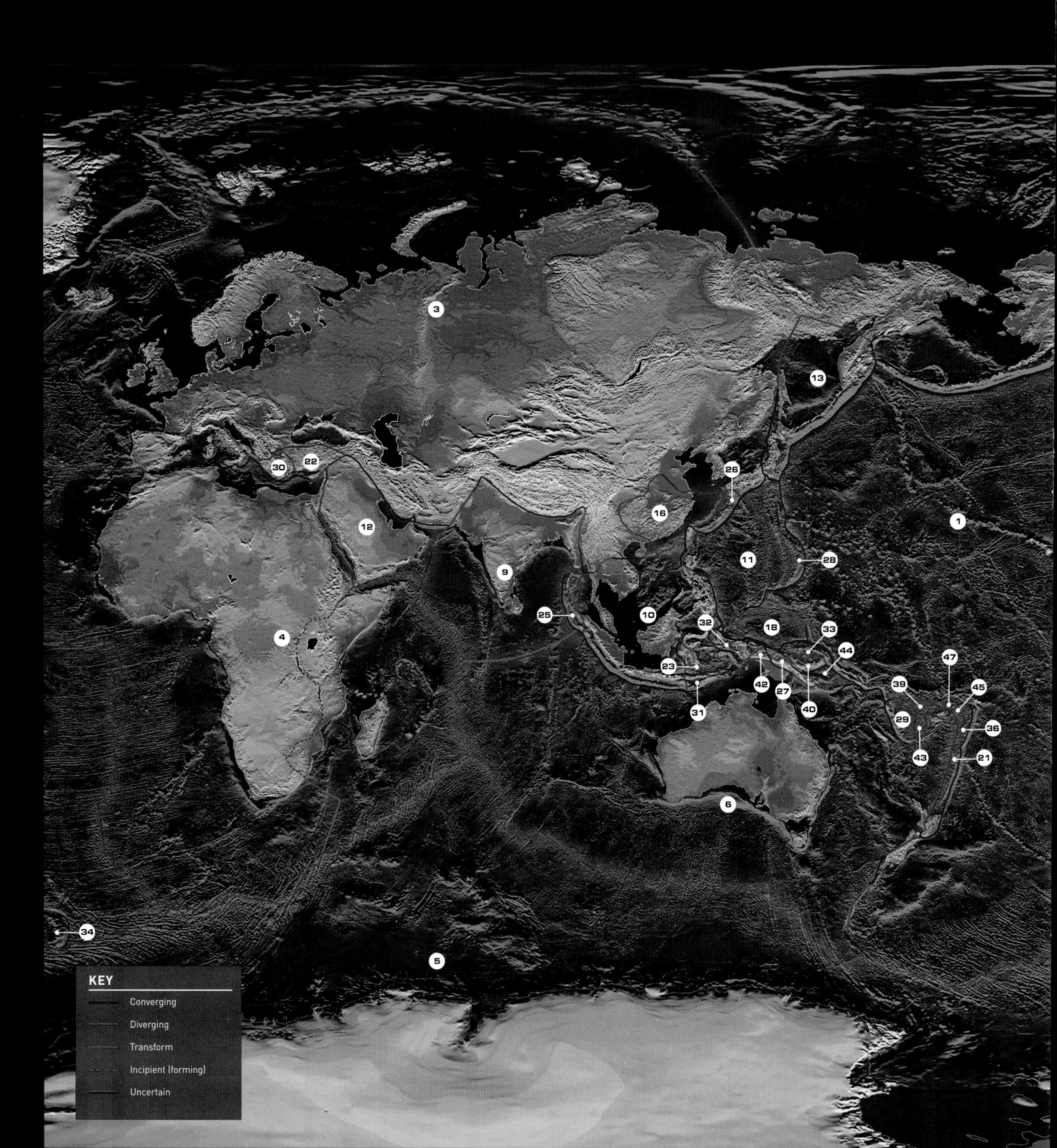

Today, Earth's surface is split up into eight to nine major plates, about six to seven medium-sized plates, and numerous much smaller plates called microplates. The boundaries between plates are of three types: divergent (where plates move apart), convergent (where they move together), and transform (where plates move past each other).

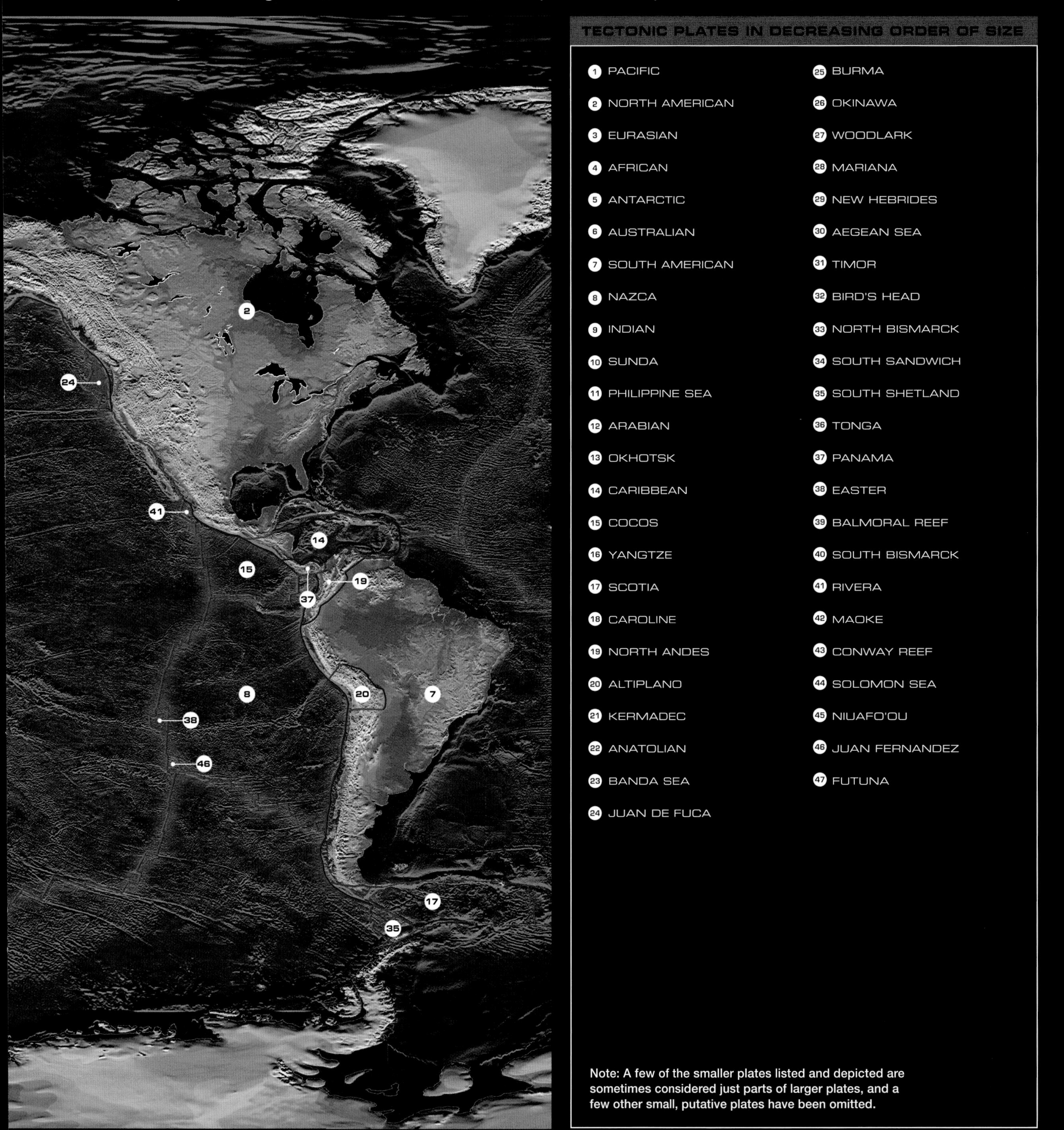

PLATE BOUNDARIES

Plate boundaries are zones around the edges of tectonic plates where a high degree of tectonic activity, such as earthquakes and volcanic eruptions, occurs. These boundaries are of three main types—divergent, transform, and convergent.

DIVERGENT BOUNDARIES

Divergent plate boundaries occur where plates slowly move apart. They are of two types: continental rifts and mid-ocean ridges (see p.246). The former are divergent boundaries that occur in the middle of landmasses, causing them to split. As a continent splits, the sea floods in to form a new ocean, changing a continental rift into a mid-ocean ridge. Both are sites of earthquakes and volcanic activity. The East African Rift, a continental rift, is a newly forming divergent plate boundary that will eventually split Africa (see p.174). The Mid-Atlantic Ridge is a mid-ocean ridge that separates the Eurasian Plate from the North American Plate in the North Atlantic and the African Plate from the South American Plate in the South Atlantic. At a mid-ocean ridge, new lithosphere (or plate) is continuously formed from magma welling up.

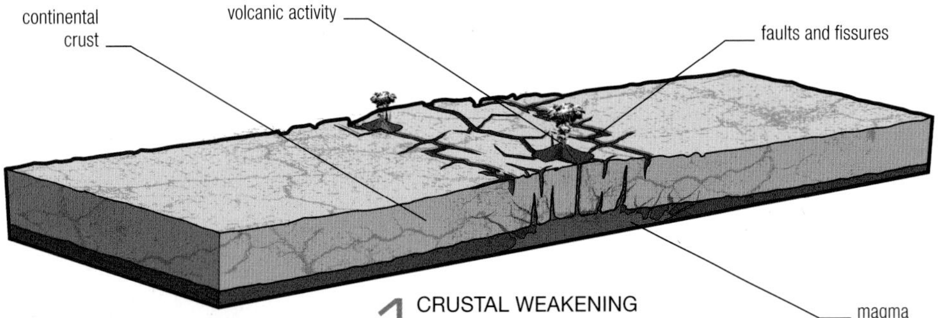

continental crust
volcanic activity
faults and fissures
magma

1 CRUSTAL WEAKENING

A new divergent boundary forms where an upflowing plume of magma rises under a section of continental crust, causing the crust to soften, weaken, stretch, and thin. Long linear faults, or fissures, appear in the crust, leading to the formation of volcanoes.

solidified lava erupted from fissure
rift valley
volcanic fissure

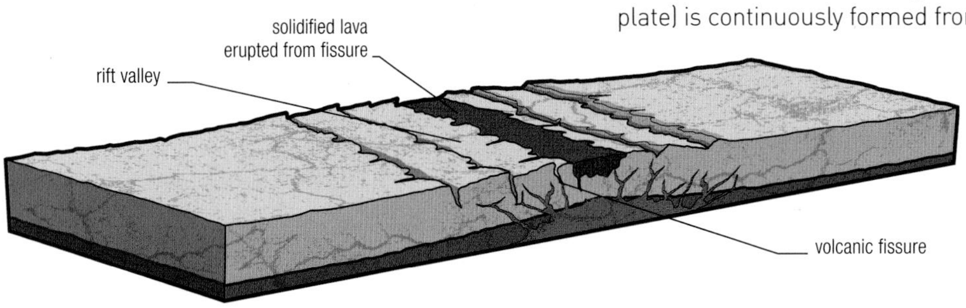

2 RIFT VALLEY FORMATION

Initially, no new lithosphere or plate is created along the rift faults. Instead, land sinks along the faults to form a rift valley, and a series of volcanoes and volcanic fissures develop in the fault zone as magma reaches the surface.

3 INFLOODING OF SEA

As the rift valley extends, the land moves apart and the sea floods in. At the rift's center at the bottom of the sea, a mid-ocean ridge develops. New oceanic crust forms here, parting land that was once joined.

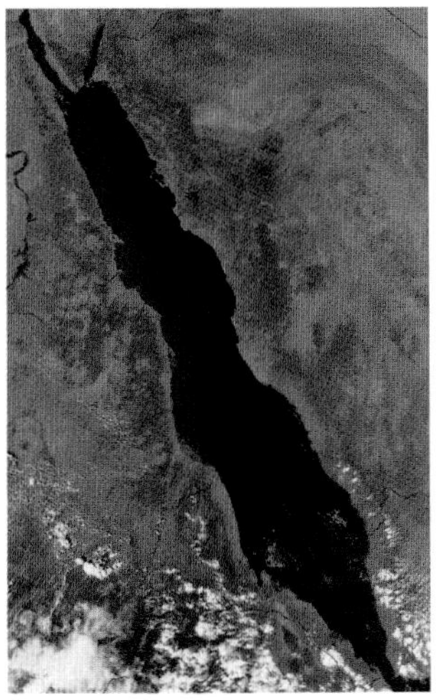

area where ridge develops
faulted continental crust
sea floods in
new oceanic crust

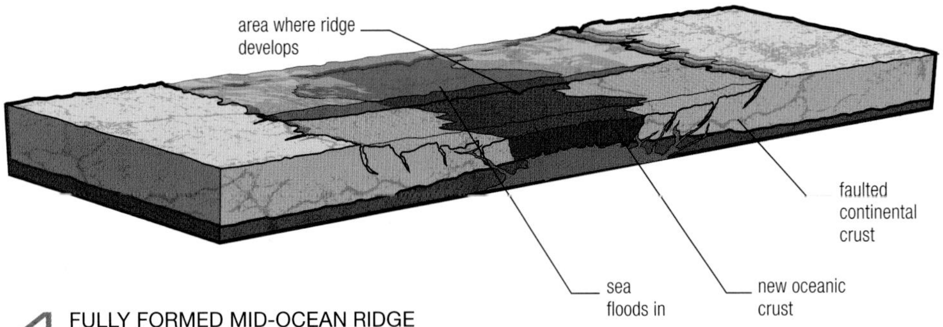

4 FULLY FORMED MID-OCEAN RIDGE

Eventually, a full-fledged mid-ocean ridge forms. As new lithosphere is created at the ridge, the plates on either side continue to move apart slowly. Ridges like this make up the world's most extensive mountain ranges.

oceanic crust
mid-ocean ridge
oceanic lithosphere

RIFT VALLEY
The Suguta Valley is part of the East African Rift where it runs through northern Kenya. This is an arid, low-lying area of salt pans, mud flats, and small volcanoes.

RED SEA
The Red Sea is a developing ocean basin, marking a divergent plate boundary where Africa split from Arabia, starting about 40 million years ago.

THINGVELLIR RIFT, ICELAND
The Mid-Atlantic Ridge rises to the sea surface in Iceland, where it is visible in a few spots as a cleft in the island.

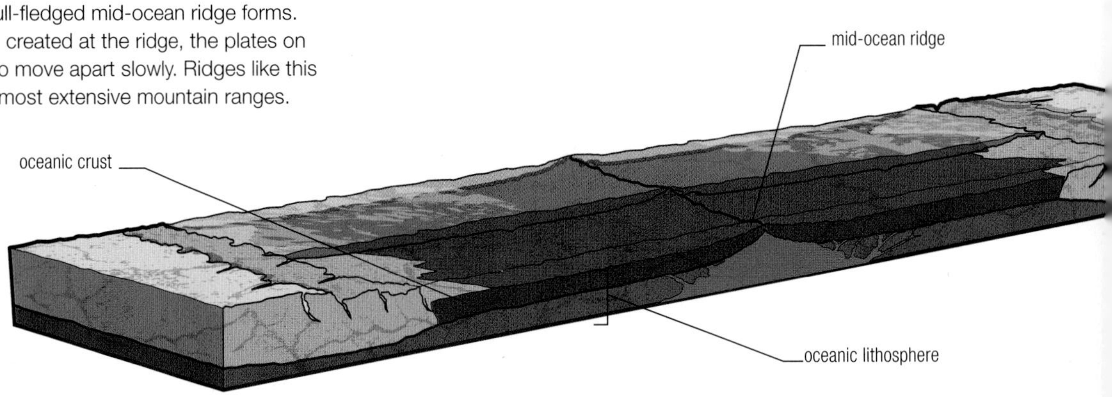

ALPINE FAULT, NEW ZEALAND
The transform boundary called the Alpine Fault is visible in this satellite image as a near-straight horizontal line running through New Zealand's South Island. The fault is a dividing line between the Australian Plate (top) and Pacific Plate (bottom).

TRANSFORM BOUNDARIES

A zone where the edges of two plates move horizontally past each other is called a transform boundary. This movement is sometimes extremely jerky and causes earthquakes. A few transform boundaries exist on land, the best known being the San Andreas Fault that runs through California. Along this boundary, chunks of continental lithosphere belonging to the Pacific and North American plates grind past each other in opposite directions, causing frequent earthquakes of high intensity. The Alpine Fault is another land-based transform boundary, which runs through New Zealand's South Island. Along this fault, slabs of continental lithosphere belonging to the Pacific and Australian plates move past each other at a rate of about 1.6in (4cm) a year, generating many high-intensity earthquakes. Another land-based transform boundary runs through north Turkey. Called the North Anatolian Fault, it is again a site of frequent, sometimes devastating, earthquakes.

The majority of transform boundaries, however, are on the ocean floor, consisting of relatively short faults in oceanic lithosphere arranged perpendicular to the mid-ocean ridges. These transform faults, as they are called, connect sections of mid-ocean ridge that are broken up into several segments and displaced, or offset, from each other. Along the transform faults, parts of oceanic lithosphere move past each other in opposite directions, causing submarine earthquakes.

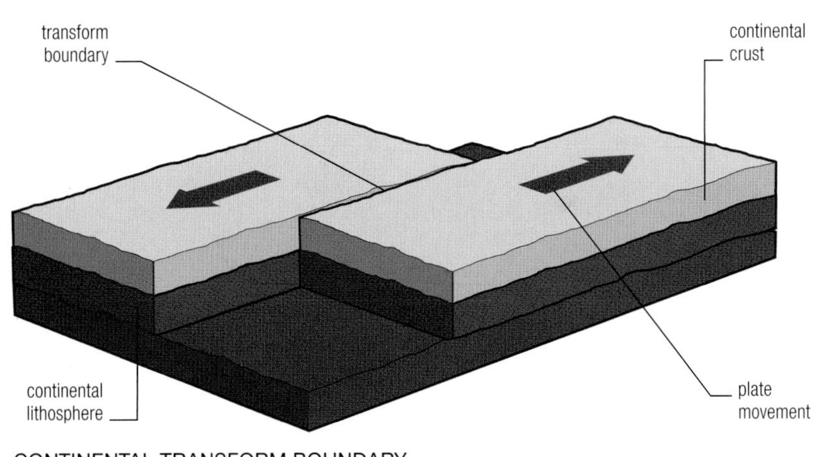

CONTINENTAL TRANSFORM BOUNDARY
At transform boundaries that run through continents, such as New Zealand's Alpine Fault, slabs of continental lithosphere move slowly past each other in opposite directions.

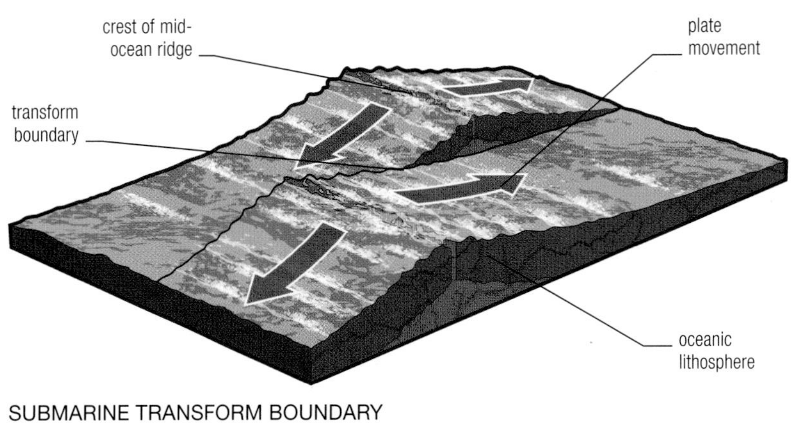

SUBMARINE TRANSFORM BOUNDARY
This type of boundary is a short fault that connects offset sections of a mid-ocean ridge. Along the boundary, slabs of oceanic lithosphere move past each other in opposite directions.

CONVERGENT BOUNDARIES

Convergent boundaries occur where two plates move toward each other, driven by heat convection in the mantle and the creation of new sea floor plates at mid-ocean ridges. At these boundaries, large sections of lithosphere (or plate) subduct or descend toward the deeper mantle and are destroyed, so these boundaries are also called destructive boundaries. They are major sites of earthquakes and volcanic activity, and fall into three types, based on whether the edges of the converging plates both consist of continental lithosphere (continent-continent collisions), or oceanic lithosphere (ocean-ocean convergence), or consist of different types of lithosphere (ocean-continent convergence).

CONTINENT-CONTINENT COLLISIONS

If the edges of two plates, both carrying continental lithosphere, collide, the mantle part of the lithosphere of one plate subducts below the other. But the continental crust, having a relatively low density, resists downward motion. Instead, the crust on either side of the boundary is compressed, folded, faulted, and pushed up to form mountains. At these boundaries, magma is produced at great depth, but rarely reaches the surface, so volcanic activity is fairly rare, although earthquakes are common.

HIMALAYAS FROM SPACE
This classic example of a continent–continent collision began about 50 million years ago, when plate motion drove the Indian Plate into the Eurasian Plate. Over millions of years, the Himalayas were pushed up and continue to grow today.

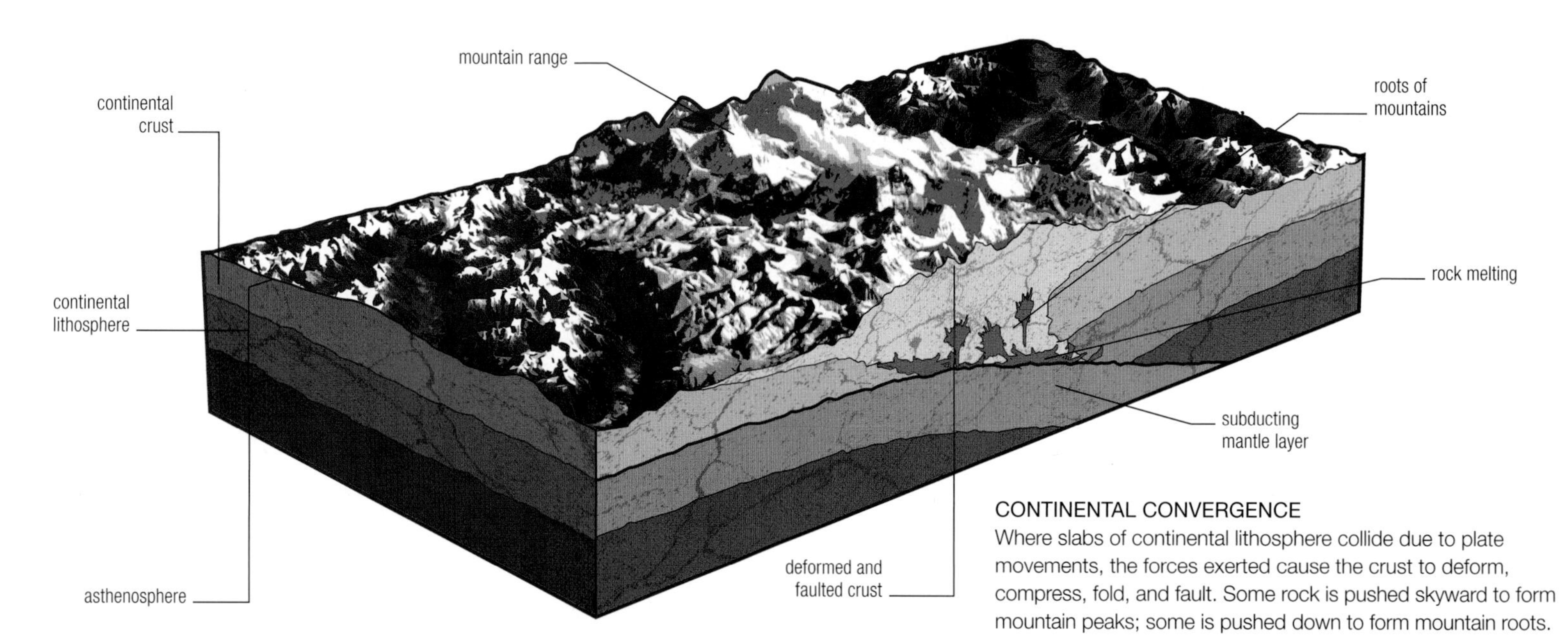

CONTINENTAL CONVERGENCE
Where slabs of continental lithosphere collide due to plate movements, the forces exerted cause the crust to deform, compress, fold, and fault. Some rock is pushed skyward to form mountain peaks; some is pushed down to form mountain roots.

APPALACHIANS
This North American mountain range, seen from space, formed by a series of collisions between landmasses 450 to 250 million years ago.

URAL MOUNTAINS
These mountains in Russia were pushed up 310 to 220 million years ago, through collisions between landmasses called Siberia, Kazakhstania, and Euramerica.

ALPS
This high mountain range in Europe formed in stages 60 to 10 million years ago, due to a collision between the African and Eurasian plates.

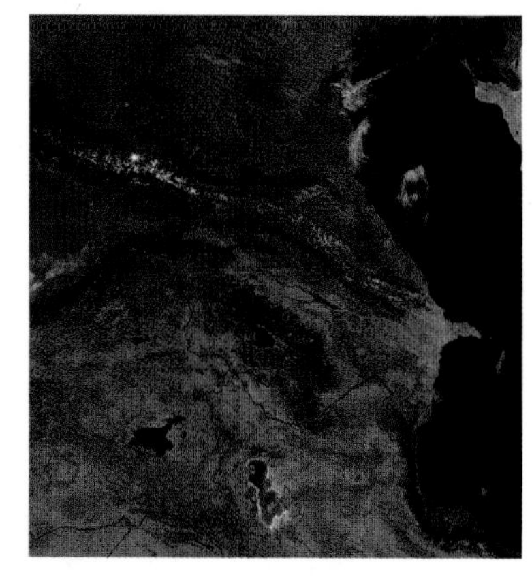

GREATER CAUCASUS
These mountains in southwest Asia formed 28 to 24 million years ago, from a continental collision involving the Arabian and Eurasian plates.

OCEAN-CONTINENT CONVERGENCE

At an ocean–continent boundary, the plate with oceanic lithosphere subducts beneath the continental plate. Along the boundary, a deep trench forms in the ocean floor. The subducting plate slides down jerkily, causing frequent earthquakes. As it descends, its temperature rises, releasing volatile substances, such as water, trapped in the oceanic crust. These act as a flux, lowering the melting temperature of rock in the mantle below the overriding plate, producing magma. This then rises through cracks in the rock above and forms deep magma chambers. The erupted magma forms a chain of volcanoes inland, called a continental volcanic arc (see pp.108–09).

KAMCHATKA
When the Pacific Plate subducted under part of the Okhotsk Plate, the result was a line of volcanoes along the Kamchatka Peninsula in eastern Russia. This volcano, with a blue crater lake, is called Maly Semiachik. It last erupted in 1952.

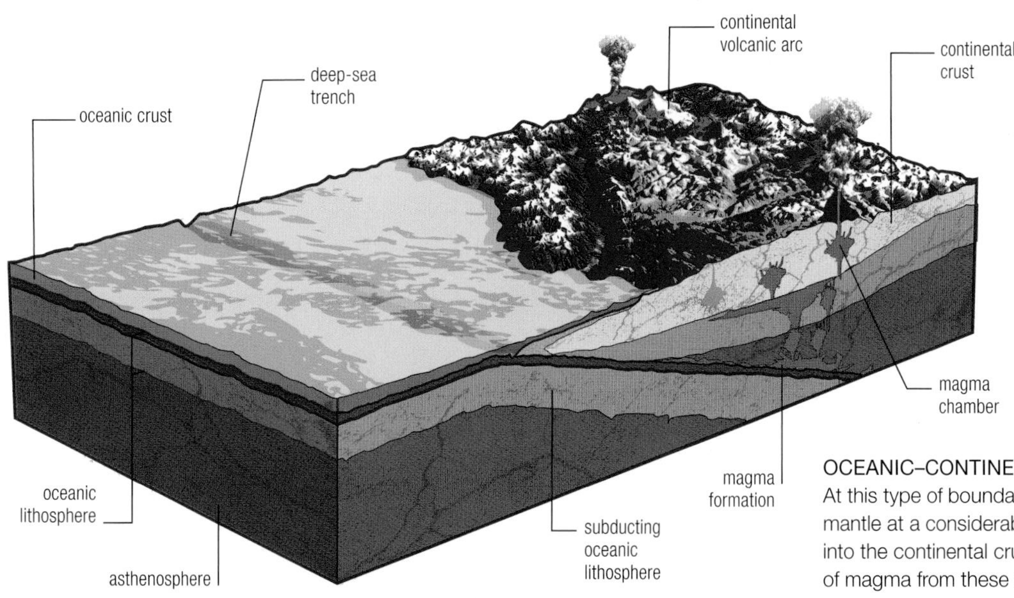

OCEANIC–CONTINENTAL CONVERGENCE
At this type of boundary, magma forms from the melting of the mantle at a considerable depth beneath the surface. It rises into the continental crust, where it forms chambers. Upflows of magma from these chambers create an arc of volcanoes.

ANDES
The Andes consists of several chains of volcanoes and other mountains formed by the Nazca and Antarctic plates descending beneath the South American Plate. This Andean volcano is called Quilotoa.

OCEAN-OCEAN CONVERGENCE

The edge of a plate carrying oceanic lithosphere slides underneath the edge of a neighboring plate also consisting of oceanic lithosphere. The sinking edge is normally the one composed of older lithosphere, because as oceanic lithosphere ages, it becomes denser and heavier. A deep trench develops along the boundary, and magma forms in the region above the subducting plate. But here the rising magma at an ocean–ocean boundary forms a line of volcanic islands arranged in a gentle curve, called a volcanic island arc (see pp.110–11). Ocean–ocean convergence zones are frequently affected by strong earthquakes.

AUGUSTINE VOLCANO
This volcano off the coast of Alaska is part of the Aleutian Arc—a long, curved chain of volcanoes, many of them on islands, in the northern Pacific, formed as the Pacific Plate has subducted under the North American Plate.

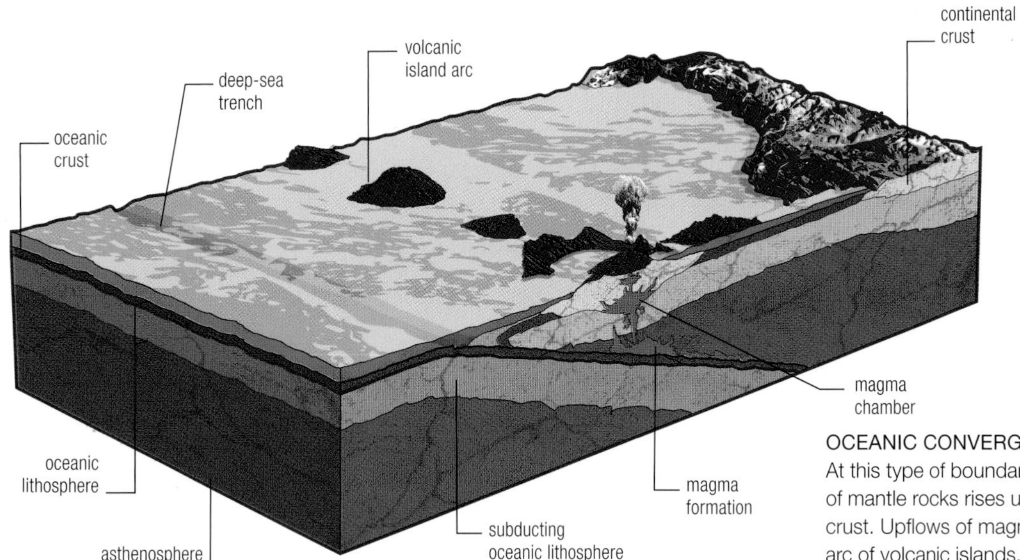

KURIL ISLANDS
The Kuril Islands are a volcanic island arc in the northwestern Pacific, created where part of the Pacific Plate subducted under the Okhotsk Plate. This one is called Yankicha.

OCEANIC CONVERGENCE
At this type of boundary, magma formed from the melting of mantle rocks rises up to form chambers under oceanic crust. Upflows of magma from these chambers create an arc of volcanic islands.

HOTSPOTS

A few locations at the top of Earth's mantle appear to be the source of peculiarly large amounts of energy. As this energy percolates to the surface, it causes a high degree of volcanic activity. These locations, many of which are far from plate boundaries, are known as hotspots.

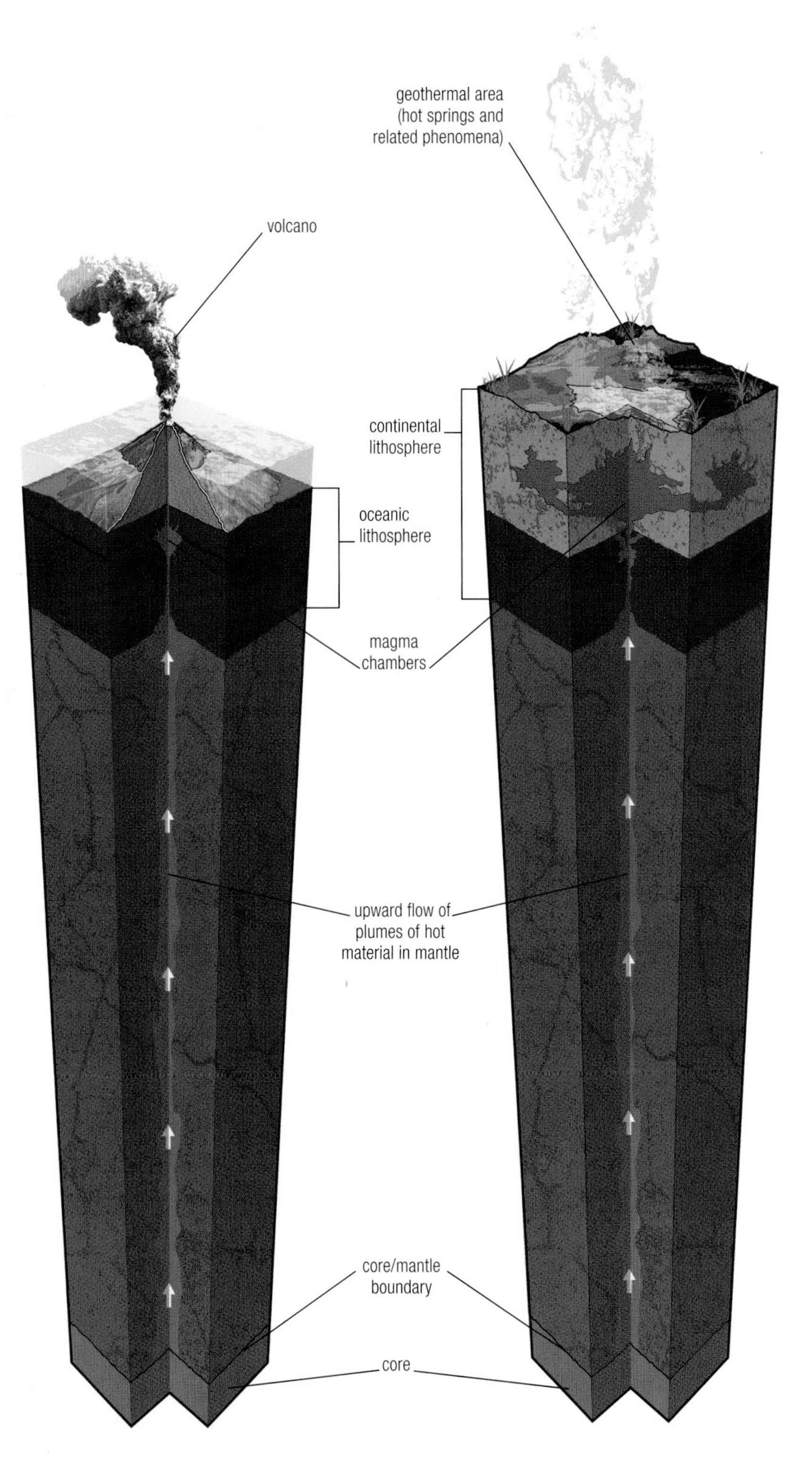

HOTSPOT UNDER OCEAN
If the mantle plume idea is correct, many volcanic islands result from upward flows of hot material within the mantle.

HOTSPOT UNDER CONTINENT
A similar mantle plume under a continent can cause a range of surface activity, from hot springs to volcanic eruptions.

HOTSPOT THEORIES

There are two main hypotheses about the cause of hotspots. The best known, and oldest, proposes that they result from mantle plumes—narrow flows of hot, semi-molten rock rising up from the core–mantle boundary to particular spots under Earth's lithosphere (outer shell). A newer hypothesis proposes that the lithosphere is being stretched in certain places by factors unrelated to such plumes, and these stretched areas allow magma (hot, molten rock) to leak up into the crust from a uniform reservoir at the top of the mantle. Regardless of the cause and exact nature, more magma is present in Earth's crust above hotspots than elsewhere. The effects vary slightly depending on whether the magma is present in continental or oceanic crust. The Yellowstone hotspot in the US (one of the best-known continental hotspots) has a large and deep magma chamber, hot springs and geysers at the surface, and large, infrequent volcanic eruptions. Other continental hotspots have created groups of small volcanoes or massive outpourings of lava. Most hotspots, however, are oceanic rather than continental.

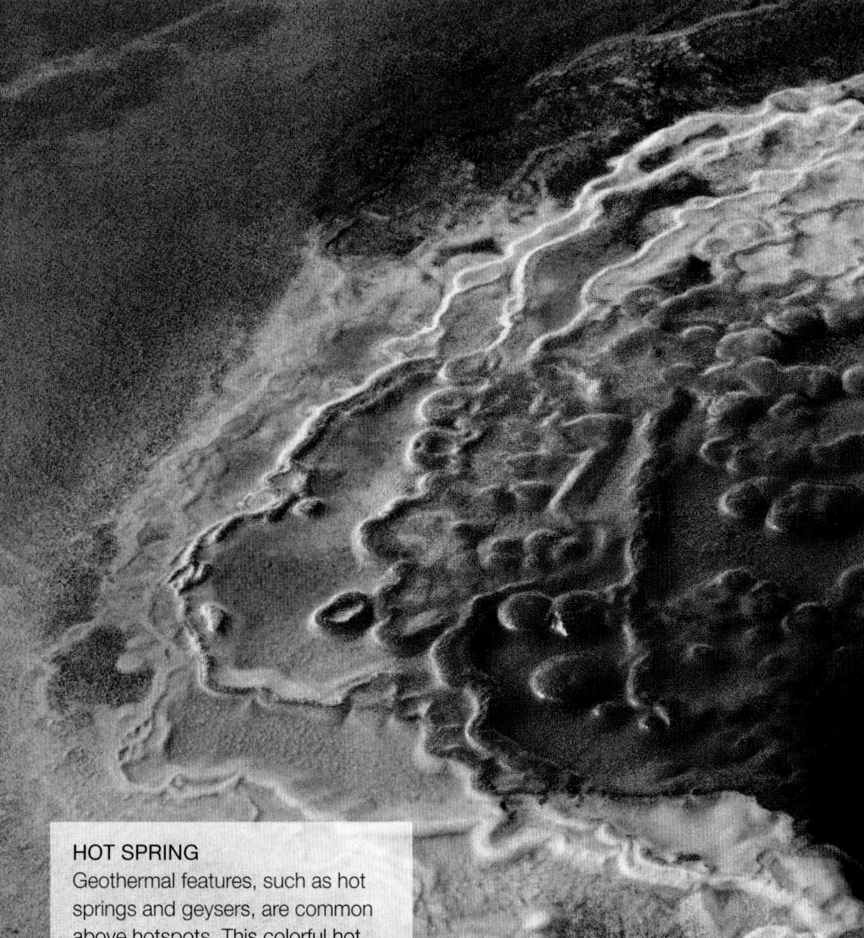

HOT SPRING
Geothermal features, such as hot springs and geysers, are common above hotspots. This colorful hot spring is located at the Norris Geyser Basin in Yellowstone Park, which sits on the Yellowstone hotspot.

SOME RECOGNIZED HOTSPOTS AROUND THE WORLD

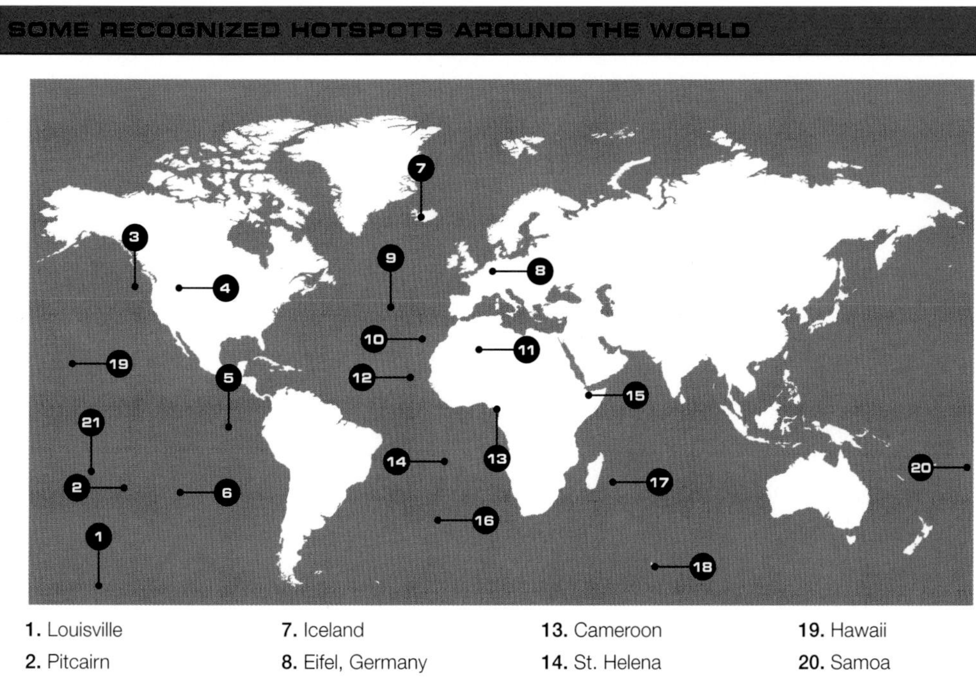

1. Louisville
2. Pitcairn
3. Cobb, US
4. Yellowstone, US
5. Galápagos
6. Easter, Pacific Ocean
7. Iceland
8. Eifel, Germany
9. Azores, North Atlantic
10. Canary
11. Hoggar, Algeria
12. Cape Verde
13. Cameroon
14. St. Helena
15. Afar, Ethiopia
16. Tristan
17. Réunion
18. Kerguelan
19. Hawaii
20. Samoa
21. Tahiti

OCEAN HOTSPOTS

About three-quarters of identified or proposed hotspots are oceanic—they exist under oceanic lithosphere, often far from plate boundaries. At these hotspots, the presence of large amounts of magma in the crust leads to volcanic eruptions on the seafloor, creating submarine volcanoes and, eventually, volcanic islands. In the 1960s, the Canadian scientist John Tuzo Wilson reasoned that plate movements that gradually push the oceanic lithosphere over such hotspots can create chains of volcanic islands and extinct submarine volcanoes called seamounts. This is still seen as the best explanation for the formation of some chains of volcanic islands (see pp.112–13), as well as linear underwater ridges. Examples include the Hawaiian Islands and the Emperor seamount chain, created by the Hawaii hotspot in the central Pacific, and the Louisville seamount chain created by the Louisville hotspot in the southwest Pacific. A hotspot in the North Atlantic, located under the Mid-Atlantic Ridge (a plate boundary), is thought to be partly responsible for the high volcanic activity that formed Iceland.

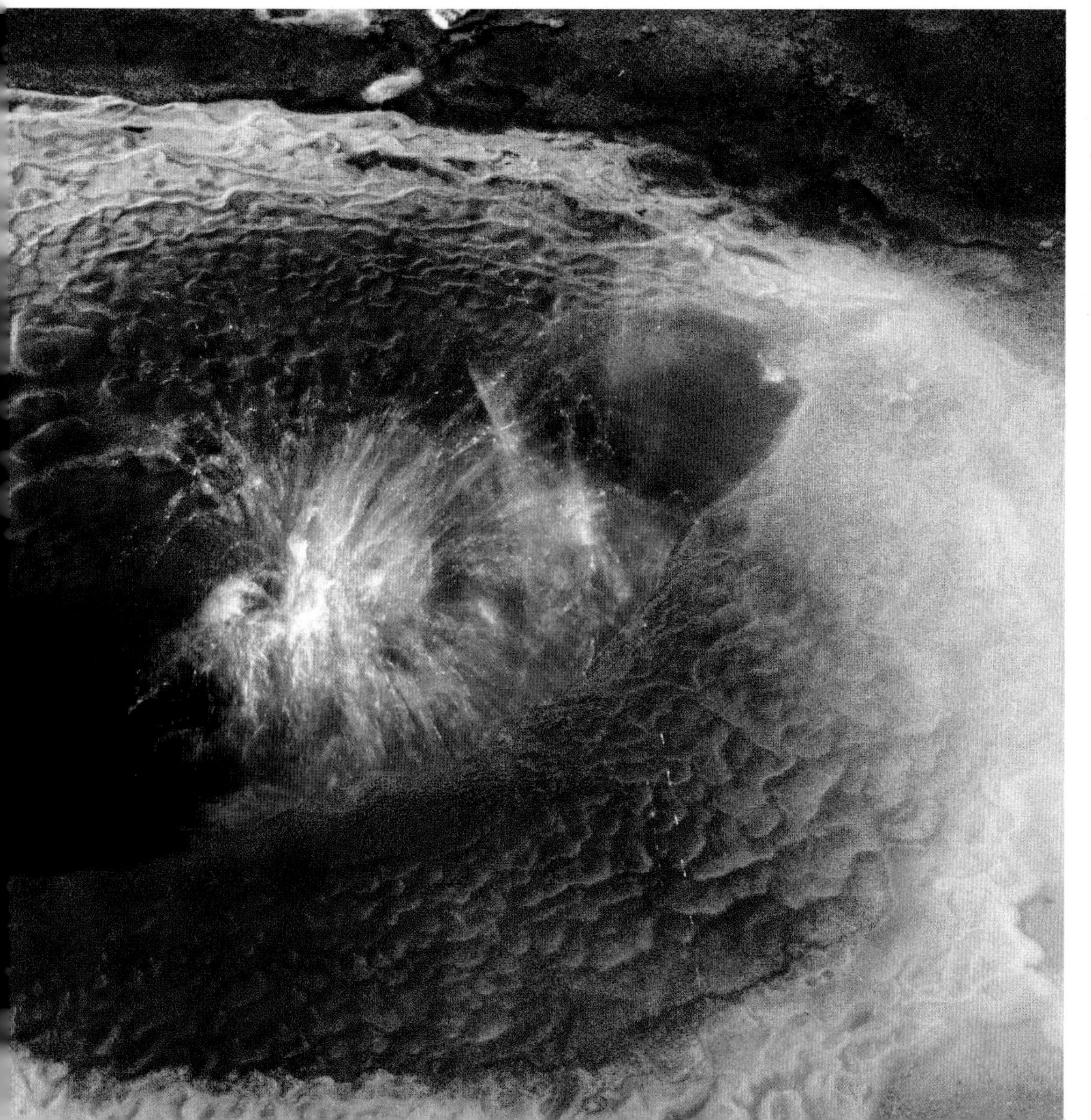

FERNANDINA VOLCANO, GALÁPAGOS ISLANDS
It is thought that the Galápagos Islands in the Pacific were created by plate movements over the Galápagos hotspot. The Fernandina volcano, which forms an entire island, is believed to lie directly over the hotspot.

60
The approximate number of volcanic regions around the world that have been suggested to be the result of hotspots. Whether some of these are truly due to hotspots, though, is fiercely debated. Most geologists only definitely agree on 20 to 30 of these.

GEOTHERMAL ENERGY

Our planet harbors a colossal amount of trapped heat energy, called geothermal energy. As this moves toward the surface and escapes, it drives plate movement, volcanic activity, and geysers.

SOURCE OF THE ENERGY

Some of Earth's internal energy is residual heat from when it first formed (see pp.8–9), but most comes from the decay of radioactive isotopes—unstable atoms of certain elements such as uranium, thorium, and potassium—scattered throughout the planet's interior. When an atom of uranium-238 decays it releases about one-trillionth of a joule of energy. This might seem negligible, but in Earth's crust alone, about 10^{24} (a trillion trillion) atoms of uranium-238 decay every second. This means that energy is generated from this source alone at a rate of about a terawatt (a trillion watts). Altogether, Earth's natural radioactivity generates heat at a rate of about 30–40 terawatts—about twice the current rate of human global energy consumption.

ENERGY FROM RADIOACTIVE DECAY

When an unstable nucleus (central part of an atom) undergoes radioactive decay, it emits both a fast-moving particle and (usually) a tiny amount of electromagnetic radiation. Energy from each of these slightly increases Earth's internal heat. Half of Earth's radioactive heat production occurs in the crust.

unstable parent nucleus

daughter product

energetic particle

photon (packet of electromagnetic radiation)

ENERGY FROM THESE ADDS TO EARTH'S INTERNAL HEAT

GEYSER ERUPTION

1 **WATER BUBBLES FORM**
The eruption of the Strokkur geyser in Iceland provides a dramatic example of geothermal energy release at Earth's surface. At first a small dome of water appears as a steam bubble rises from below.

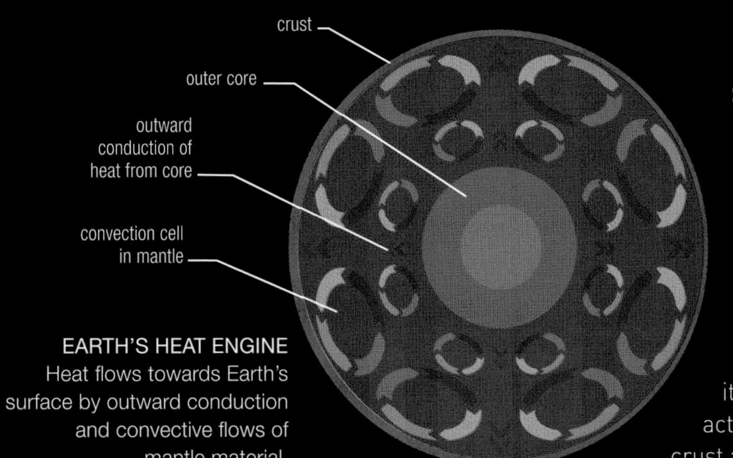

EARTH'S HEAT ENGINE
Heat flows towards Earth's surface by outward conduction and convective flows of mantle material.

SPREADING THE ENERGY

The heat energy in Earth's interior moves slowly outward from its core toward the cooler crust by two main mechanisms. One of these is conduction—the direct particle-to-particle transfer of energy through stationary matter. The other, more predominant, mechanism of heat transfer is convection—slow, circular movements of the semi-molten rocks of Earth's mantle. On reaching the crust, the energy dissipates [illegible]ous ways, mainly through activity at plate [illegible]aries, where it is released in earthquakes and volcanic eruptions, and at hotspots where it produces phenomena such as geyser eruptions as well as volcanic activity. Some of the energy is simply conducted through the crust and radiated at the surface.

2 THE GEYSER ERUPTS
Suddenly the boiling hot water is forced violently upward by the steam, with an accompanying explosion of sound. More steam and hot water rapidly follow, pushing the geyser fountain upward.

3 THE ERUPTION REACHES ITS FULL HEIGHT
The Strokkur geyser reaches a maximum height of about 65½ft (20m). During the eruption, some geothermal energy is released as sound energy and as heat, since the escaping steam warms the nearby air.

HARNESSING THE ENERGY

For more than 100 years, people have used power plants to capture and make use of geothermal energy. These plants generally work by drilling wells to tap into the hot water deep underground, which also can convert to steam as it reaches the surface. The resulting hot water is used to heat homes and the steam produced drives electricity-generating turbines. Worldwide, today's geothermal plants produce useful, non-polluting energy at a rate of about 40 gigawatts, supplying about 0.25 percent of worldwide energy needs. This represents only a tiny fraction (about 0.1 percent) of the total energy flowing out of Earth. Much of the energy flow is concentrated in particular regions—around the boundaries of tectonic plates and at hotspots. These are the places where geothermal plants are built.

BLUE LAGOON
At the Svartsengi plant in Iceland, bathers can enjoy geothermal energy directly, since some of the plant's output of warmed water is pumped into a nearby lagoon.

MEASURING PLATE MOVEMENTS

Scientists have developed several methods to measure how fast Earth's plates are shifting around and how they moved in the past. They can also predict how these plates will move in the future.

PRESENT-DAY MOVEMENTS

Interferometry is one method used to measure current movements. By comparing times when radio telescopes, positioned on either side of a plate boundary, receive signals from distant galaxies, the distance between the telescopes can be calculated. Repeated over many years, it can estimate relative motion between the plates on which the telescopes stand. Other methods compare distances by reflecting laser beams off satellites or use the Global Positioning System (GPS). To estimate motions relative to Earth's mantle, plate positions are measured relative to fixed hotspots (see pp.32-33).

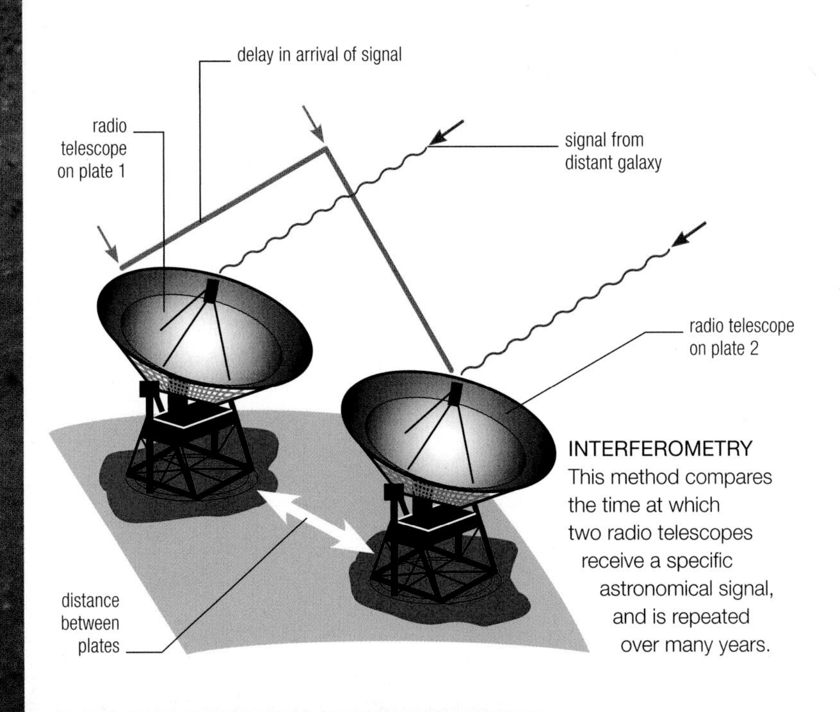

INTERFEROMETRY This method compares the time at which two radio telescopes receive a specific astronomical signal, and is repeated over many years.

SAN ANDREAS FAULT The average rate of movement along this plate boundary where it runs through California is 1.4in (3.5cm) a year. But the movement is episodic, with no movement most years and considerable shifts in others as a result of earthquakes.

CLOSER VIEW Earthquakes can cause shifts of up to several feet along a fault. However, these shifts can also cause vertical misalignments, as on the San Andreas Fault seen here. Monitoring the pattern of shifts can help predict the location of the next earthquake.

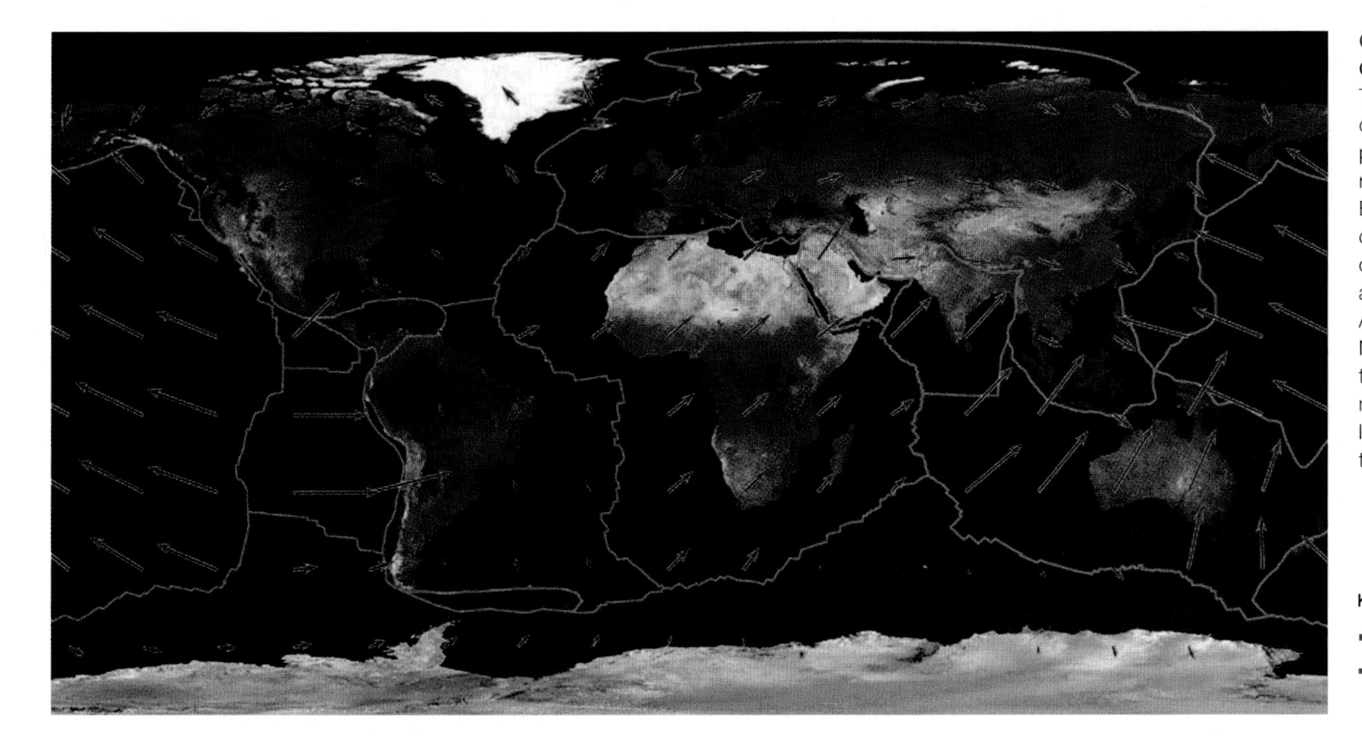

CURRENT RATES OF MOVEMENT
This map shows the direction and rate of plate movements today, relative to the average Earth. Overall, plates comprising largely oceanic lithosphere (such as the Pacific Plate, the Australian Plate, and the Nazca Plate) move faster than those composed mainly of continental lithosphere (such as the Eurasian Plate).

KEY
- Plate movements
- Plate boundaries

PAST MOVEMENTS

The measurement of plate motions can be used not only to calculate present-day movements, but past movements as well. One method is to analyze the magnetic properties of rocks on the ocean floor. When these rocks first form at mid-ocean ridges, Earth's magnetic field gives them a magnetic "signature," which depends on the strength and direction of the magnetic field at the time. By compiling and analyzing magnetic maps of the sea floor, the rates at which plates have moved away from mid-ocean ridges can be ascertained. Another method involves studying chains of islands that form as plates pass over hotspots in Earth's mantle—these have shown, for example, that the Pacific Plate has been moving in a northwesterly direction at the rate of about 3in (7cm) per year for tens of millions of years. Unfortunately, no sea floor is older than 200 million years, so to work out how Earth's plates moved before then, continental rocks have to be analyzed. As plate movements have rotated the continents, ancient rocks have turned with them, and magnetic minerals in those rocks, which pointed north when they first formed, may now point elsewhere. So magnetic measurements on these rocks can be used to give an estimate of past rotational motions.

South America
Has moved north and parts of it are now close to Florida

North Africa
Has joined with Europe

Horn of Africa
Has split from Africa but joined with Arabian Peninsula

Australia
Has moved north and crashed into Japan

Antarctic Peninsula
Has moved north and now has a temperate climate

Eastern Siberia
Lies partly in the tropics

FUTURE PLATE MOVEMENTS
Based on current rates of plate motion, it is possible to estimate where various landmasses will be at various times in the future—this map looks forward 100 million years. Some projections suggest that all the continents will come back together again in about 250 million years to form a new supercontinent.

GIANT'S CAUSEWAY
In the late 18th century, scientists arguing over what natural process had caused this collection of 40,000 hexagonal basalt columns in Northern Ireland—and how old they are—helped initiate the modern science of geology. Today, they are known to have formed from cooled lava extruded in vast quantities from Earth some 60 million years ago.

THE GEOLOGIC TIME SCALE

The development of a system for dividing up Earth's history—a geologic time scale—and the establishment of how old Earth is, has a history going back centuries. But serious scientific work on the matter started about 200 years ago.

ZION CANYON
This canyon has rock layers extending from the Permian to Triassic periods.

DIVIDING UP GEOLOGIC TIME

By the late 18th century, scientists had already realized that layers of sedimentary rocks had been deposited at different times in Earth's history, with the oldest rocks at the bottom. By comparing rock layers, or formations, in different locations, an understanding was gradually built up of the sequence in which the formations across large regions of the world had been laid down. This process was assisted by studies on fossils, which could often help establish that rocks from different sites were the same age, and on overlaps between the sequences observed at different sites. For example, comparison of the rock formations in three canyons in the western US showed overlap, with some formations in Zion Canyon also present in the Grand Canyon and others present in Bryce Canyon. This meant that a sequence of deposition covering all three sites could be compiled. As knowledge about the overall deposition sequence built up, geologists started to ascribe names to particular parts of it. Thus, the name Devonian was given to a section fairly low down in the sequence, containing some formations visible in Devon, England. A section higher up, which included formations seen in the Swiss Jura Mountains, was called Jurassic, and a younger sequence of rocks that contained a lot of chalk was called Cretaceous. The scientists who devised these names had only a vague idea how old each layer was.

HORSESHOE BEND, GRAND CANYON
Rock layers in the Grand Canyon, Arizona, range in age from Precambrian schists in its very deepest parts to sedimentary rocks of the Permian Period at the top.

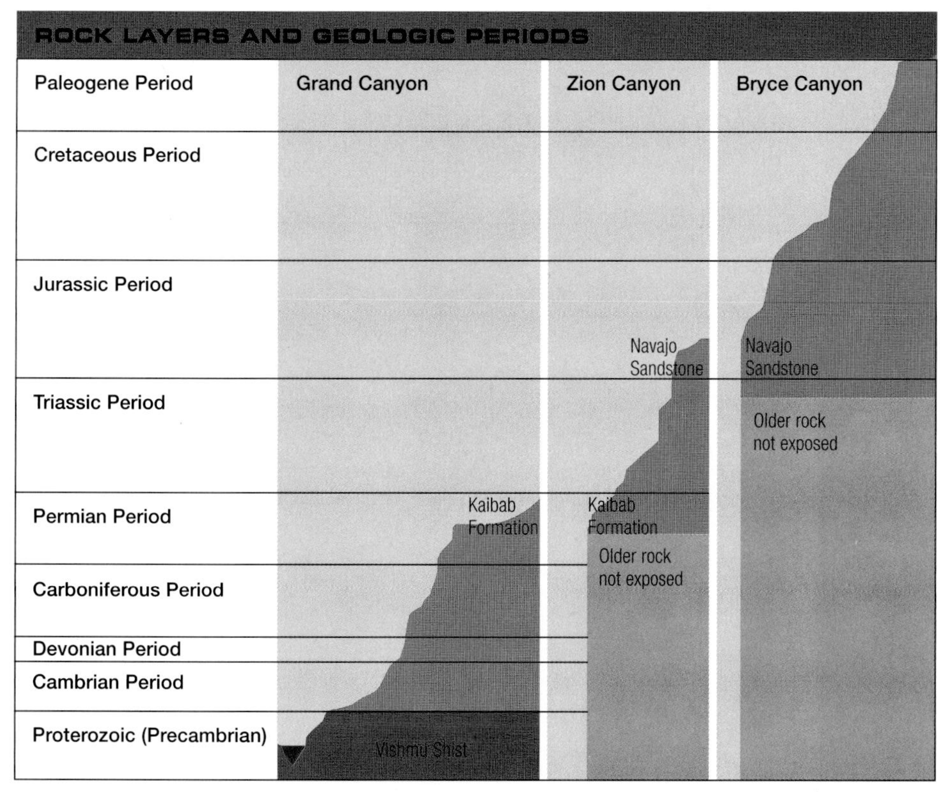

OVERLAPPING ROCK SEQUENCES
The overlapping sequence of rock layers in the Grand Canyon, Zion Canyon, and Bryce Canyon extend back through many geologic periods, omitting a few (such as the Silurian Period) since no new rock layers were being deposited or have survived within the area at the time.

GEOLOGIC PERIODS

As more parts of the sedimentary rock sequence were named, geologists began to formalize the way in which geologic time should be divided up. Substantial sections of the rock sequence were called periods. Later, periods were subdivided into epochs. Going up in the hierarchy, some periods were grouped into longer intervals called eras, and these in turn into yet longer ones, called eons. Today, four eons of Earth's history are recognized—the Hadean, the Archaean, the Proterozoic, and the Phanerozoic. The three eons before the Phanerozoic are commonly called the Precambrian.

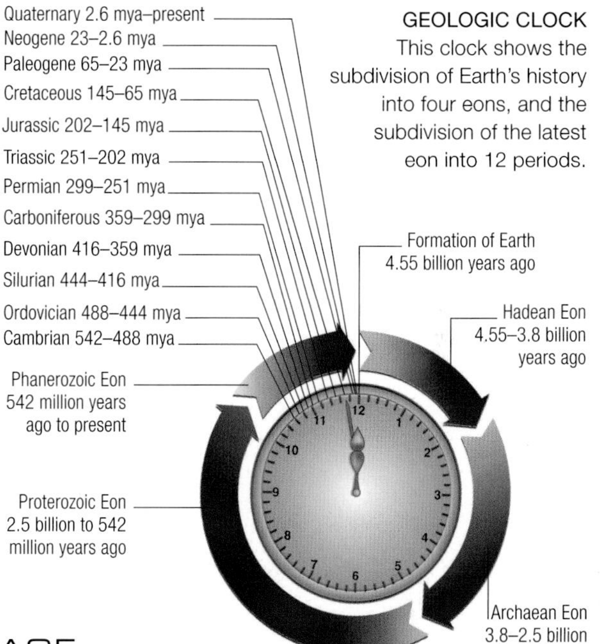

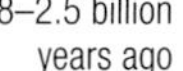

GEOLOGIC CLOCK
This clock shows the subdivision of Earth's history into four eons, and the subdivision of the latest eon into 12 periods.

MEASURING EARTH'S AGE

By the mid-19th century, although they understood the relative ages of different rocks, geologists were still unclear about their absolute ages or the age of Earth itself. Eventually, in the early 20th century, an accurate method for dating rocks was devised, based on measuring radioactive decay processes within the rocks. This method, called radiometric dating, was soon being applied to put figures on the ages of particular geologic periods. Earth's age was trickier, since no rocks still exist from when our planet first formed. In the 1920s, Arthur Holmes, a pioneer in radiometric dating, calculated the age of Earth to be as much as 3.0 billion years old. Eventually, in the 1950s, modern techniques were used to calculate the age of some meteorites—rocks formed at the beginning of the Solar System that had fallen to Earth. On the assumption that their age and Earth's must be the same, this gave a figure for Earth's age of 4.55 billion years, a figure that remains accepted today.

BRYCE CANYON
The exposed rock layers in the walls of Bryce Canyon, Utah, range in age from the Triassic to the Paleogene period.

CANYON DIABLO METEORITE FRAGMENT
The age of our planet was eventually established by measuring the age of the Solar System, in particular the age of meteorites, such as the Canyon Diablo meteorite of which a fragment is shown here.

2 MOUNTAIN BUILDING

<< La Sal Mountains
The snowcapped La Sal range, part of the Rocky Mountains, can be seen in the distance behind the sandstone formations of Arches National Park, Utah.

MOUNTAINS OF THE WORLD

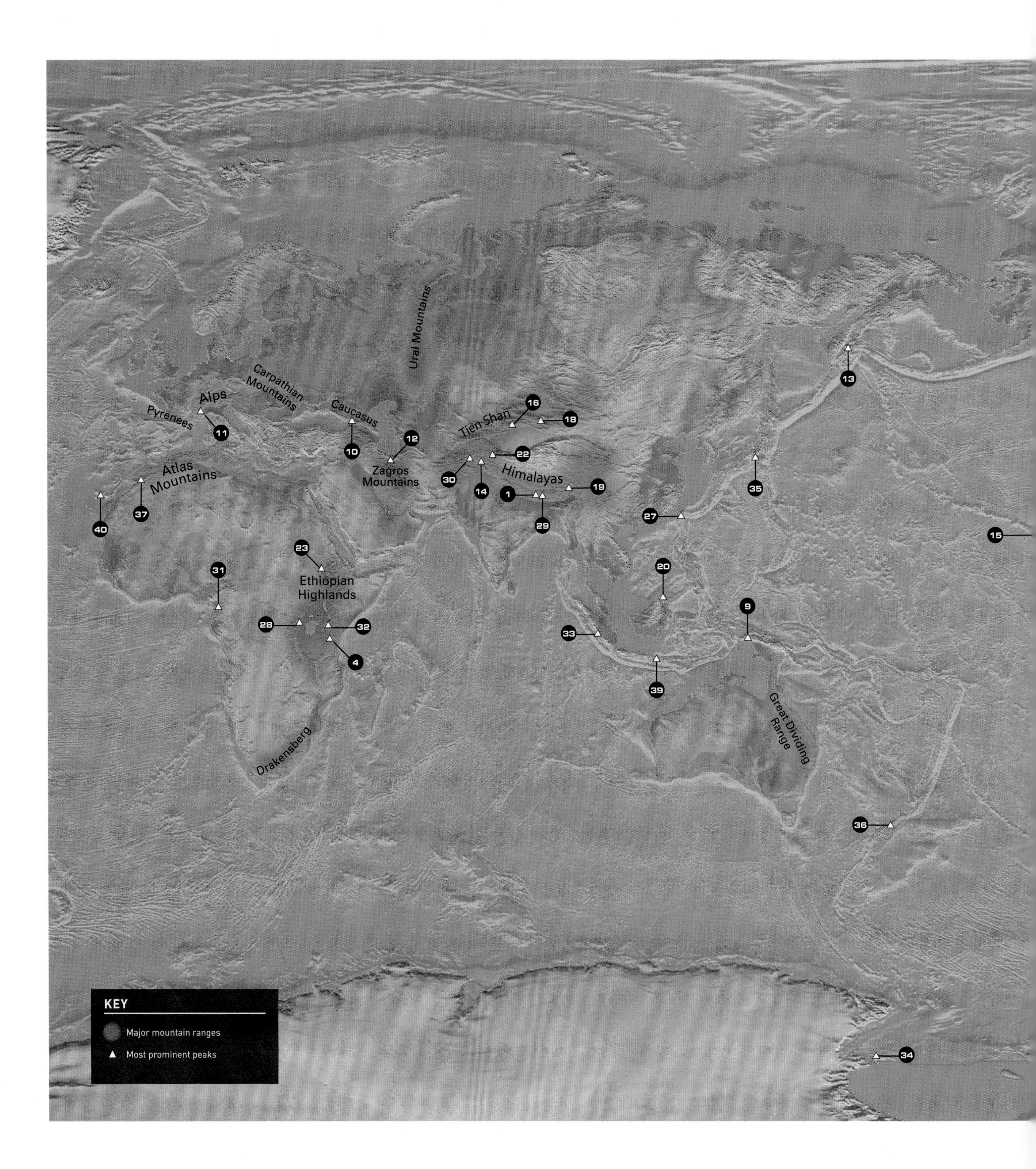

High mountain peaks form pinpricks on the surface of our planet, reaching nearly 6.25 miles (10km) above sea level. Some of these are volcanoes, built up by volcanic eruptions. But many have been pushed up where Earth's crust is being squeezed.

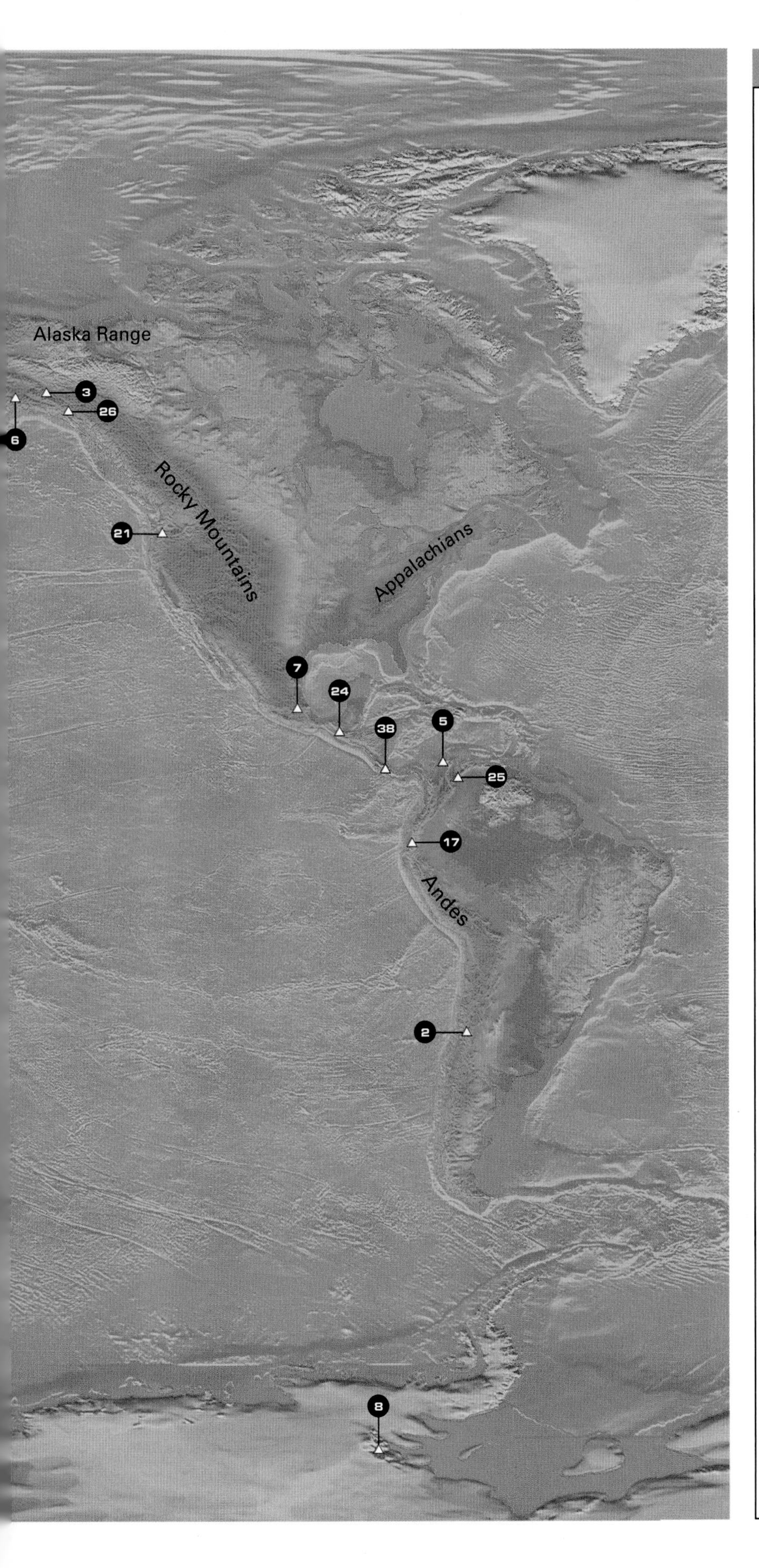

40 MOST PROMINENT PEAKS ON EARTH

1. MT. EVEREST, NEPAL/TIBET
Elevation 29,029ft (8,848m)

2. ACONCAGUA, ARGENTINA
Elevation 22,841ft (6,962m)

3. MT. MCKINLEY (DENALI), US
Elevation 20,321ft (6,194m)

4. KILIMANJARO, TANZANIA
Elevation 19,340ft (5,895m)

5. CRISTOBAL COLON, COLOMBIA
Elevation 18,700ft (5,700m)

6. MT. LOGAN, CANADA
Elevation 19,551ft (5,959m)

7. PICO DE ORIZABA, MEXICO
Elevation 18,619ft (5,675m)

8. VINSON MASSIF, ANTARCTICA
Elevation 16,049ft (4,892m)

9. PUNCAK JAYA, INDONESIA
Elevation 16,023ft (4,884m)

10. GORA ELBRUS, RUSSIA
Elevation 18,510ft (5,642m)

11. MT. BLANC, FRANCE/ITALY
Elevation 15,777ft (4,809m)

12. DAMAVAND, IRAN
Elevation 18,405ft (5,610m)

13. KLIUCHEVSKOI, RUSSIA
Elevation 15,863ft (4,835m)

14. NANGA PARBAT, PAKISTAN
Elevation 26,656ft (8,125m)

15. MAUNA KEA, US
Elevation 13,795ft (4,205m)

16. JENGISH CHOKUSU, CHINA
Elevation 24,406ft (7,439m)

17. CHIMBORAZO, ECUADOR
Elevation 20,702ft (6,310m)

18. BOGDA SHAN, CHINA
Elevation 17,864ft (5,445m)

19. NAMCHA BARWA, CHINA
Elevation 25,531ft (7,782m)

20. KINABALU, MALAYSIA
Elevation 13,435ft (4,095m)

21. MT. RAINIER, US
Elevation 14,409ft (4,392m)

22. K2, PAKISTAN/CHINA
Elevation 28,251ft (8,611m)

23. RAS DEJEN, ETHIOPIA
Elevation 14,836ft (4,533m)

24. VOLCAN TAJUMULCO, GUATEMALA
Elevation 13,845ft (4,220m)

25. PICO BOLIVAR, VENEZUELA
Elevation 16,342ft (4,981m)

26. MT. FAIRWEATHER, US/CANADA
Elevation 15,325ft (4,671m)

27. YU SHAN, TAIWAN
Elevation 12,966ft (3,952m)

28. NGALIEMA, CONGO
Elevation 16,762ft (5,109m)

29. KANGCHENJUNGA, NEPAL
Elevation 28,169ft (8,586m)

30. TIRICH MIR, PAKISTAN
Elevation 25,289ft (7,708m)

31. MT. CAMEROON, CAMEROON
Elevation 13,435ft (4,095m)

32. MT. KENYA, KENYA
Elevation 17,057ft (5,199m)

33. GUNUNG KERINCI, INDONESIA
Elevation 12,484ft (3,805m)

34. MT. EREBUS, ANTARCTICA
Elevation 12,448ft (3,794m)

35. FUJI SAN, JAPAN
Elevation 12,388ft (3,776m)

36. MT. COOK, NEW ZEALAND
Elevation 12,320ft (3,755m)

37. JEBEL TOUBKAL, MOROCCO
Elevation 13,671ft (4,167m)

38. CERRO CHIRRIPO, COSTA RICA
Elevation 12,533ft (3,820m)

39. GUNUNG RINJANI, INDONESIA
Elevation 12,224ft (3,726m)

40. PICO TEIDE, SPAIN
Elevation 12,198ft (3,718m)

THE ROOTS OF MOUNTAINS

Mountains are not just features of Earth's surface. One of the great geological discoveries of the 19th and early 20th centuries was that mountain ranges have deep roots, where Earth's crust is much thicker than normal.

EARTH'S CRUST

The rocks in the continents, exposed at Earth's surface, are made predominantly of the minerals quartz and feldspar. These form grains of sand in sandstones, or interlocking crystals in once molten granites, which are typical of Earth's crust. Rocks from deep within the crust can be found in ancient mountain belts, where they have been exposed by erosion. However a very different type of rock called peridotite, largely made up of the dark colored minerals olivine and pyroxene, is brought up from depths of several tens of miles in volcanic eruptions. These rocks come from Earth's mantle, the part of the interior that lies beneath the crust. The boundary between the crust and mantle is called the Mohorovičić Discontinuity, or Moho, named in honor of the Croatian geophysicist Andrija Mohorovičić, who first discovered it while studying how earthquake vibrations move through the crust and mantle.

MAPPING THE ROCKS

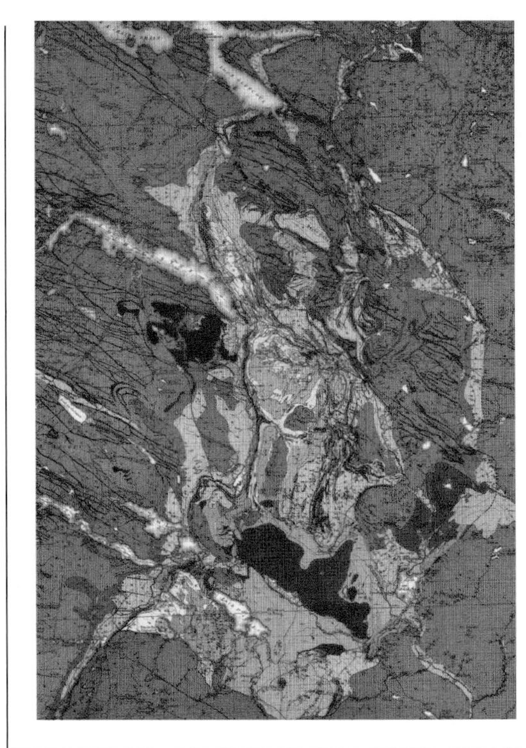

GEOLOGICAL MAP
Geologists have carefully recorded the rocks in the continents, making geological maps in which the different rock types are shown as different colors. This map of the Assynt region of Scotland was made by geologists at the end of the 19th century, and revealed for the first time the detailed nature of the rocks in the heart of an ancient mountain range. The pink colors represent rocks that formed billions of years ago deep in Earth's crust. The yellow, blue, and green colors show rocks that have been displaced by geological movements during mountain building in this part of Scotland more than 400 million years ago.

ANCIENT ROOTS
The deeply eroded landscape of the Isle of Lewis in Scotland's Outer Hebrides is underlain by rocks that crystallized at high temperatures and pressures in the deep roots of an ancient mountain range, nearly 3 billion years ago.

ICEBERG MOUNTAINS

One of the most important ideas in geophysics is that rocks in Earth's crust and mantle sink or rise depending on their density. Thus, the relatively light rocks that make up Earth's crust are effectively floating on the underlying denser mantle rocks. Just like an iceberg floats above the sea because it has deep icy roots, so mountain ranges must have deep roots in the crust. As a result, the elevation of a mountain range is determined by the thickness of the underlying crust—a theory referred to as "isostasy." English astronomer George Airy first proposed this idea in 1854, after studying the pull of gravity at the edges of the high Himalayas. The presence of a thick root of low density crust beneath these mountains has reduced sideways the pull of gravity here compared to that in the lowland plains of India, where there are higher density mantle rocks at relatively shallow depth.

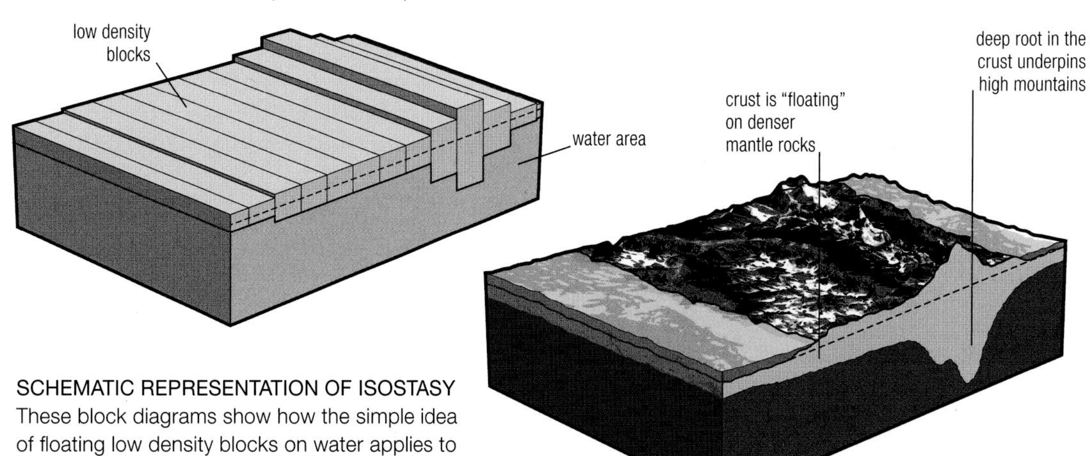

SCHEMATIC REPRESENTATION OF ISOSTASY
These block diagrams show how the simple idea of floating low density blocks on water applies to the crust beneath a mountain range. Just like the floating blocks, the mountains rise higher where their crustal roots are deeper.

GRAVIMETER
Gravity surveys are a routine way of looking inside Earth. Measurements from a gravimeter reveal minute variations in the pull of gravity, caused by variations in the density of the rocks.

PROBING THE EARTH'S CRUST

Geophysicists can look deep into the crust by measuring small changes in the pull of gravity. Gravity is affected by the density of rocks, and so if rocks are less dense than normal, the local force of gravity is reduced. These measurements are made with extremely delicate instruments, called gravimeters, which consist of carefully balanced weights and springs. In this way, very small changes in gravity can be measured. Measurements in major mountain belts, such as the Andes or Himalayas, show that the pull of gravity here can only be explained if the crust has a deep root, extending up to 50 miles (80km) beneath the surface, compared with a more usual depth of about 22 miles (35km) beneath the adjacent lowland plains.

HOW MOUNTAINS FORM

Mountains are not simply a feature of Earth's surface. They are the result of processes that have acted to deform the Earth's crust at a variety of depths below the surface, especially near converging boundaries between plates.

THRUST FORMATION

Ever since geologists first started studying the rocks in mountainous regions, they realized that they were twisted and fractured, cut by great fault lines and rucked up to form corrugations or folds. The world's highest mountains, in the Andes and Himalayas, show evidence for intense squeezing of the crust, pushing the mountains up along reverse or thrust faults, with rock layers stacked on top of each other. The thrust formation process causes a profound change to the nature of crust: by squeezing the layers in a horizontal direction, the crust thickens in a vertical direction. It is this thickening that is the key to forming high mountain ranges.

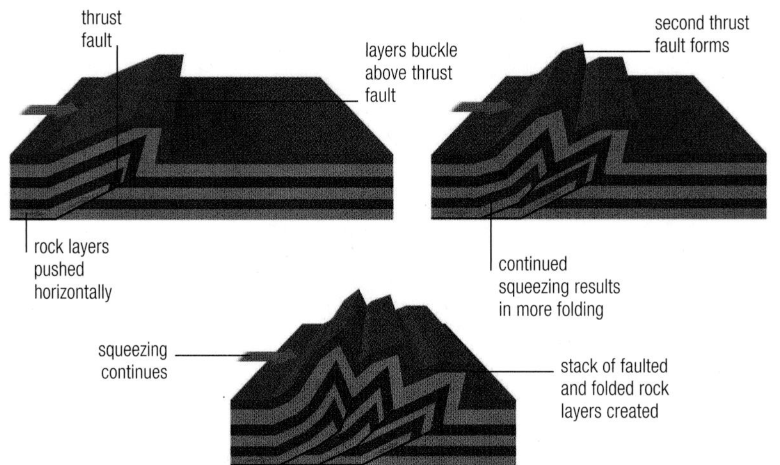

THRUST FORMATION
When layers of rock are compressed, they will tend to break along gently inclined faults, called thrust faults, as well as ruck up to form folds. As the squeezing continues, the rocks are progressively stacked up. Due to erosion, mountain ridges do not always coincide with the crests of folds.

THE PALMDALE CUT
Folded and faulted layers of sandstone and shale are spectacularly exposed on the site of the Antelope Valley Freeway, close to the San Andreas Fault in California.

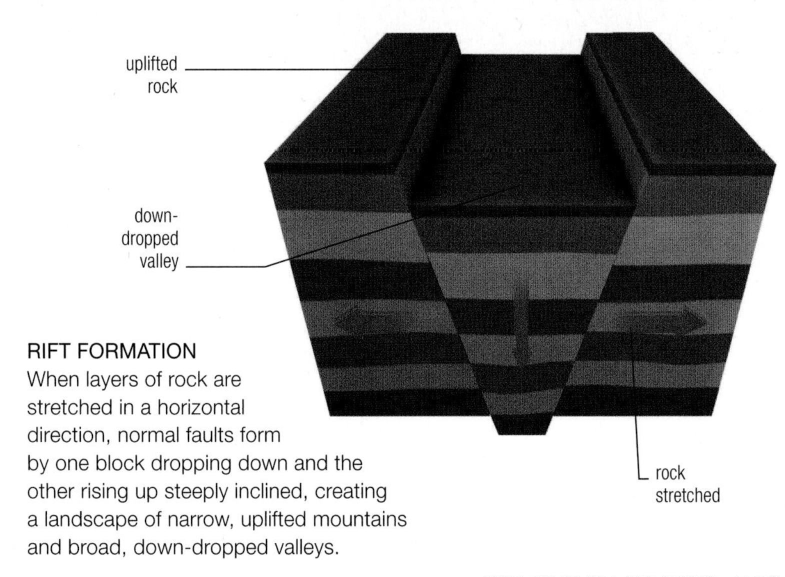

RIFT FORMATION
When layers of rock are stretched in a horizontal direction, normal faults form by one block dropping down and the other rising up steeply inclined, creating a landscape of narrow, uplifted mountains and broad, down-dropped valleys.

RIFTING

Narrow uplifted mountains may also form when the crust is stretched and rifted, as the tectonic plates move apart. Here, movement on steeply inclined faults, called normal faults, results in one block rising up, but with the other block dropping down to form a deep valley or rift. Because this process stretches and thins the crust, the region overall subsides, eventually forming a drowned landscape such as the Aegean region of Greece.

TRE CIME DI LAVAREDO, ALPS
These jagged limestone peaks in the Italian Dolomites have been pushed up by movement on thrust faults as part of the formation of the European Alps.

FOLDING

When the rock layers are squeezed, they are bent to form folds, defining what geologists called synclines (forming troughs) and anticlines (forming arches or crests). Folding usually occurs at the same time as movements on the faults. The horizontal distance between neighboring fold crests and troughs defines the fold wavelength, which can vary in a typical mountain belt from less than an inch to hundreds of miles. The very long wavelengths result in an almost imperceptible tilting or broad doming of the surface. Geologists believe this is often caused by upwelling of hot and buoyant mantle rocks at depth.

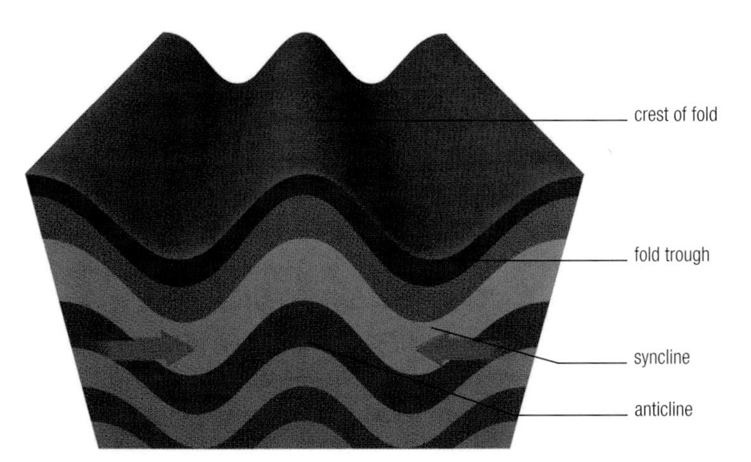

FOLD FORMATION
If layers of rock are underlain by weak rocks at depth, they will form simple corrugations when squeezed laterally. Sometimes the land surface is folded in the same way, although erosion will wear it down.

SIERRA NEVADA
The snow-capped Sierra Nevada Mountains formed where the crust is rifting. It has been suggested that these mountains are underlain by hot and buoyant mantle rocks, allowing them to reach their present high elevation.

EROSION AND WEATHERING

The creation of a mountain range leads directly to profound interactions between the rocks and atmosphere. At high elevations, snow accumulates to eventually form glaciers. As these slide downhill they scrape the bedrock, carving out valleys. At lower elevations, the water drains off the mountains in raging torrents, carrying some of the bedrock with it to the surrounding plains, and eventually the sea. These processes expose once deeply buried rock at the surface, which starts to weather, undergoing chemical alteration. Planes of weakness in the rock layers are worn away, creating a landscape of ridges in arid regions, where the rocks are devoid of vegetation.

KUQA CANYON, CHINA
These steeply inclined layers of sandstone form part of major folds in the crust that are being rapidly worn away by the forces of erosion, creating a landscape of planes and ridges.

MOUNTAINS ON THE MOVE

The boundaries between Earth's plates are alive with geological activity, rocked by earthquakes and shifting along fault lines. This results in profound changes to the landscape, creating deep depressions or towering mountain ranges.

ACTIVE TECTONICS

Created by the forces that drive the tectonic plates, mountain ranges are pushed up where adjacent plates converge. The mountains themselves are also on the move, which causes pervasive straining of the rocks. Eventually the rocks break along fault lines, creating earthquakes. Geologists can observe these movements by studying the evidence in the landscape, such as the displacement of river channels or beaches uplifted over thousands of years.

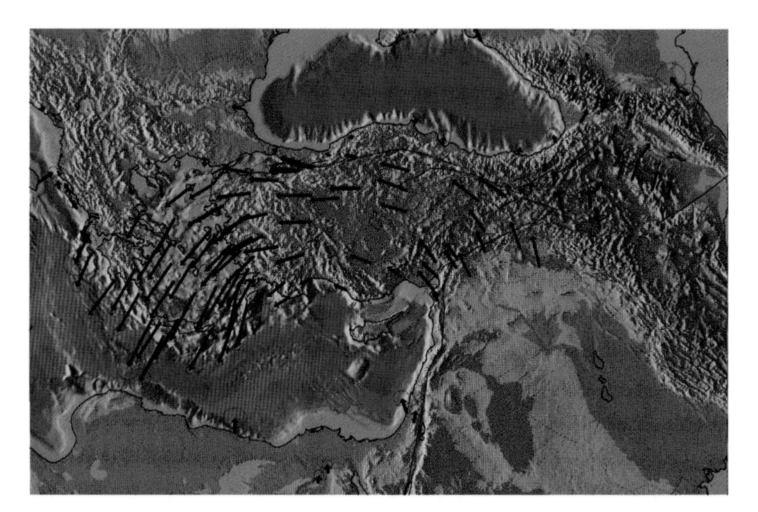

MEASURING PLATE MOVEMENT
Measurement instruments in Turkey and the Aegean Sea reveal movement of the crust in a vast region between the Eurasian, Arabian, and African plates.

NEW ZEALAND COASTLINE
This flat-topped headland is the remains of an old shoreline that existed tens of thousands of years ago. Since then, the land has been uplifted by 66ft (20m) during a succession of earthquakes, creating the rugged landscape in this part of New Zealand.

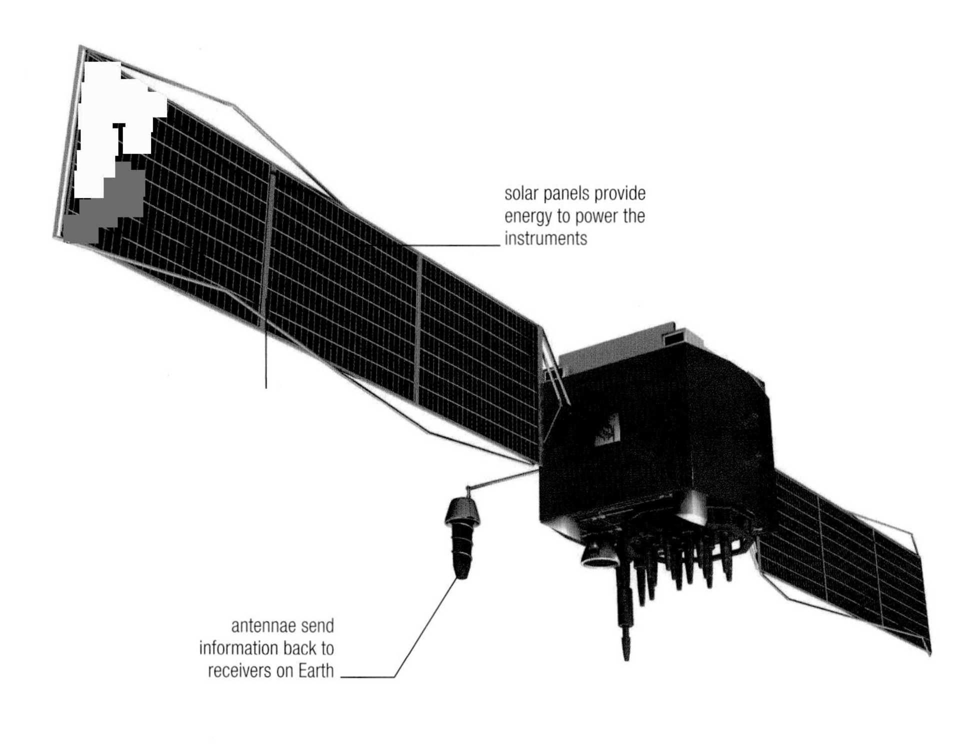

GPS SATELLITE
The global positioning system is based on a network of satellites that orbit Earth, sending radio signals down to receivers on the planet's surface. A GPS measurement is made by timing the radio signal to determine the distance between the receiver and several satellites.

GPS SURVEYING

Since the 1980s, accurate measurements of Earth's surface have been made using the global positioning system (GPS). GPS is routinely used today by walkers and motorists to navigate, and with only a small handheld instrument it is possible to locate oneself to within a few feet. But by taking many measurements over a period of hours, using an antenna and a more sophisticated instrument, it is possible to work out positions of survey points much more accurately. These measurements, repeated over several years, are now used to record the longer-term motion of Earth's surface, revealing the flow of the crust in the tectonically active regions. Following earthquakes and tsunamis, areas can be surveyed to see how far the land has shifted and to help predict future movement.

MEASURING A FAULT
This geophysicist is setting up the antenna for a GPS receiver so that she can accurately monitor the position of the ground station and measure the vertical and horizontal movement of the land after the 1994 Northridge earthquake in California.

MEASURING MOVEMENT

Developed in the 1990s, interferometry is a new method of accurately measuring the movement of the land surface using radar images from space. A satellite orbits Earth, sending and receiving accurately timed radar signals. This way, an image of the shape of the landscape can be created. By comparing radar images taken over a period of years it is possible to detect shifts in the landscape as small as a few inches, and to show these shifts as an interferogram image.

CALIFORNIA INTERFEROGRAM
The colors in this image represent the amount of displacement of the crust (increasing from blue to red) across the Hayward Fault in California (marked by the thin red line) over a period of several years.

MEASURING THE GROWTH OF MOUNTAINS

One way geologists measure the rise of a mountain range over millions of years is to study fossil plants preserved in sedimentary rocks high up in the mountains. By observing how plants cope with high altitudes today, it is possible to estimate the elevation at which the fossil plants originally grew. Temperature and humidity generally decrease with elevation, and plants have responded by modifying the shapes and sizes of their leaves. Geologists can use a fossil leaf to determine the change in elevation of a mountain range through time. Studies like this have shown that the Central Andes of South America have risen about 1.2 miles (2km) in the last 10 million years.

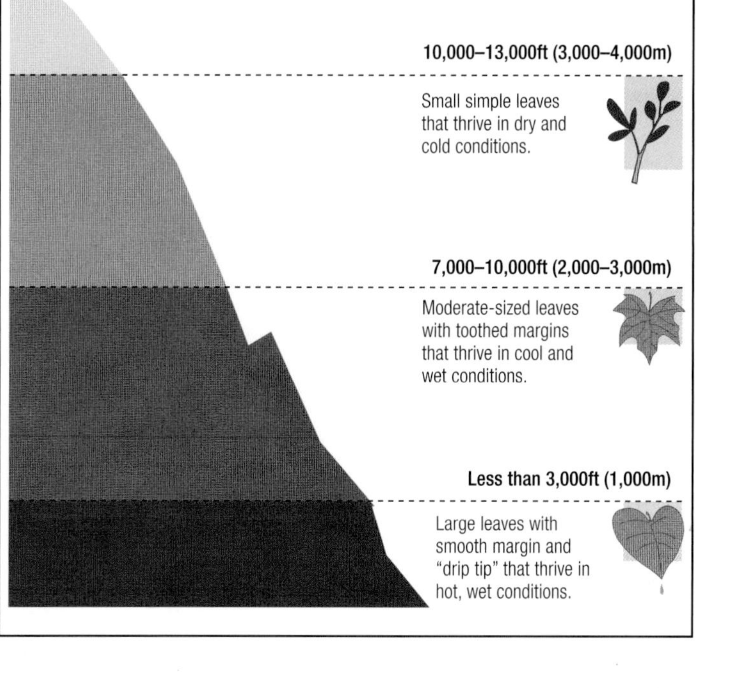

LIFECYCLE OF MOUNTAINS

Mountain ranges are not fixed and static features of our planet. Almost like living organisms, they are born, grow old, and then die. The typical lifespan of a mountain range is tens of millions of years, forming part of a more fundamental cycle, controlled by the movements of the tectonic plates.

MOUNTAINS OVER TIME

The lifecycle of a mountain range is part of the longer history of the birth and death of an ocean, controlled by the movements of the tectonic plates. This has been called the Wilson cycle after the Canadian geophysicist J. Tuzo Wilson who first described it. The continents are part of the plates and move as one with them. So, the Wilson cycle also involves the rifting and then collision of continents, thickening the crust and pushing up mountain ranges.

AUSTRALIAN OUTBACK
Ayers Rock is the deeply eroded roots of an ancient and long dead mountain range. The relatively flat landscape of western Australia has formed over hundreds of millions of years, unaffected by mountain building and smoothed out by the forces of erosion.

NEW ZEALAND'S SOUTHERN ALPS
The rugged peaks of South Island, New Zealand, are part of a very young mountain range that has formed in the last 10 million years. These mountains are active today, rocked by earthquakes along the many fault lines in the landscape.

1 RIFTING OF CONTINENTS
Divergent plate motion can sometimes split a continent apart. This begins with a rift zone, often where the continent has been warped into a broad dome by a warm plume rising in the mantle. In the early stages, a deep rift valley, like the East African Rift Valley, is created in the stretched crust.

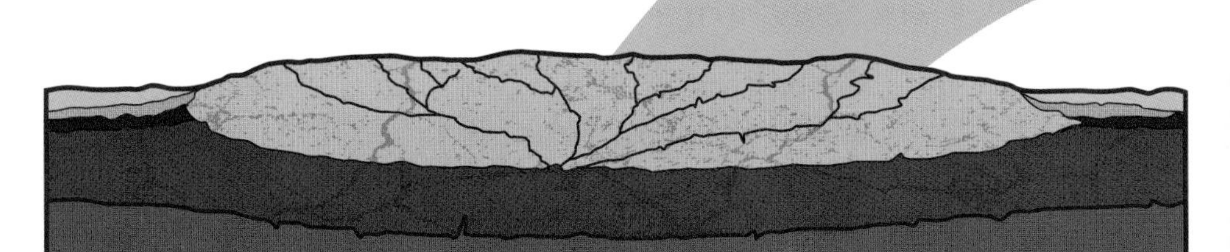

6 MATURE OCEAN
Finally, a new continent is created as the continental margins are welded together, and relative plate motions cease. Over geological time, erosion will wear down the mountain range to create a stable continental interior close to sea level.

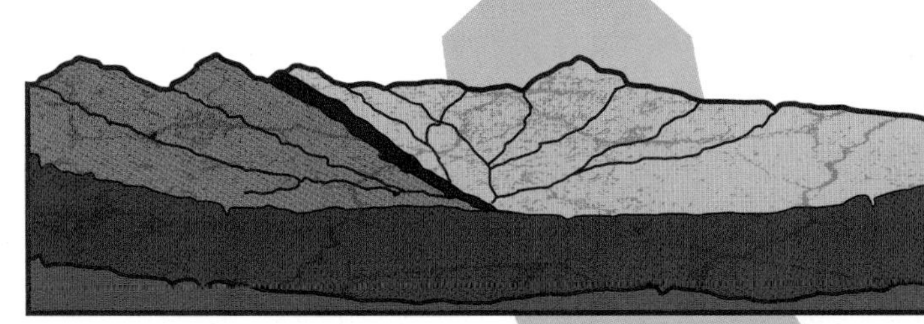

5 CONTINENTAL COLLISION
When no more ocean floor is left, the two continental margins collide in the early stages of continental collision, squeezing and thickening the crust over a wide region, creating a large mountain range, such as the Himalayas.

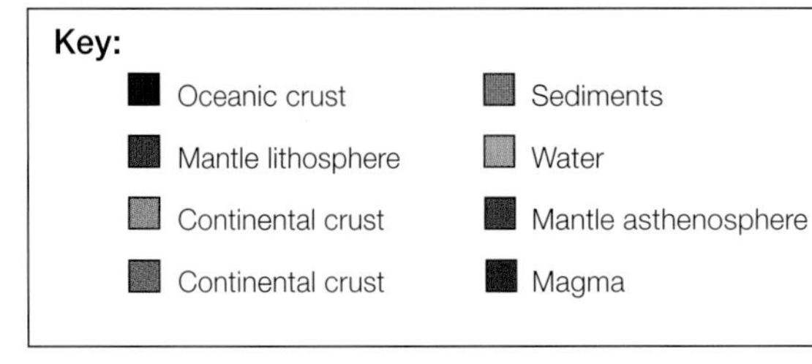

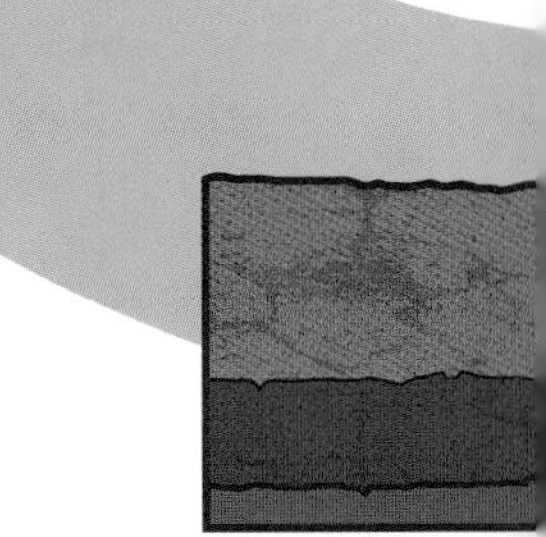

AFAR REGION
The volcanically active region of the Afar in Ethiopia reveals a landscape created during the very early stages of the birth of an ocean, as the tectonic plates begin to rift apart. However, the unusually large amount of volcanic rock is most likely because this region is underlain by a deep mantle plume rising up beneath the crust in the mantle.

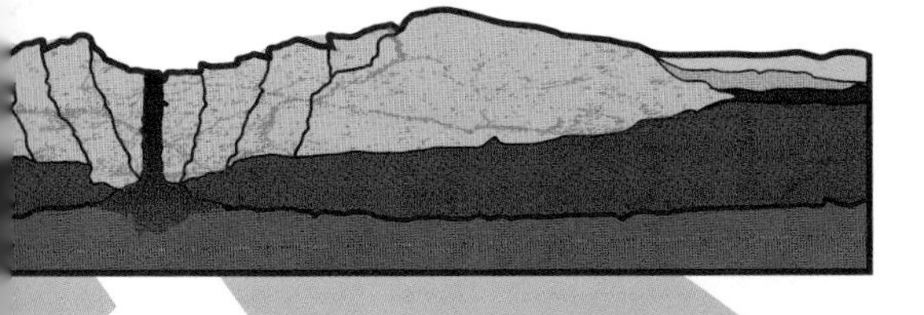

2 YOUNG OCEAN
As the plates continue to move apart, the floor of the rift system will eventually sink below sea level, to be flooded by ocean. Ocean floor is created by the intense volcanic activity, as the underlying mantle wells up and erupts at a mid-ocean ridge. Over time, the ocean will widen from a narrow ocean basin like the Red Sea to a wide ocean like the Atlantic, with the rifted halves of the continent on either side.

3 MATURE OCEAN
Eventually, as the ocean floor cools, it breaks free from one or both continental margins, to make a subduction zone. Here, the subducting ocean floor releases water back into the mantle, causing it to melt and triggering volcanic activity. The subducting plate also rubs against the overlying continent, squeezing and compressing, triggering earthquakes and starting mountain building, such as in the Andes.

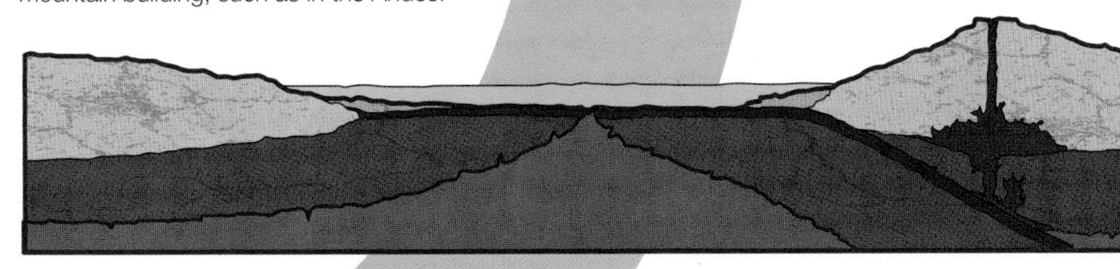

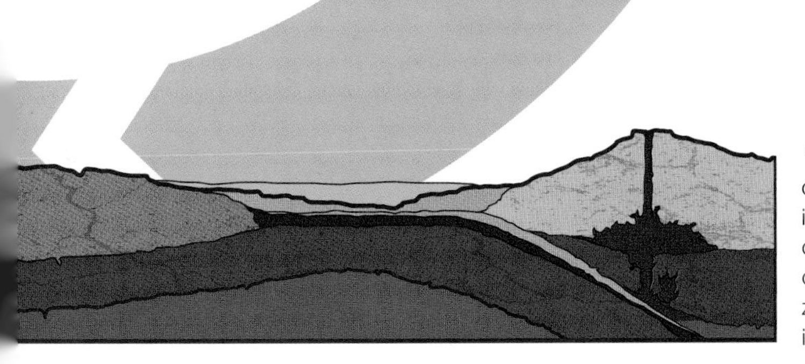

4 SUBDUCTION
If the plate motions change, driven by deep forces inside Earth, the two sides of the ocean will begin to converge as the subduction zone consumes the ocean floor, including the mid-ocean ridge.

THE ROCK CYCLE

Over the billions of years of geologic time, repeated Wilson cycles have helped to create the continents. The ancient continental interiors, well away from today's high mountain ranges, formed deep in the roots of long dead and ancient mountains that have been completely worn away by the forces of erosion—washed out to sea to be deposited in the oceans. Later plate convergence may eventually push these sediments up above sea level, to be eroded and carried to the sea once more. And so the Wilson cycle is also the cycle of rocks in the outer part of Earth, as they form and are then carried deep into the crust, to eventually be uplifted and eroded once again.

RED SEA COAST
The Red Sea is a long, narrow body of water, created by the rifting of the Arabian Plate with Africa. As the plates move apart, the rift zone subsides, eventually to be flooded by the neighboring body of water. The sea floor is created when mantle rocks rise up and melt, erupting in undersea volcanoes. So far, the movements of the plates have resulted in a narrow sea only 220 miles (355km) at its widest.

NORTH AMERICA, EAST COAST
Over time, if the tectonic plates continue to move apart, a wide ocean such as the Atlantic will form. The margin of this ocean slowly subsides, as rock detritus brought down by rivers from the interior of the continent is deposited on the continental shelves and carried out to sea.

ROSY HUE

China's Gansu Province is home to landscapes of reddish sandstone, eroded over time to form rolling cliffs and unusual freestanding pillars and towers. These landforms occur throughout southern China and are called danxia, meaning "rosy clouds," in reference to the rich hues of the exposed rock.

HIMALAYAS

The Himalayas are home to the planet's highest peaks. This mountain range makes up the southern fringe of a vast uplifted area that extends for thousands of miles through Tibet and much of central Asia, pushed up during the last 55 million years by the northward collision of India with Asia.

BIRTH OF A MOUNTAIN RANGE

One hundred and twenty-five million years ago the Indian continent lay thousands of miles further south, on the other side of the Tethys Ocean, where it formed a small piece of the supercontinent of Gondwana. When Gondwana rifted apart, the Tethys Ocean began to shrink as India drifted northward. By 55 million years ago, India made contact with the southern margin of what is today central Asia. Since then, India has pushed its way a further 1,240 miles (2,000km) northward, squeezing and thickening the Asian continental crust. This is how the vast ranges of central Asia were created. The Himalayas are the southern margin of this collision zone, where the Indian subcontinent is colliding with and sliding underneath the Asian continent along a gigantic fault, pushing the mountains above ever higher. Glaciers have carved out many of the distinctive ranges in the broken crust above this fault.

MOUNTAIN PEAKS

The mountains of the Great Himalayas soar to more than 29,000ft (8,800m) above sea level. They are formed of Upper Paleozoic rocks, slices of the Tethys Ocean floor, and granite intrusions that have been thrust over early Cenozoic sediments. The highest peaks are covered with permanent snow and glaciers.

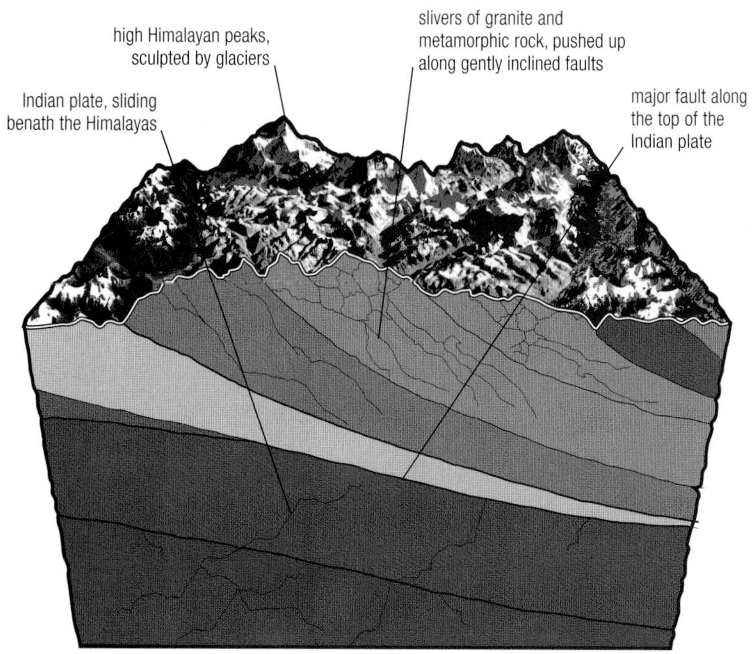

CONTINENTS COLLIDING
A gigantic fault line runs along the southern edge of the Himalayas. The fault slopes gently down toward the north beneath the highest peaks.

VIEW FROM THE TOP
Makalu, the fifth-highest mountain in the Himalayas, rises above the clouds on the border between Nepal and China. The third-highest peak, Kangchenjunga, can be seen on the horizon.

TIBETAN PLATEAU

North of the Himalayas lie the Tibetan Plateau and ranges of central Asia, such as the Karakoram and Tien Shan mountains, which contain many of the world's highest peaks. These mountain ranges were created in the same continental collision that pushed up the Himalayas. In the process, rocks that lay deep within the crust were uplifted and are now exposed at the surface. The plateau covers an area of 965,000 sq miles (2.5 million sq km) and has a remarkably uniform elevation of about 3 miles (5km). The region is cut by vast, strike-slip faults that extend for hundreds to thousands of miles.

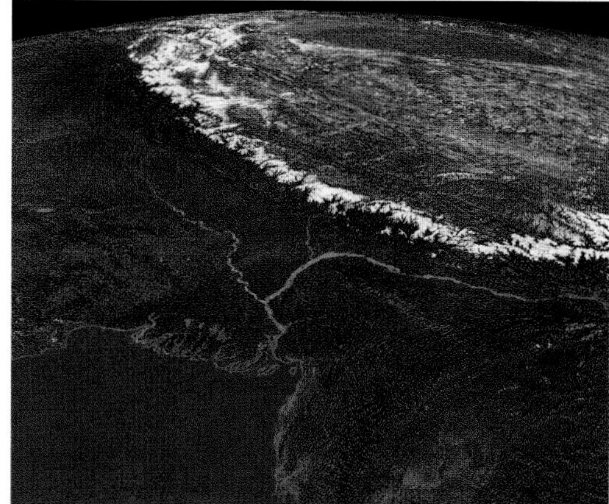

HIMALAYAS FROM SPACE
This satellite view of the Himalayas (in white) shows part of the huge Tibetan Plateau that lies north of the mountain range.

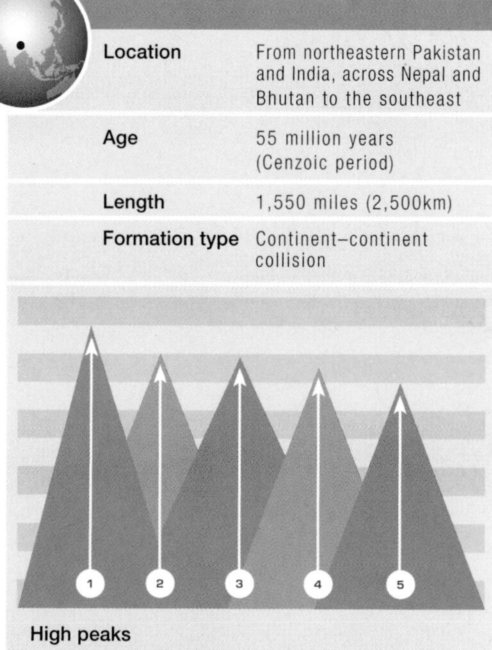

Location	From northeastern Pakistan and India, across Nepal and Bhutan to the southeast
Age	55 million years (Cenzoic period)
Length	1,550 miles (2,500km)
Formation type	Continent–continent collision

High peaks

1. Mount Everest 29,035ft (8,850m)
2. K2 28,251ft (8,611m)
3. Kangchenjunga 28,169ft (8,586m)
4. Lhotse 27,939ft (8,516m)
5. Makalu 27,762ft (8,462m)

THE SOUTHERN ALPS

Running along the western side of South Island, New Zealand's Southern Alps are relatively young. Formed only 10 million years ago, these mountains are rapidly growing where two tectonic plates are converging and sliding past each other.

RAPID RISE

The islands of New Zealand lie on the boundary between two giant tectonic plates—the Pacific and Australian plates—that are moving toward each other by 1.5in (4cm) every year. The landscape here is shifting along many fault lines, which extend for hundreds of miles.

In South Island, the plates comprise two different continents that are sliding past each other along the major Alpine Fault, which runs right down the west coast. As this happens, the rocks to the east of the Alpine Fault are being rapidly pushed up at the rate of about 0.4in (10mm) per year, creating the snow-capped peaks of the Southern Alps. A deep mountain root is also forming here as the crust is being squeezed. In the very wet climate, rivers are cutting deep into the mountains, carrying vast amounts of rock debris to the west coast, to be deposited offshore in the Tasman Ocean. In this way, rocks that were once buried 12 miles (20km) below the surface are now exposed.

To geologists, the Southern Alps are a mountain range in their early childhood, created by changes in the motions of the tectonic plates in the last 10 million years. In that time, the tectonic plates have slid past each other several hundred miles. The mountains' rugged peaks reveal the ever-present battle between the forces deep in the Earth that are pushing them up, and the forces of erosion wearing them down.

MOUNT COOK
The highest mountain in the Southern Alps is Mount Cook, which rises 12,316ft (3,754m) above sea level. In Maori it is known as Aoraki.

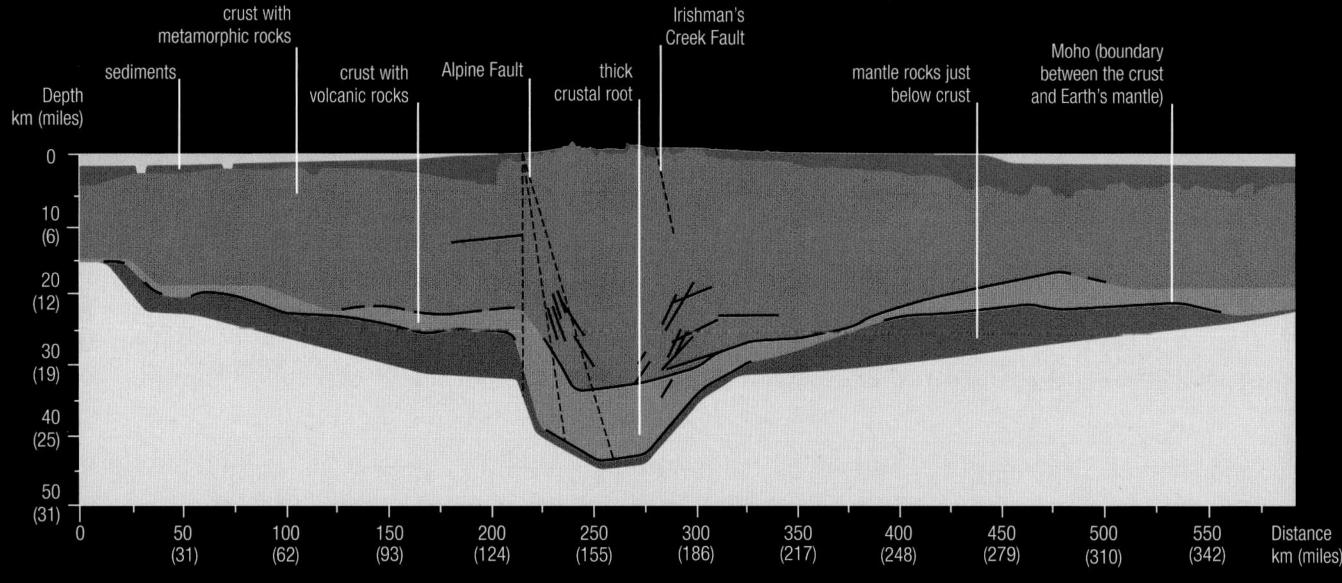

THICK CRUST

Geophysicists have probed the crust beneath New Zealand's Southern Alps by listening to the seismic vibrations created by earthquakes and explosions. By accurately timing the arrival of these vibrations at various points in New Zealand, it is possible to work out the thickness of the crust the vibrations have traveled through. This has shown that the crust here has deep roots, reaching about 30 miles (45km) beneath the highest mountains.

LAYERS IN THE CRUST
The speed of seismic vibrations, as they travel through the crust, show that there are layers of sediments and metamorphic and volcanic rocks (shown here by the different colors), revealing both a detailed picture of the crust and the complex geological history of this region.

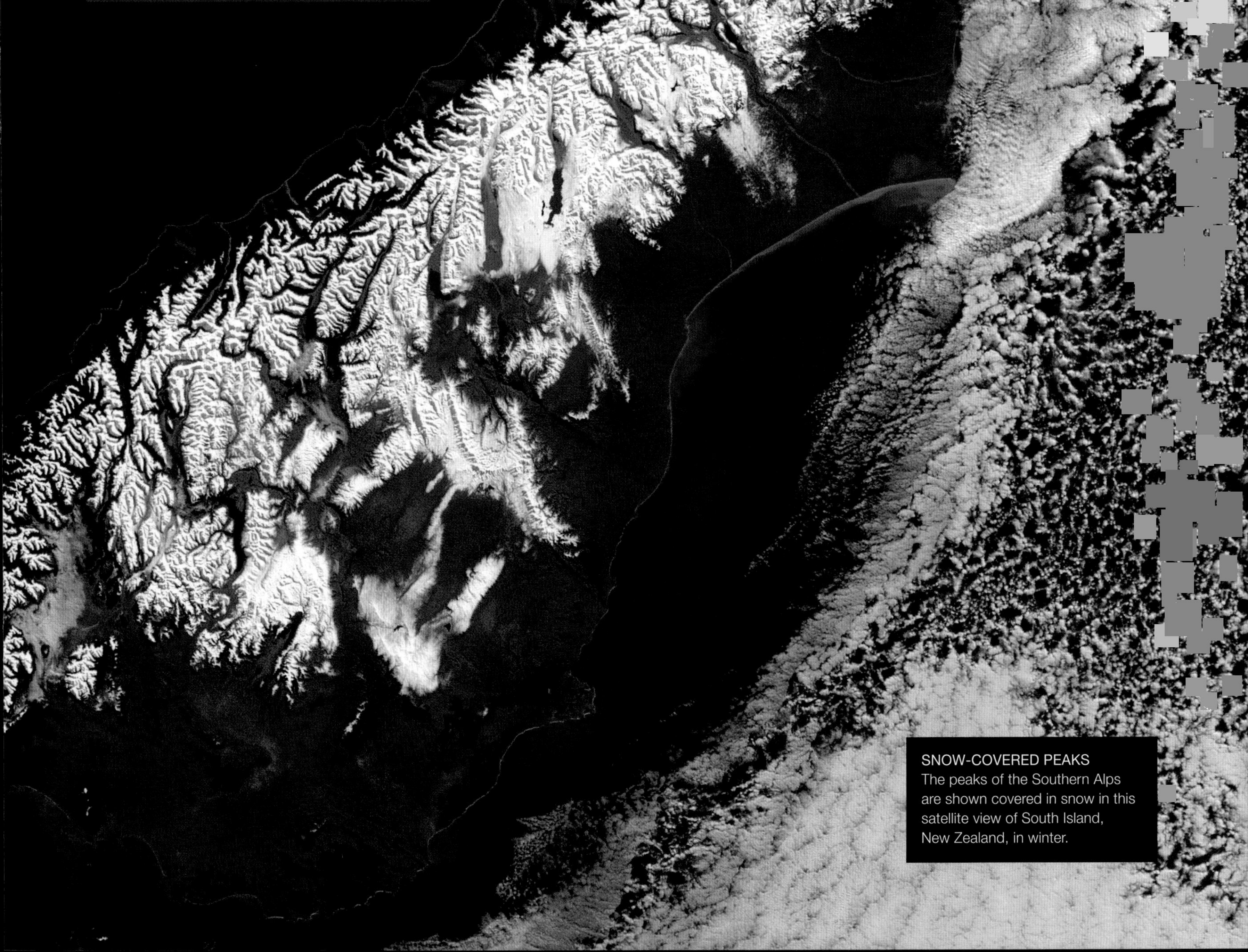

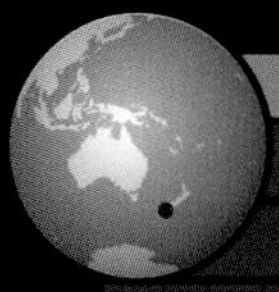

SOUTHERN ALPS

Location	South Island, New Zealand
Age	10 million years
Length	435 miles (700km)
Formation type	Continental collision

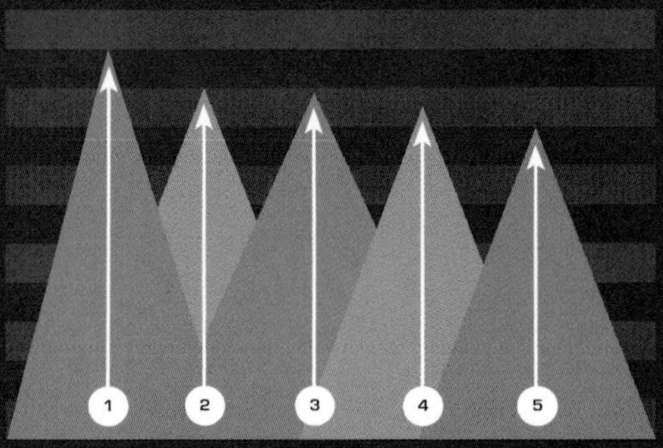

High peaks

1. Mount Cook 12,316ft (3,754m)
2. Tasman 11,473ft (3,497m)
3. Dampier 11,286ft (3,440m)
4. Silberhorn 10,758ft (3,279m)
5. Lendenfeld 10,502ft (3,201m)

SNOW-COVERED PEAKS
The peaks of the Southern Alps are shown covered in snow in this satellite view of South Island, New Zealand, in winter.

THE ANDES

Extending down the western side of South America, the Andes are Earth's longest continuous mountain range. Here, earthquakes and intense volcanic activity show that immense geological forces are at work, continuing to build up the mountains today.

FORMATION

In geological terms, the Andes are relatively young. The region was still a vast inland sea or lake when the last of the dinosaurs were alive. Since then, over the past 25 million years, the mountains have gradually risen. They are still growing, nearly doubling their height in the last 10 million years. They owe their existence to the action of the Pacific sea floor, as it rubs against the South American plate in a subduction zone, slipping back into Earth's interior. This plate convergence has squeezed and thickened the crust, pushing up the mountains along gently-inclined thrust faults. The intense volcanic activity shows that there is molten rock at depth, and this is also adding to the crust when it cools and crystallizes.

183
The number of active volcanoes in the Andes, which extend over seven coutries and are divided into the Southern Andes, Central Andes, and Northern Andes.

THE ANDES	
Location	Western side of South America
Length	4,500 miles (7,200km)
Age	25 million years
Type	Subduction

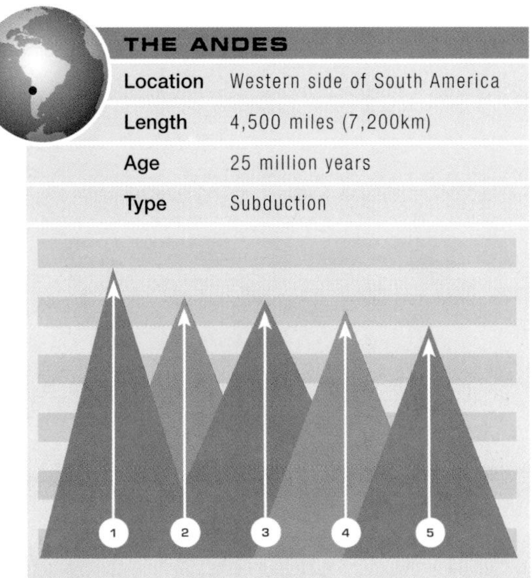

High peaks
1. Aconcagua, Argentina 22,831ft (6,959m)
2. Ojos del Salado, Argentina–Chile 22,595ft (6,887m)
3. Pissis, Argentina 22,300ft (6,795m)
4. Bonete, Argentina 22,175ft (6,759m)
5. Tres Cruces Sur, Argentina–Chile 22,139ft (6,748m)

ATACAMA DESERT
The barren and dry Atacama region lies on the western flanks of the high Andes in northern Chile. It is so dry because of the "rain shadow" caused by the high Andes.

PLATEAU IN THE ANDES
The Altiplano, a high plateau within the widest part of the Andes, is a 372-mile (600-km) long, 80-mile (130-km) wide region. Lake Titicaca (seen in this satellite image as dark blue and cloud covered), is the world's highest navigable lake.

CORDILLERA REAL
Snow-capped Illimani rises more than 21,000ft (6,400m) tall, towering above La Paz, the capital of Bolivia. It is part of the Cordillera Real, a subrange of the Andes, and is made of once-molten granite that cooled about 25 million years ago.

ALTIPLANO PLATEAU

The Andes span great extremes of climate, flanked by the lush Amazon jungle in the east and the arid Atacama desert along the coast. The interior of the Central Andes is a vast plateau, about 13,000ft (4,000m) high, hemmed in by the snow- and ice-capped peaks of the surrounding cordilleras. This is a harsh environment with only half as much oxygen as at sea level. Yet here can be found the sparkling blue waters of Lake Titicaca, the highest navigable lake in the world.

MINERAL WEALTH

The Andes are part of the "Ring of Fire" (see pp.86–87). Numerous hot springs and fumaroles testify to the presence of very hot rock not far below the ground. The volcanic activity is caused by seawater, taken deep into Earth's mantle down the subduction zone offshore, melting the rocks, rather like adding salt to ice. Eventually the water-rich magma rises to erupt at the surface. But there is a valuable dividend, because as the molten magma cools, water is driven off into the surrounding rocks, carrying gold, silver, tin, and copper. This way, the Andes have been endowed with immense mineral wealth.

HOT SPRINGS
Steam rising from hot water at El Tatio, heated by volcanic activity deep beneath Andean volcanoes in Chile.

> "THE PRESENT IS THE KEY TO THE PAST."
>
> **CHARLES LYELL**, GEOLOGIST, IN *PRINCIPLES OF GEOLOGY*, 1830–33

TRANSVERSE RANGES

Just to the north of the vast sprawling city of Los Angeles, complex motions of the crust are pushing up mountains. Here, the Transverse Ranges are caught up in the boundary between the Pacific and North American plates.

CALIFORNIAN RANGES

California lies on the boundary between two tectonic plates: the Pacific Plate is sliding approximately to the northwest, at about 2in (45mm) per year relative to the North American Plate. Most of this motion is taken up along great fault lines that run the length of California, such as the San Andreas Fault, where the Pacific Plate is slipping sideways on right lateral strike-slip faults. However, just north of Los Angeles, where the San Andreas Fault shows a pronounced kink, fault lines have a different orientation, running nearly east–west. Here, in the last 20 million years, slivers of crust that straddle the major strike-slip faults further north and south have been squeezed, pushed up, and twisted around. These movements have created the Transverse Ranges, such as the San Gabriel and San Bernardino Mountains. Much of the bedrock is made of granitic rocks intruded into the crust more than 65 million years ago, but sandstones and shales, laid down a few million years ago, are also being folded and forced upward. Notable recent evidence of this occurred during the 1994 Northridge earthquake in the northern suburbs of Los Angeles itself.

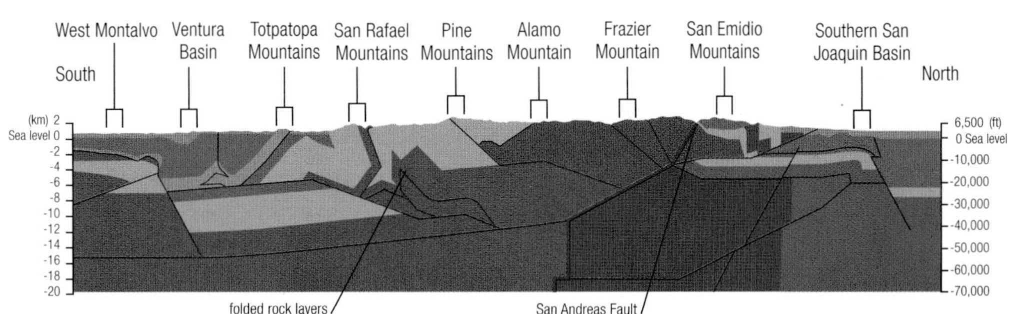

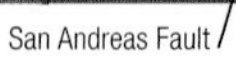

GEOLOGY OF THE RANGE
A cross-section view of the rocks in the Transverse Ranges reveals the gently inclined reverse faults along which the mountains are rising. The rocks themselves are mainly sandstones, shales, and granitic rocks. The oldest sedimentary rocks in the cross section are about 100 million years old, and the igneous units are mostly older.

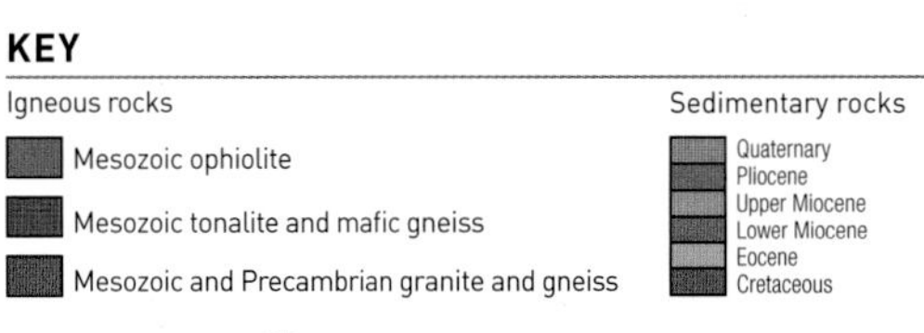

CUTTING ACROSS THE GRAIN

Geologists have long speculated why the Transverse Ranges cut across the grain of the major northwest fault lines in California. One idea is that an old subduction zone underlies the mountains, forming a sort of rudder. Twenty million years ago this rudder was aligned more parallel to the other major fault lines. But the movement of the tectonic plates has twisted this rudder around, taking blocks of the overlying crust with it into their present east–west orientation.

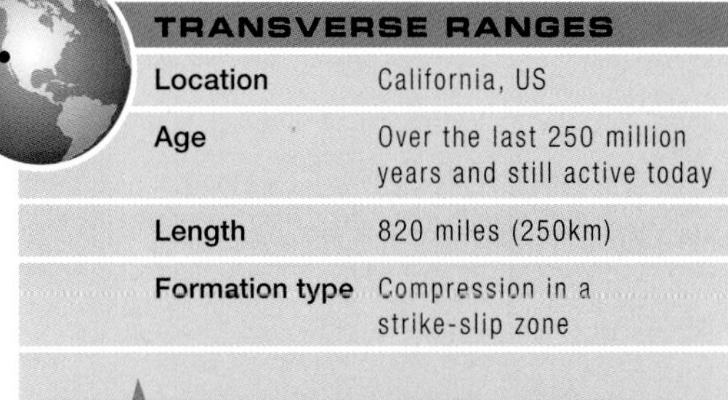

TRANSVERSE RANGES	
Location	California, US
Age	Over the last 250 million years and still active today
Length	820 miles (250km)
Formation type	Compression in a strike-slip zone

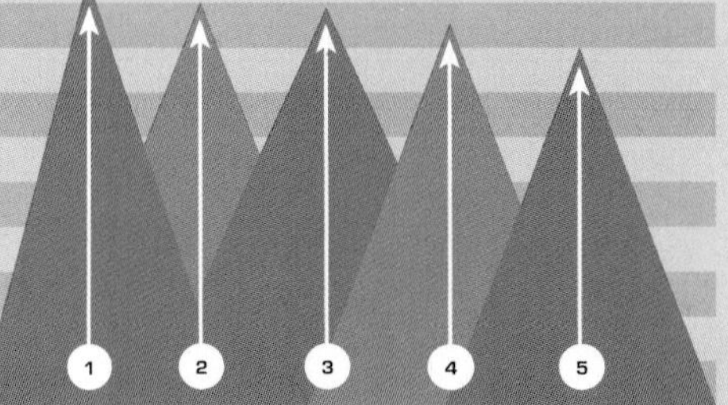

High peaks

1. San Gorgonio Mountain 11,499ft (3503m)
2. San Bernadino Peak 10,650ft (3,250m)
3. Mount San Antonio 10,065ft (3,068m)
4. Mount Pinos 8,830ft (2,692m)
5. Frazier Mountain 8,026ft (2,446m)

GRANODIORITE ROCK
This rock is typical of much of the San Gabriel Mountains, made mainly of interlocking crystals of quartz and feldspar. It formed during events long before the Transverse Ranges were born, when the crust in this area was hot enough to melt.

SAN BERNARDINO MOUNTAINS
With the Mojave Desert to their north, the San Bernardino Mountains span a distance of about 60 miles (100km). Most of the range lies within a National Forest of the same name.

MOUNT MCKINLEY

Also known as Denali, Mount McKinley is the only mountain in North America over 20,000ft (6,000m), towering above the Alaskan landscape. Like the Transverse Ranges, it is being pushed up today by local squeezing, caught like a seed in the huge active Denali strike-slip fault line. The Denali Fault runs across Alaska in a broad arc, slipping about 0.3in (10mm) per year. It last ruptured in 2002 during the magnitude 7.9 earthquake. These movements are rapidly raising the underlying granite bedrock of Mount McKinley, which was intruded as magma into the crust about 60 million years ago.

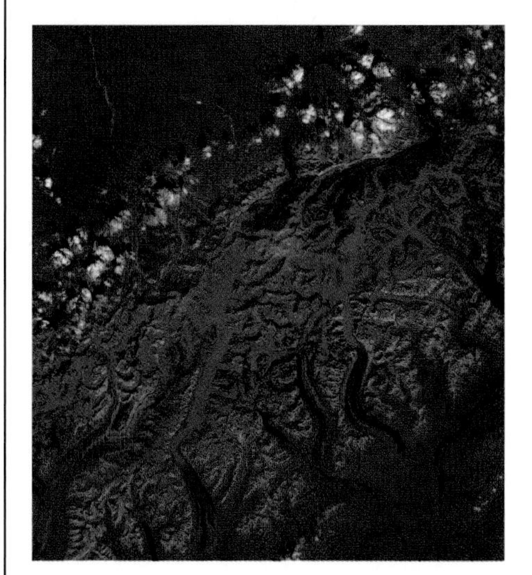

Mount McKinley from space
Standing out near the middle of this satellite image is Mount McKinley, flanked to the south by the glaciated peaks of the Alaska Ranges. At 20,320ft (6,194m) above sea level, it is the highest mountain in North America.

Denali National Park and Preserve
The Polychrome Hills form a stark contrast to the snow-covered peaks of the Alaska Range, in Denali National Park and Preserve. The striking colors are due to weathered volcanic rocks that make up much of the bedrock of this part of the Alaska Ranges.

ERODED ROCK STRATA
Forced upward by tectonics, the rock that forms this mountain in Southern Australia's Northern Flinders Ranges was then eroded, leaving the ridges of different rock layers seen here. The central oval shape is the eroded dome, all that remains of the mountain's core.

BASIN AND RANGE

The Basin and Range province forms an extraordinary and unique landscape in western North America, consisting of hundreds of narrow mountain ranges, interspersed with wide, flat valleys that in places lie below sea level. Active fault lines and volcanoes show that Earth's crust is being pulled apart in this vast region, triggering melting at a greater depth.

BASIN AND RANGE	
Location	North America
Age	20 million years
Length	500 miles (800km)
Formation type	Rifting

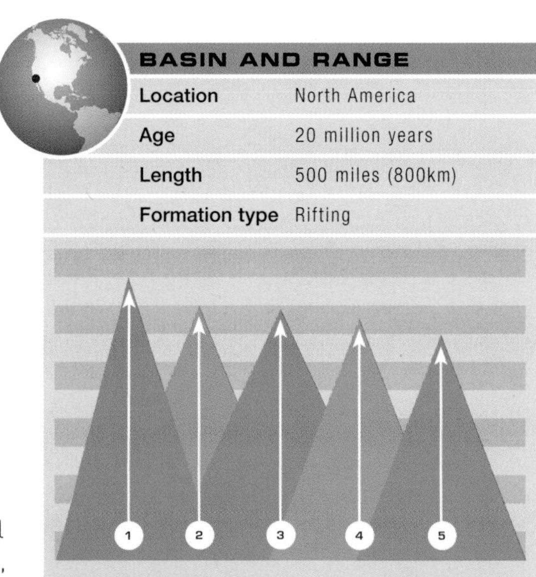

High peaks

1. White Mountain Peak 14,245ft (4,342m)
2. Wheeler Peak 13,064ft (3,982m)
3. Mount Jefferson 11,942ft (3,640m)
4. Charleston Peak 11,916ft (3,632m)
5. Arc Dome 11,771ft (3,588m)

CREATING THE LANDSCAPE

The Basin and Range province lies between the Pacific coastal ranges and the Colorado Plateau, centered on the state of Nevada. Here, the crust is actively being stretched in an approximately east–west direction, broken up along north–south normal fault lines. This has created normal faulting characterized by a fault block dropping down on one side, creating deep valley floors, and another rising up on the other side, forming the high, narrow mountain ranges. In places, the mountains rise to more than 13,000ft (4,000m) above sea level. They are made up of late Precambrian and Paleozoic rock, which is being rapidly eroded, filling up the valleys between with fresh sediment. The mountains themselves are still rising along the great fault lines at their edges, which are inclined steeply away from the valleys.

The presence of small basalt and rhyolite volcanic cones shows that there is molten rock at depth, some of which is reaching the surface. This geological activity started about 20 million years ago and is, geologists believe, driven by the rise of hot and buoyant mantle rocks at great depth below the crust. This is uplifting the surface, causing the crust here to rift apart as it collapses into the surrounding regions.

BLOCK FAULTING
Steeply inclined normal faults break up the crust in the Basin and Range into a series of parallel raised mountain blocks separated by flat valleys.

down-dropped valley with sediment layers

steeply inclined normal fault

upthrown eroded mountain block

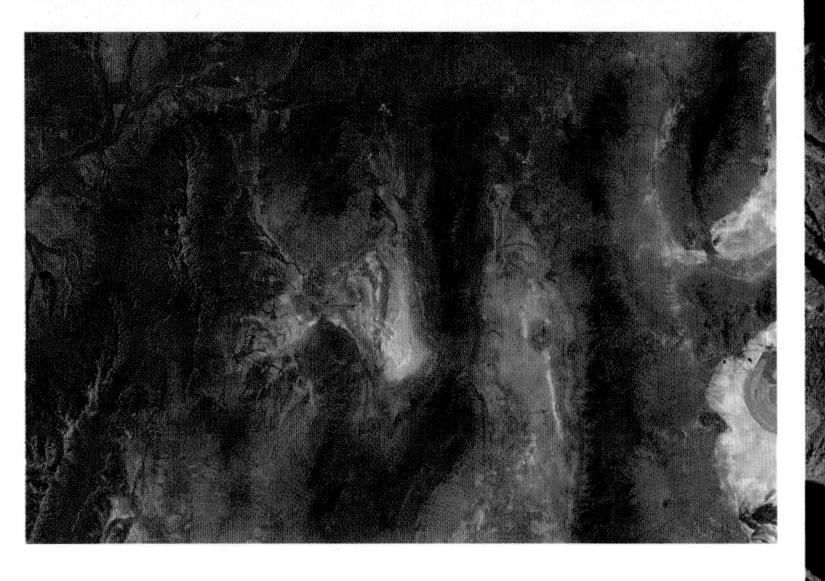

BASIN AND RANGE FROM SPACE
This color satellite image shows the classic "basin and range" landscape, with narrow mountain ranges (sometimes cloaked in green vegetation) separated by wider desert valleys, often with dry salt lakes.

PANAMINT RANGE
This view of the Panamint Range on the edge of the Mojave Desert looks across Death Valley, revealing some of the greatest topographic relief in the United States, from more than 260ft (80m) below sea level at the valley floor to nearly 11,480ft (3,500m) above sea level at the mountains' peaks.

REGION OF EXTREMES

The Basin and Range contains some of the greatest extremes of elevation contrast in North America. Death Valley is the lowest place in North America, at 282ft (86m) below sea level. But only a few miles to the west, on the upthrown side of one of the fault lines that runs along the edges of the valley, the Panamint Range rises up to 11,049ft (3,368m) above sea level. This huge elevation contrast has also created extremes in climate. During the summer, nighttime temperatures in Death Valley can reach 86°F (30°C), while at the same time falling to below 32°F (0°C) on the summit of Mount Telescope.

DEATH VALLEY
Unusually cold winter temperatures in Death Valley have resulted in a coating of frost on the valley floor and surrounding hills.

BASIN AND RANGE MOUNTAINS OF GREECE

In the Cretaceous and early Tertiary periods, much of the crust in Greece and the Aegean was pushed up by the same northward movement of Africa into Europe that created the European Alps (see pp.74–75). However, in the last 20 million years, Greece has subsided as the uplifted landmass collapsed southward toward the Hellenic subduction zone. The drowned landscape of the Aegean and southern Greece has been created as a series of uplifted blocks and grabens (depressed blocks of land), moving along steeply inclined faults. In northern Greece, the crust is also rifting apart, but here the uplifted blocks form high mountains, such as Mount Olympus.

Mount Olympus
This image of Mount Olympus was taken from the International Space Station. At 9,570ft (2,917m) above sea level, it is the highest peak in Greece.

THE ROCKIES

The western side of North America is dominated by the great cordilleras or Rocky Mountains, rising up to 14,433ft (4,400m) above sea level and extending for nearly 3,000 miles (5,000km) from northwestern Canada to New Mexico. To 19th-century explorers and settlers, they formed a major barrier to reaching the west coast. But they have also long excited the curiosity of geologists, trying to understand the origin of our planet's mountains.

RAISING THE ROCKIES

The Rocky Mountains were mainly raised between 80 to 50 million years ago, and the intense folding and faulting show that they are the result of compression. Layers of sedimentary rock, laid down over the previous several hundred million years, have been piled up into a series of slices along huge, gently inclined thrust faults. However, the region to the west of the Rockies has a more complex and drawn out history over a period of more than 100 million years, when slivers of crust, or terranes, were dragged sideways and juxtaposed along great strike-slip fault lines. All these movements are the result of the forces generated by the subduction of the ocean floor beneath the western margin of North America.

A RUGGED CONTINENT
The Rocky Mountains of western North America stand out clearly in this shaded-relief map.

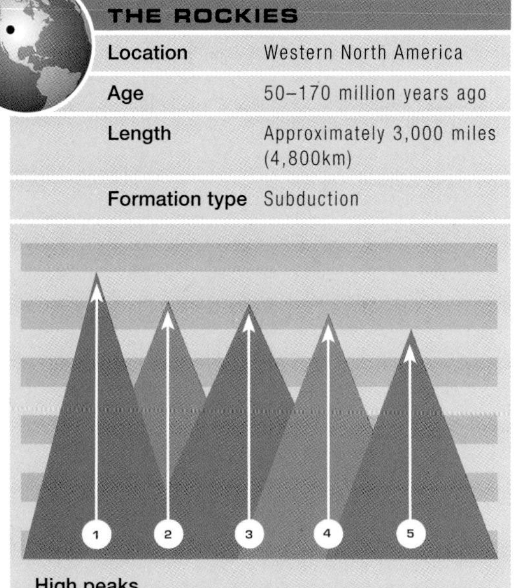

THE ROCKIES	
Location	Western North America
Age	50–170 million years ago
Length	Approximately 3,000 miles (4,800km)
Formation type	Subduction

High peaks

1. Mount Elbert, Colorado 14,439ft (4,401m)
2. Mount Massive, Colorado 14,429ft (4,398m)
3. Mount Harvard, Colorado 14,422ft (4,396m)
4. La Planta Peak, Colorado 14,366ft (4,379m)
5. Blanca Peak, Colorado 14,350ft (4,374m)

CASTLE MOUNTAIN, BANFF, CANADA
These fortresslike peaks are made of 500-million-year-old limestones and shales, pushed up by movement on thrust faults during the creation of the Canadian Rockies.

SURROUNDING LANDSCAPE

To the east of the Rocky Mountains are the Great Plains of North America. These lie only a few hundred feet above sea level, underlain by the detritus eroded and carried by rivers from the Rocky Mountains themselves. In Canada, this region was covered by the great ice sheets that advanced southward during the last glaciation. The weight of the ice pushed down the land surface, creating Hudson Bay, as well as sculpting and smoothing the landscape.

TETON MOUNTAINS, WYOMING
The Tetons are bounded by a young fault zone, and rise spectacularly from the plain without any intervening foothills. There are about a dozen peaks within this range rising to more than 12,000ft (3,650m).

THE GRANDE RONDE RIVER, WASHINGTON
West of the Rocky Mountains, this stripey landscape has been created where rivers have cut deeply through layers of basalt, erupted about 16 million years ago.

GEOLOGIC CROSS-SECTION
This cutaway view of the rocks in the Idaho and Wyoming part of the Rocky Mountains shows the ages of the rocks (see key) and the giant thrust faults. Movement on the faults between 120 and 50 million years ago has stacked up and folded the rocks to build the mountains—a classic example of a "thin skinned" fold and thrust belt. The most prominent faults and folds have been given individual names.

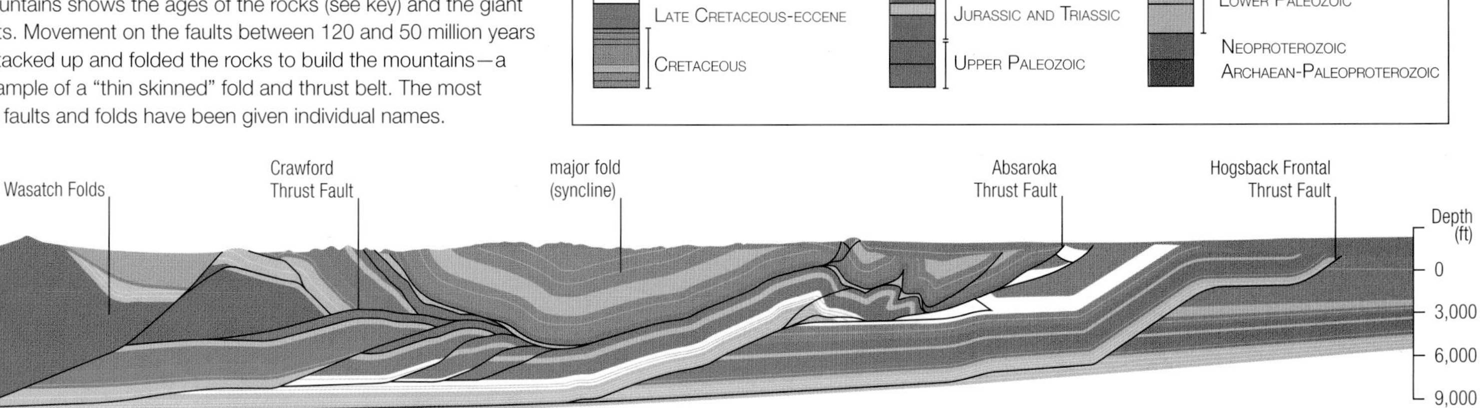

EAST AFRICAN RIFT

When the continental crust begins to split, long grooves or depressions form, creating rift valleys. The flanks of the rifts also tend to warp upward, creating rift mountains bounded by steep escarpments. The largest of these fault systems, the East African Rift, is clearly visible from space as a great scar in the continent.

GREAT RIFT VALLEY OF EAST AFRICA

Down the eastern side of Africa, from the Afar region in the north to the Kalahari Desert in the south, runs a great rift in the continent. Many of the great lakes of Africa fill this rift, which in places splits in two. Earthquakes and volcanic activity show that process of rifting is active today, splitting apart up to 0.3in (10mm) per year. In fact, the rift is a failed ocean, which started to form about 35 million years ago. Away from the Rift, in the highlands of east Africa, there are the remains of a landscape that has been elevated into a great dome, up to 7,000ft (2,000m) above sea level. Geologists believe that this dome was pushed up by a plume of hot and buoyant mantle rocks that rose from deep in the interior. As the dome bulged upward it weakened the overlying crust, which began to rift apart.

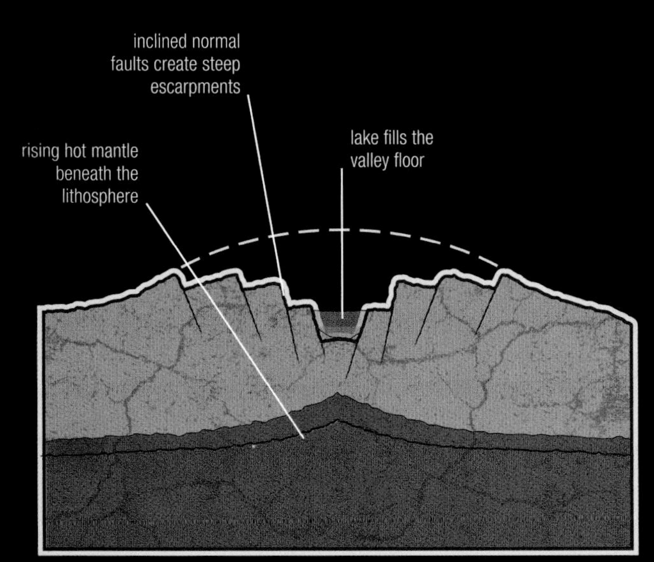

RIFT VALLEY FORMATION
A rift often forms where the lithosphere is warped up into a broad dome, splitting at its crest along steeply inclined normal faults. Movement along the faults results in subsidence on one side, and the uplift of rift mountains on the other. Rivers carry rainwater from the flanking highlands into the valleys, which rapidly fill, forming lakes.

AERIAL VIEW OF GREAT RIFT VALLEY
This image shows three of the great lakes of east Africa—Lake Edward, Lake Kivi, and Lake Tanganyika. They are located in the Democratic Republic of the Congo, Rwanda, Burundi, Uganda, and Tanzania, and fill the western arm of the East African Rift system. The flanks of the rifts have been warped upward, while further away the relatively flat highlands form part of a vast doming of Earth's surface.

ETHIOPIA RIFT VALLEY, ETHIOPIA
The horizontally bedded rocks exposed in the steep, eroded sides of gorges are the remains of a vast volcanic outpouring that occurred about 30 million years ago above a mantle plume when the crust was just beginning to rift apart.

GREAT RIFT VALLEY	
Location	East Africa
Age	35 million years old and still active today
Length	2,175 miles (3,500km)
Formation type	Continental rifting

30 THE AVERAGE WIDTH IN MILES (50KM) OF THE GREAT RIFT VALLEY

CRADLE OF MANKIND

The East African Rift seems to have provided the right environment for our evolutionary ancestors. Here, there is fossil evidence that hominid populations thrived in a landscape of lakes and plains, with rock shelters in the steep valley sides and a climate that was protected from extremes. However, it is the volcanic activity along the Rift that has helped to bury and preserve the fossil evidence for this. One of the most famous is the well-preserved skeleton of Lucy, or *Australopithecus afarensis*, found in the Ethiopian sector of the Rift. Lucy shows that hominids were walking on two legs about 3.2 million years ago. About the same time, but much farther south in Tanzania, footprints show that a family of hominids walked across a muddy plain near the Olduvai Gorge.

> "OUT OF AFRICA ALWAYS SOMETHING NEW."
>
> **PLINY THE ELDER**, ROMAN AUTHOR AND NATURALIST 23–79CE

OTHER RIFT MOUNTAINS

Lake Baikal is the deepest place on a continental plate—the lake bed is up to 5,370ft (1,637m) deep. It also is a rift valley, formed where the Siberian region started to split apart about 25 million years ago. Some of this motion has been caused by movement on both strike-slip faults and rift (normal) faults, carrying crust eastward and out of the way of advancing India.

Lake Baikal from space
The straight edges of the ice-covered Lake Baikal are clearly visible in this satellite view. The lake is more than 390 miles (600km) long and 50 miles (80km) wide, following the fault lines where the crust is splitting apart. The flanking mountains on either side rise up to about 9,200ft (2,800m) above sea level.

Dead Sea Rift Valley
A series of rifts along the Dead Sea transform fault, where the Arabian Plate slips northward past the African Plate, have created deep lake-filled valleys like the Dead Sea itself. Today, the lake level is dropping, exposing the remains of old lake beds.

GRANITE GIANT
The Brandberg Massif towers over the surrounding Namib Desert. This granite intrusion formed more than 120 million years ago. It stretches 9.3 miles (15km) across and rises 8,441ft (2,573m) above sea level. The dark ring seen in this satellite image is the steep-sided rock that circles the mountain.

THE ALPS

To many, the snow-capped peaks of the European Alps represent the typical mountain range. However, it has taken geologists nearly 100 years to make sense of the complex sequence of geological events that led to their formation.

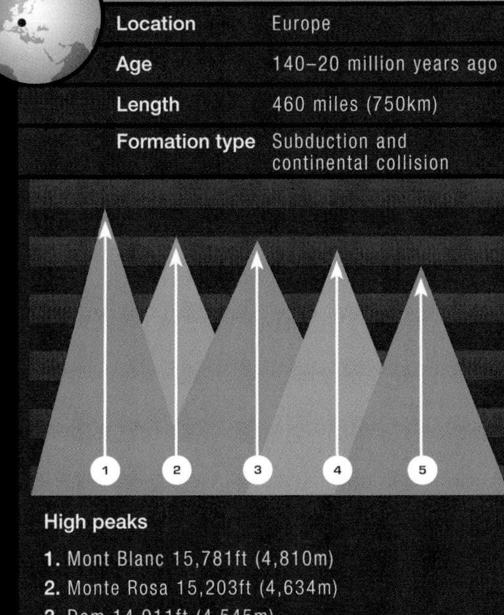

THE ALPS	
Location	Europe
Age	140–20 million years ago
Length	460 miles (750km)
Formation type	Subduction and continental collision

High peaks

1. Mont Blanc 15,781ft (4,810m)
2. Monte Rosa 15,203ft (4,634m)
3. Dom 14,911ft (4,545m)
4. Weisshorn 14,780ft (4,505m)
5. Matterhorn 14,700ft (4,478m)

RAISING THE ALPS

The European Alps were created by the convergence of the European and African tectonic plates, which resulted in the closing up and subduction of the ancient Tethys Ocean that once separated them. The mountains are built of rocks that originally formed part of both the European and African crust, and fragments of the sea floor itself.

Mountain building occurred in at least two stages, with an early collision between Africa and the Apulian micro-plate in the Late Cretaceous about 100 million years ago, creating the Austroalpine mountains in the eastern Alps. Subsequently, the narrow Penninic Ocean to the north was subducted, leading to the main phase of mountain building in the Tertiary period between 60 and 20 million years ago. All these movements pushed up a series of slivers—called nappes—from great depth in the crust of both plates. This brought up rocks that have been transformed at either high or low temperatures or pressures, as well as creating the marked curvature of the mountain range. In addition, great masses of Mesozoic limestone and other sediments, that were deposited on the northern margin of the ancient Tethys Ocean, were thrust up to build what is today the northern and southern Alps. Erosion by rivers and glaciers has cut deeply into the rocks in the central parts of the Alps, depositing the debris on its margins in vast sheets that now hold important oil and gas deposits.

ALPINE FORMATION
The final stages in the formation of the Alps occurred 30 million years ago, creating a distinctive landscape of uplifted mountains.

Foothills
Folded and faulted rock layers underlie hilly land near the mountains

Mountain peaks
Central uplifted zone underlain by metamorphic rocks

Fault line
Scar in the landscape follows major fault line

Molten crust
Rises beneath mountain range

Metamorphic rocks
Subjected to high temperatures and pressures

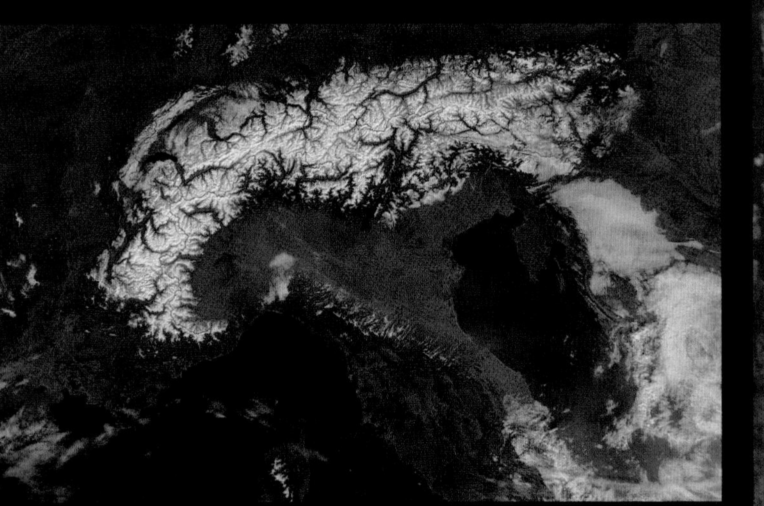

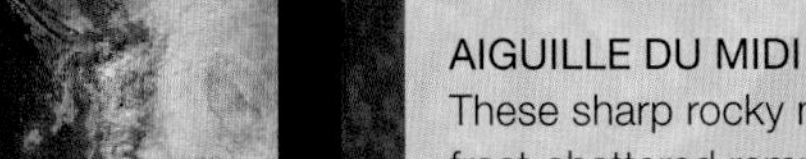

THE ALPS FROM SPACE
This image from space shows the distinctive shape of the Alps. The high parts have been deeply eroded, exposing metamorphic rocks that formed tens of miles deep in the crust.

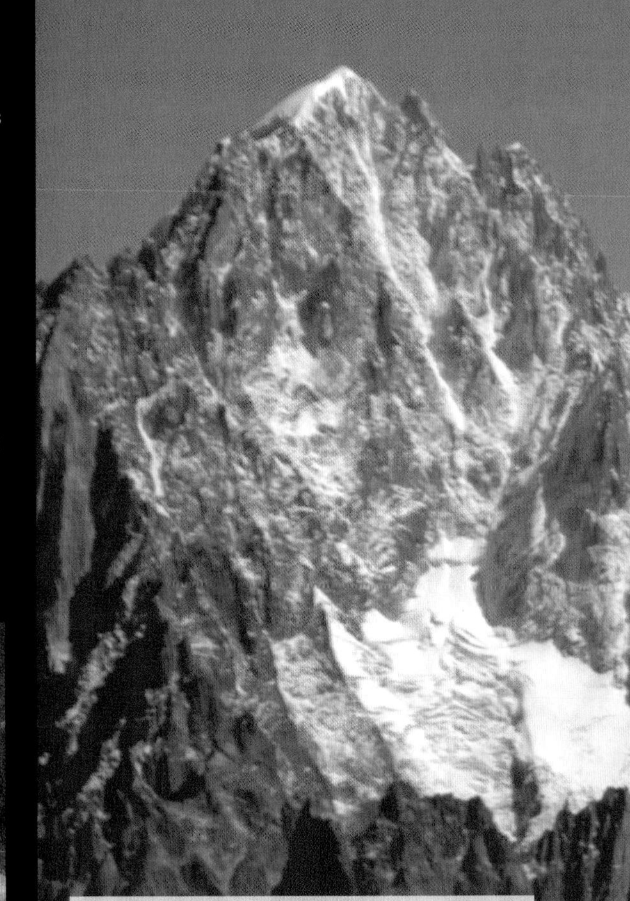

AIGUILLE DU MIDI
These sharp rocky needles are the frost-shattered remnants of once-molten crystalline rocks that formed deep in the continental crust. They were brought to the surface along thrust faults during the earth movements that created the Alps.

ROCK TYPES

The most famous of the metamorphic rocks that make up the Alps is blueschist, which is full of the minerals glaucophane, lawsonite, and epidote. This rock formed when sediment or volcanics on the margins of the ancient Tethys Ocean were carried to depths of tens of miles in a subduction zone. Here, at very high pressures, but relatively low temperatures, lawsonite and glaucophone crystallized and the rock was transformed. During later Alpine earth movements, these rocks were carried upward, to be exposed at the surface. Other metamorphic rocks in the region contain the mineral chlorite and garnet crystals.

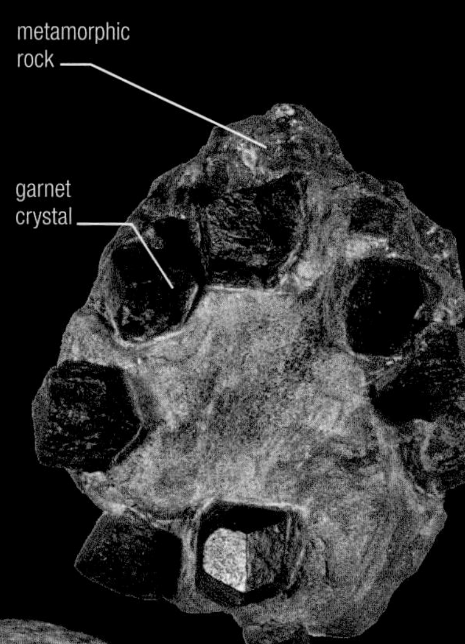

GARNETS
The Austrian Alps are famous for these large red garnet crystals, with perfect crystal shapes that have grown slowly in the rock.

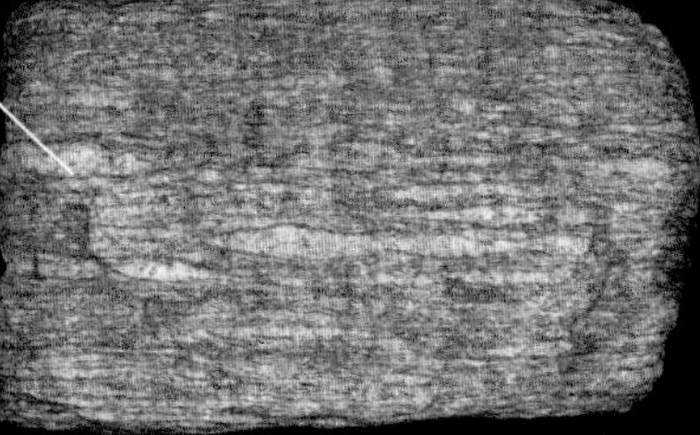

BLUESCHIST
This metamorphic rock has marked foliation where it has been compressed by the huge forces deep inside Earth.

CHLORITE
Chlorite is a green platy mineral often found in metamorphic rocks. This sample is from the Matterhorn, Zermatt, Switzerland. It formed at a later stage in the uplift of these rocks.

AUSTROALPINE NAPPES
A great fold is visible in this mountain, where the rocks on the southern margin of Europe have been rucked up and twisted round by tectonic forces during the early stages of plate collision in the Alps, displacing the crust northward.

THE URALS

The Urals are often seen as the natural boundary between Europe and Asia. They extend for about 1,550 miles (2,500km) in a north–south direction all the way from the Arctic Ocean to near the Caspian Sea, marking the boundary between Europe and Asia. They are also a major source of Russia's mineral wealth.

FORMATION

The creation of the Ural Mountains records the final stages in the closing up of an ocean that was created by rifting of the supercontinent of Pangaea 400–500 million years ago. Fragments of this ocean, including chains of volcanic islands, were caught up in the final collision between what is now Europe and Siberia 220–300 million years ago, pushing up the mountain chain. Geologically speaking, little has happened since then except erosion by rivers and glaciers, reducing the once high mountains down to their relatively low present elevations. This has exposed a bedrock of quartzites, schists, and gabbros, and rich deposits of gold, platinum, chromite, magnetite, and coal. Precious and semiprecious stones, such as emeralds and diamonds, are also found there.

THE URALS	
Location	Russia
Age	250–300 million years ago
Length	1,550 miles (2,500km)
Formation type	Continental collision

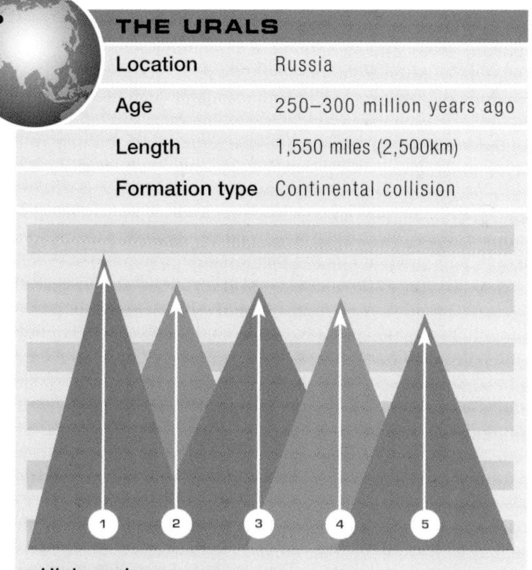

High peaks

1. Mount Narodnaya 6,200ft (1,895m)
2. Mount Karpinsky 6,160ft (1,878m)
3. Mount Manaraga 5,450ft (1,662m)
4. Mount Yamantau 5,380ft (1,640m)
5. Mount Telposiz 5,305ft (1,617m)

SATELLITE VIEW OF NOVAYA ZEMLYA
Looking like a northern offshore continuation of the Urals, the mountainous island of Novaya Zemlya records continental collision event, dating from about 220 million years ago. Erosion has exposed rocks from deep within the crust.

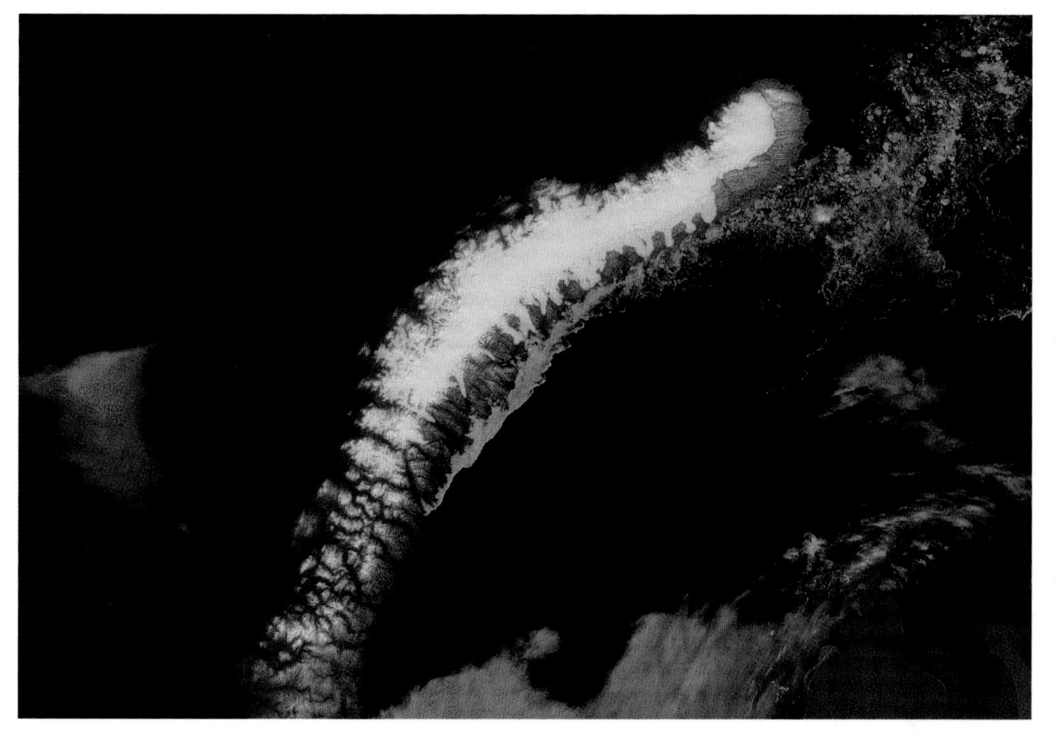

SUBPOLAR URALS
The Subpolar Urals form the highest parts of the Ural Mountains. When the ice sheets retreated at the end of the last ice age they left an eroded and bleak landscape of exposed sedimentary and metamorphic rock.

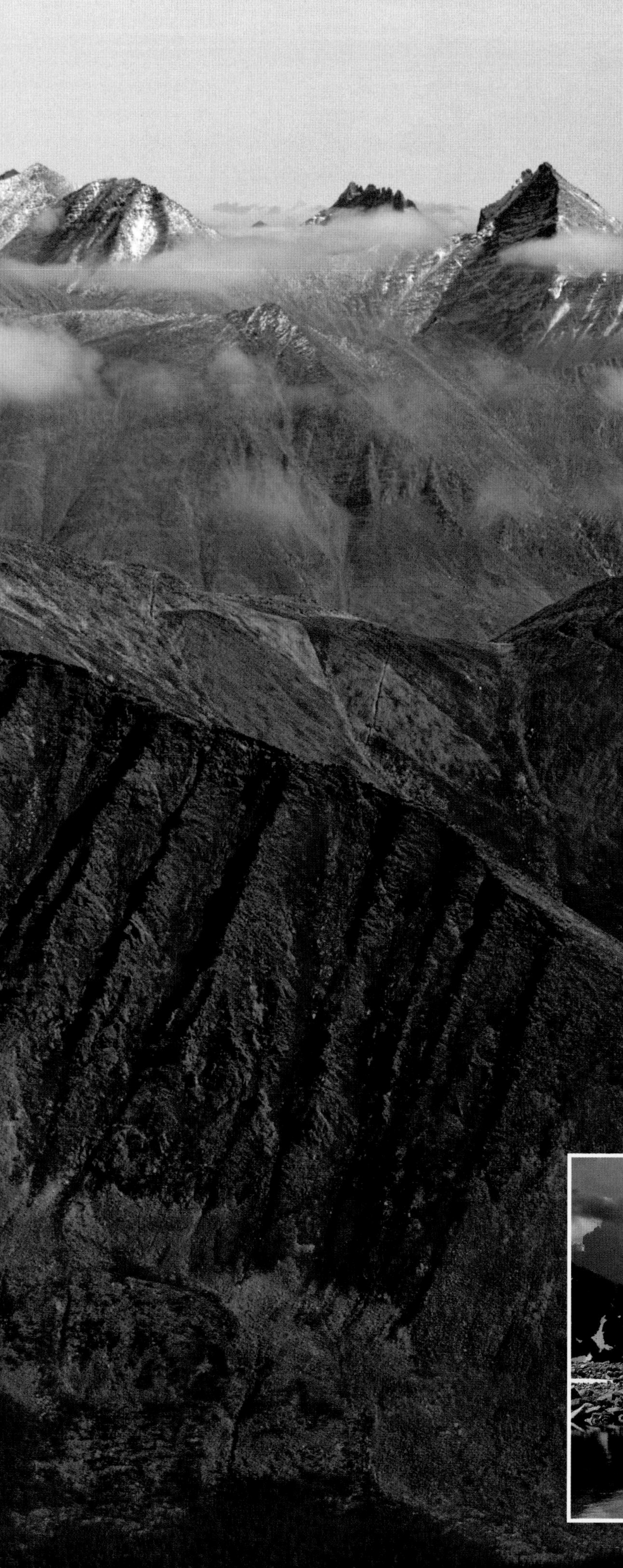

MOINE THRUST BELT

Another ancient mountain range can be found in the hills of northwest Scotland, which are underlain by rocks that date from the Proterozoic Eon and the Cambrian and Ordovician periods of geologic time. These were pushed up to form a great mountain range along thrust faults, such as the Moine Thrust Belt, which runs near the western edge of the Scottish Highlands. This occurred when the ancient continents of Laurentia, Baltica, and Avalonia collided about 400–500 million years ago, sealing the final fate of the ancient Iapetus Ocean. Today, the mountains have been largely eroded away by the action of rivers and glaciers, revealing sedimentary rocks that were once deposited on the edge of this ancient ocean or laid down by rivers.

The Moine Thrust Belt was discovered by two Victorian geologists—Ben Peach and John Horne—toward the end of the 19th century. Their work revealed the colossal scale of movements in Earth's crust that lead to the formation of a mountain range.

The thrust plane is gently inclined to the southeast, separating the underlying Torridonian sandstones and Lewisian metamorphic rocks from the overlying cooked-up remains of sedimentary rock deposited on the flanks of the Iapetus Ocean. Movement of the rocks on the thrust plane was up to several tens of miles.

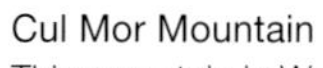

Cul Mor Mountain
This mountain in Western Ross, Scotland, is made up of distinctive Torridonian red sandstones, laid down by rivers in the Proterozoic Period. About 400 million years ago these rocks were deeply buried when metamorphic rocks, which now make up much of the highlands of Scotland, were pushed along the Moine Thrust Belt. Erosion has revealed the Torridonian sandstones once again.

Glencoul Thrust
The effects of thrust formation can clearly be seen at Glencoul Thrust in Scotland, where older rocks have stacked above newer layers.

MOUNT NARODNAYA
The highest mountain in the Urals, Mount Narodnaya, is made up of weakly metamorphosed sedimentary rocks, such as quartzites and slates, originally laid down in the Proterozoic and Cambrian periods of geological time.

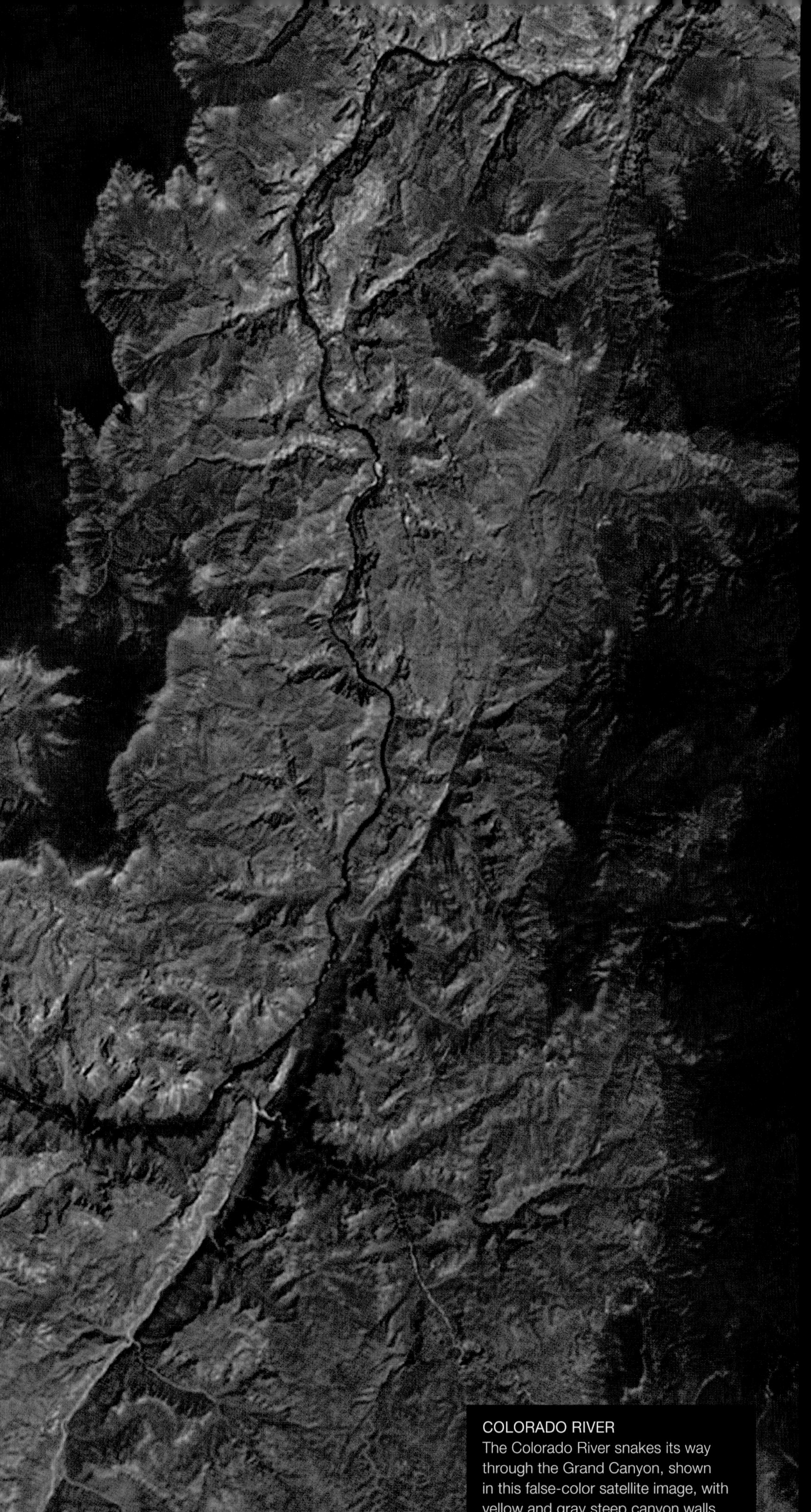

COLORADO RIVER
The Colorado River snakes its way through the Grand Canyon, shown in this false-color satellite image, with yellow and gray steep canyon walls and red vegetation-covered plateaus.

BRIGHT ANGEL POINT
Within the canyon, the Colorado River has cut more than 270 miles (4,40km) of deep cliffs through red sandstone, as seen here from a viewpoint called Bright Angel Point.

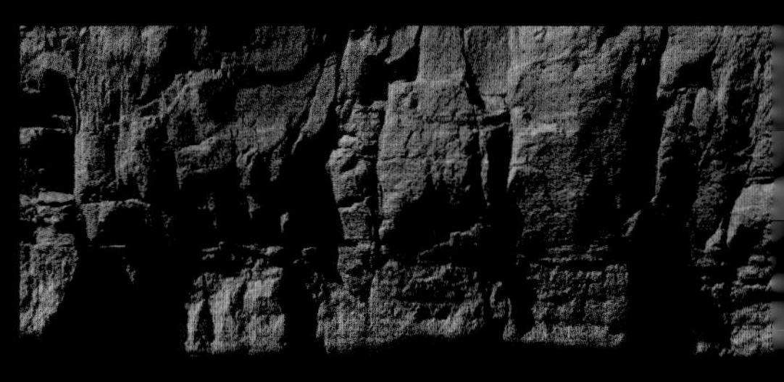

SEDIMENTARY ROCK
Many of the Grand Canyon's rocks are sedimentary, laid down when the area was below sea level. This picture shows the joint between two sedimentary formations.

GRAND CANYON
Toroweap Point rises 3,000ft (914m) above the Colorado River. From this vantage point, 1.7 billion years of geological history is visible in the exposed rock strata.

TRANSANTARCTIC MOUNTAINS

At the southern end of the world, and partly buried under ice, the Transantarctic Mountains form the planet's remotest mountain chain, dividing Antarctica in two. Discovered near the turn of the 20th century, their origin still remains a puzzle to geologists.

A GREAT DIVIDE

About 100 years ago, the early Antarctic explorers found their route to the South Pole blocked by the Transantarctic Mountains. Captain Robert Scott, on his fateful expedition in 1911–12, crossed these mountains by laboriously making his way up the deeply crevassed Beardmore Glacier, where the ice flowing down from the high polar plateau has carved a great valley through the mountains. Amazingly, Scott's party found time to collect rock samples from the sides of the glacier, picking up fragments of coal, sandstone, and volcanic rocks. These rocks are now recognized to be from the Paleozoic to Mesozoic eras, laid down when Antarctica was ice free. They form part of the backbone of much of the Transantarctic Mountains. Beneath them lie older metamorphic rocks that form the deep roots of ancient mountain ranges that are long gone, worn away by the forces of erosion.

TRANSANTARCTIC MOUNTAINS	
Location	Antarctica
Age	About 60 million years
Length	2,000 miles (3,200km)
Formation type	Rift

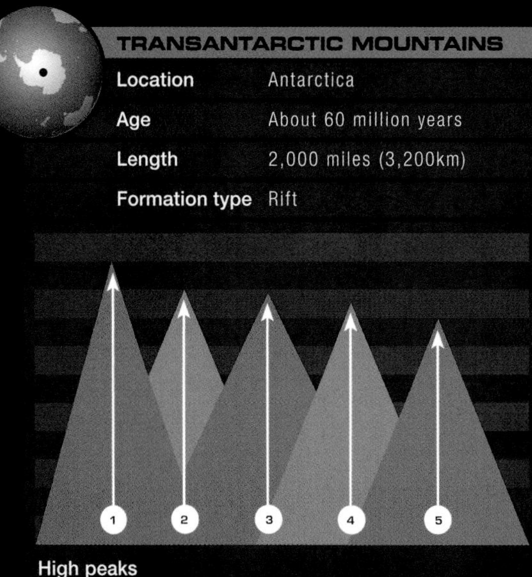

High peaks

1. Mount Kirkpatrick, Queen Alexandra Range 1,486ft (4,528m)
2. Mount Kaplan, Queen Maud Mountains 1,388ft (4,230m)
3. Mount Minto, Victoria Land 1,367ft (4,166m)
4. Mount Lister, Prince Albert-McMurdo 1,320ft (4,023m)
5. Faure Peak, Horlick Mountains: 9,219ft (2,810m)

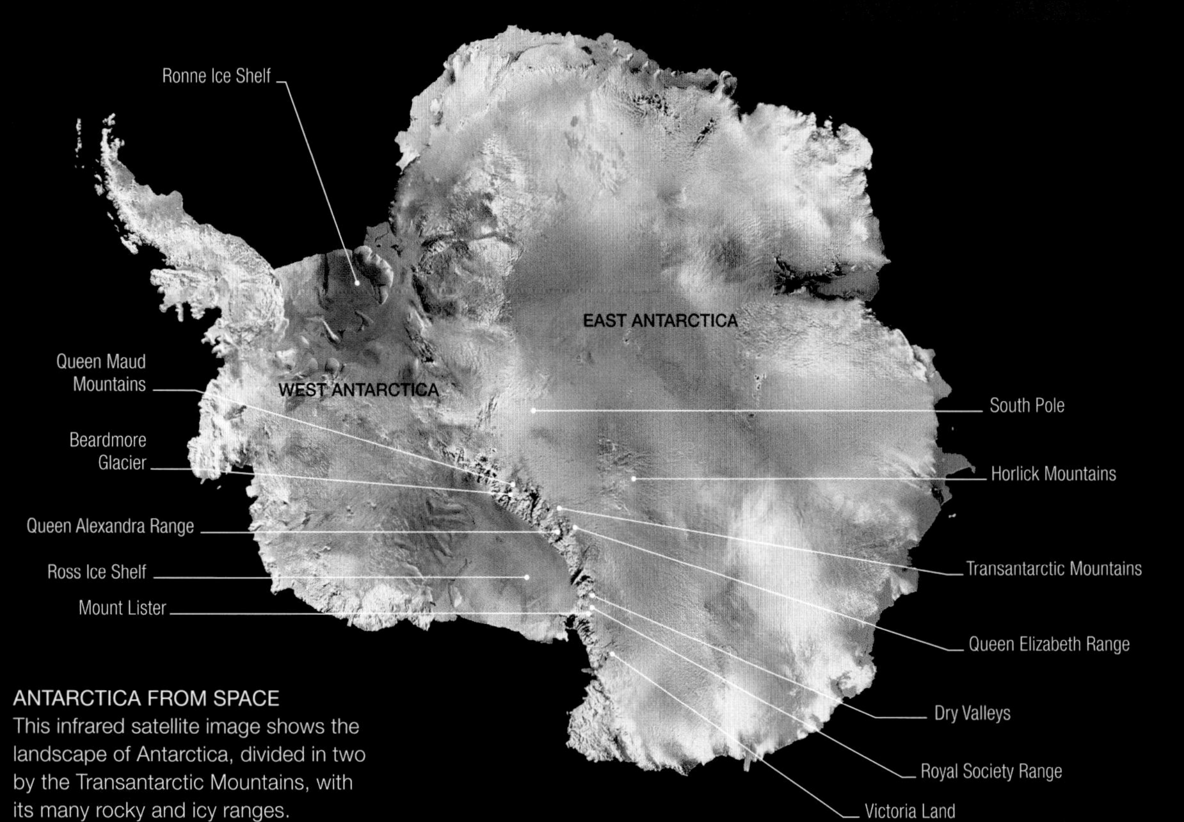

ANTARCTICA FROM SPACE
This infrared satellite image shows the landscape of Antarctica, divided in two by the Transantarctic Mountains, with its many rocky and icy ranges.

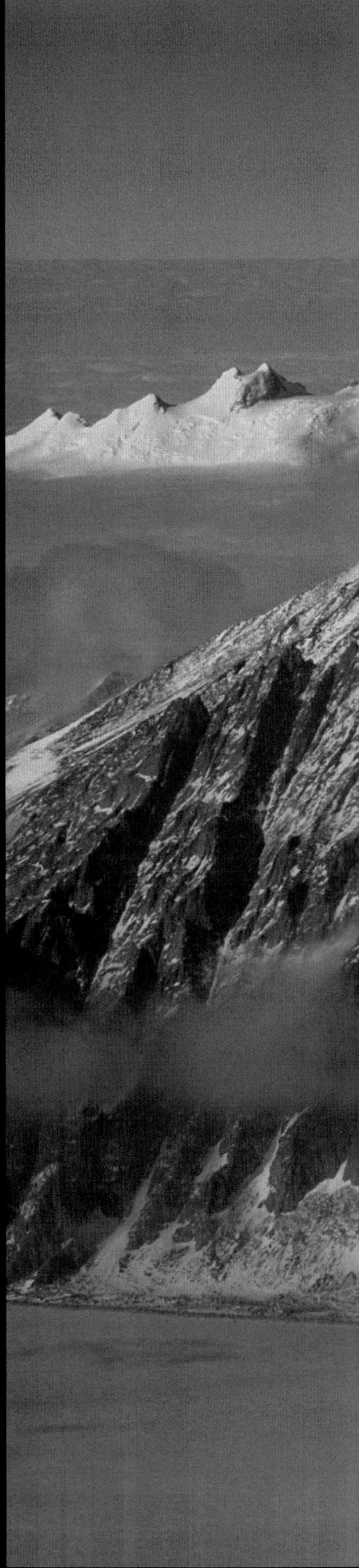

QUEEN ALEXANDRA RANGE
Rising above the snow and ice is the Queen Alexandra Range, forming the highest peaks in the Transantarctic Mountains.

ROYAL SOCIETY RANGE
In the Ross Sea segment of the Transantarctic Mountains lies the Royal Society Range—named by Captain Scott after the scientific sponsors of his expedition.

DRY VALLEYS
Ancient valleys carved into the Transantarctic Mountains expose the nearly flat-lying sedimentary and volcanic layers of rock that formed in the Paleozoic and Mesozoic periods.

QUEEN MAUD MOUNTAINS
The steep mountain front of the Transantarctic Mountains reveals a rift scarp. Here, glaciers flowing off the polar plateau—such as the Beardmore Glacier—have carved deep valleys.

FORMATION

The Transantarctic Mountains also mark the boundary between two blocks of crust with very different geological histories at the edge of the supercontinent of Gondwana. During the Paleozoic and Mesozoic periods, rivers flowing from highlands in both East and West Antarctica deposited sediments in a vast inland basin, now preserved in the Transantarctic Mountains. Even today, the uplifted layers of this sedimentary rock are still virtually flat, and show none of the folding and thrust faulting commonly associated with mountain belts. However, modern geological mapping has revealed steeply inclined normal faults cutting the rocks, showing that the Transantarctic Mountains are rift mountains that first started to rise more than 140 million years ago as part of the break-up of Gondwana. Rifting in Antarctica continued into the Tertiary, when the mountains rose higher on the flanks of a rift valley. But this does not fully explain the great height of these mountains today, and it has been suggested that there is unusually hot and less dense mantle rock beneath the mountains, which is helping to buoy them up more.

3

VOLCANOES

<< Eruption
This dramatic eruption occurred on the flank of Piton de la Fournaise, a shield volcano on Réunion Island in the Indian Ocean.

WHAT IS A VOLCANO?

A volcano is an opening in Earth's crust where magma—a mixture of red-hot molten rock, mineral crystals, rock fragments, and dissolved gases—from inside the planet erupts onto the surface. Magmatic volcanoes like this are by far the best known type, but there is a second, less well known type that erupts mud rather than magma (see pp.196–97).

CREATION OF A VOLCANO

Magma is produced by the melting of rock in the Earth's upper mantle and lower crust. This occurs only at certain places, notably at convergent and divergent plate boundaries (see pp.28–31) and at hotspots or mantle plumes (see pp.32–33). Magma is usually less dense than the surrounding rock because it is hotter (heating matter causes it to expand), so it rises up, traveling through weaknesses or fractures in the crust all the while incorporating small to large amounts of the surrounding bedrock called lithic fragments. Eventually it collects in large cavities called magma chambers, several miles below Earth's surface. From there, the magma rises through channels called conduits or pipes until it reaches the surface or, in the case of a submarine volcano, the ocean floor. There, the magma escapes either through an opening, called a vent, or a crack, called a fissure. The escape of the magma is known as an eruption, and it can vary from a quiet outpouring—in the form of a fountain or stream of lava—to a highly explosive event in which the magma and contained gases are blown violently into the air and can travel down the slopes of the volcano at great speeds in the form of a pyroclastic flow, a mixture of hot rock, gas, and ash.

GROWTH OF A VOLCANO

Volcanoes grow mainly from the accumulation of their own eruptive products—solidified lava, cinders, and ash. Lava is the name given to magma when it flows out of a volcano. This molten material eventually cools and solidifies to form solid rock. Cinders and ash are magma that has been blown into the air, cooled then deposited as solid fragments. Different types of volcanoes build up either by the accumulation of one main material, such as cinders of lava flows, or from a combination of several products. Volcanoes can also grow partly by intrusion—when magma moves up within the volcano and solidifies internally, pushing overlying rock upward to form a bulge. As they grow, many volcanoes develop a classic volcano shape, a steep-sided cone. However, not all volcanoes are cones: some are broad, gently sloping shield-shaped structures while others consist of enormous shallow craters or water-filled depressions in the ground. Volcanoes vary considerably in their activity, so their growth is very intermittent. Some can continue growing for millions of years before the supply of magma runs out and they become extinct.

THE SEQUENCE OF GROWTH

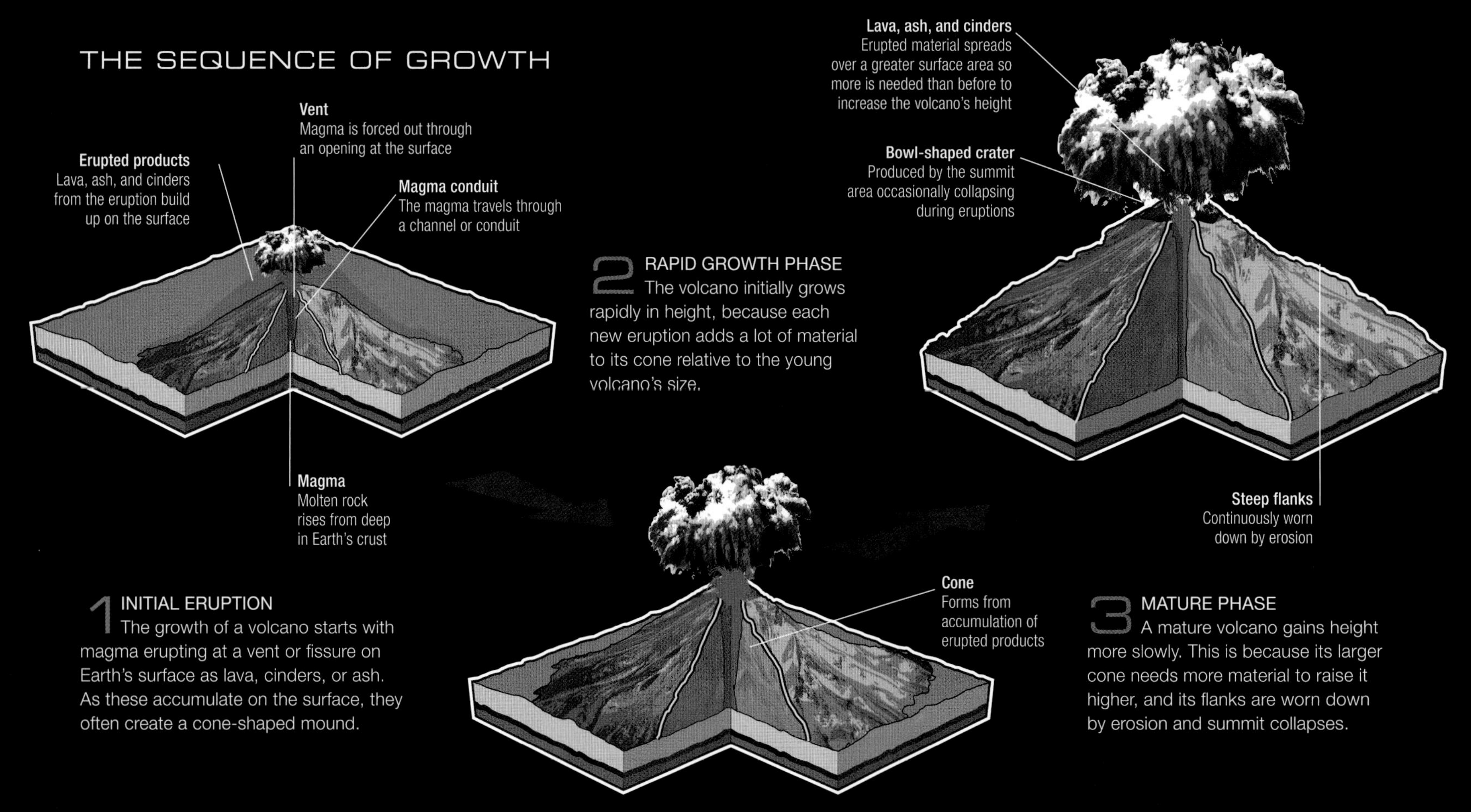

1 INITIAL ERUPTION The growth of a volcano starts with magma erupting at a vent or fissure on Earth's surface as lava, cinders, or ash. As these accumulate on the surface, they often create a cone-shaped mound.

2 RAPID GROWTH PHASE The volcano initially grows rapidly in height, because each new eruption adds a lot of material to its cone relative to the young volcano's size.

3 MATURE PHASE A mature volcano gains height more slowly. This is because its larger cone needs more material to raise it higher, and its flanks are worn down by erosion and summit collapses.

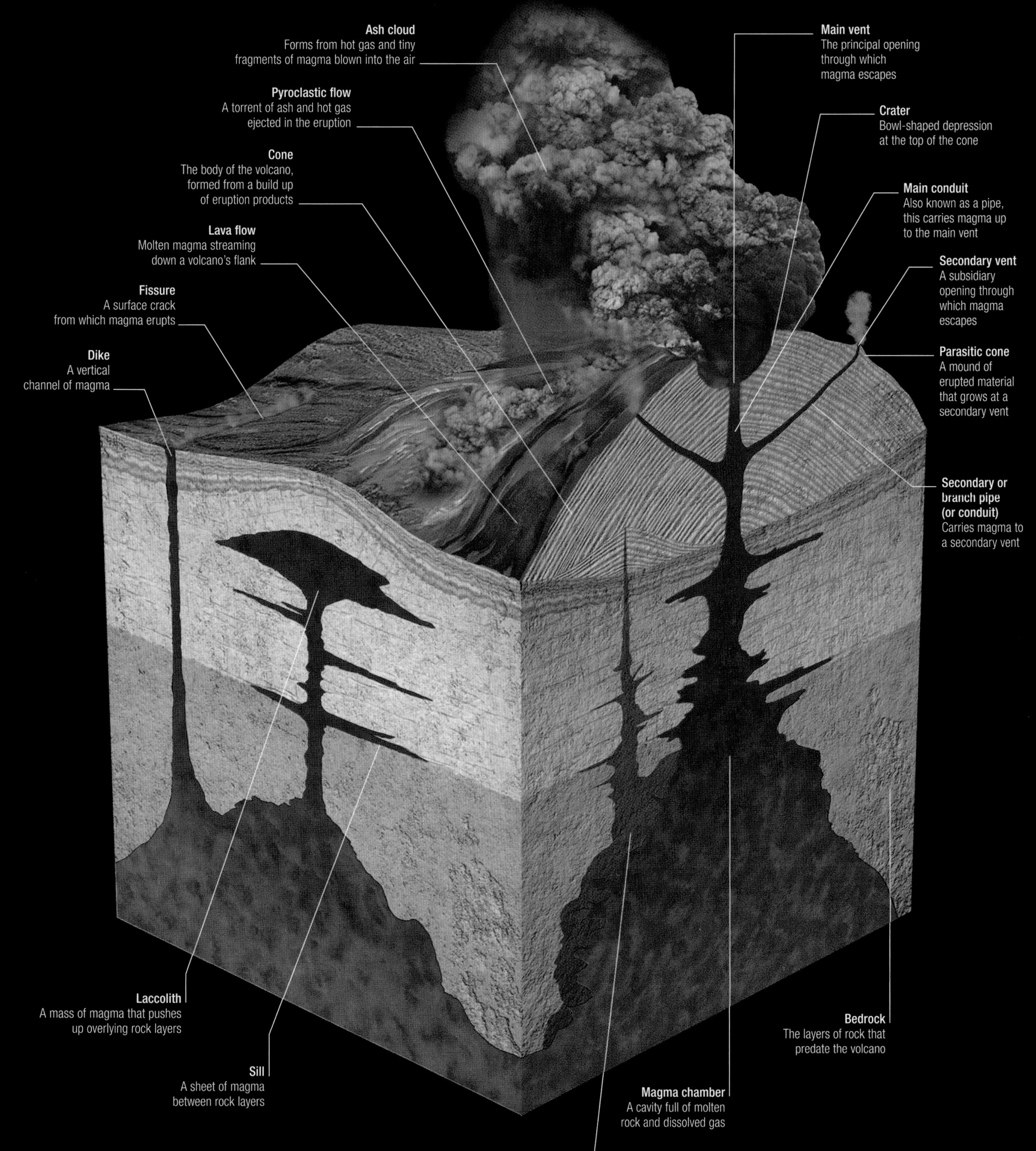

INSIDE A STRATOVOLCANO

The source of a volcano's activity is a magma chamber—a cavity containing molten rock and gas that lies 0.6–6 miles (1–10km) below Earth's surface. A volcano usually has one main conduit or pipe through which magma from this chamber reaches the surface, but it can also erupt from secondary or side vents, forming parasitic cones, or from surface fissures. Magma may also intrude into the surrounding rock without reaching the surface, forming subterranean structures such as dikes, sills, and laccoliths, which eventually cool to form solid bodies.

THE WORLD'S VOLCANOES

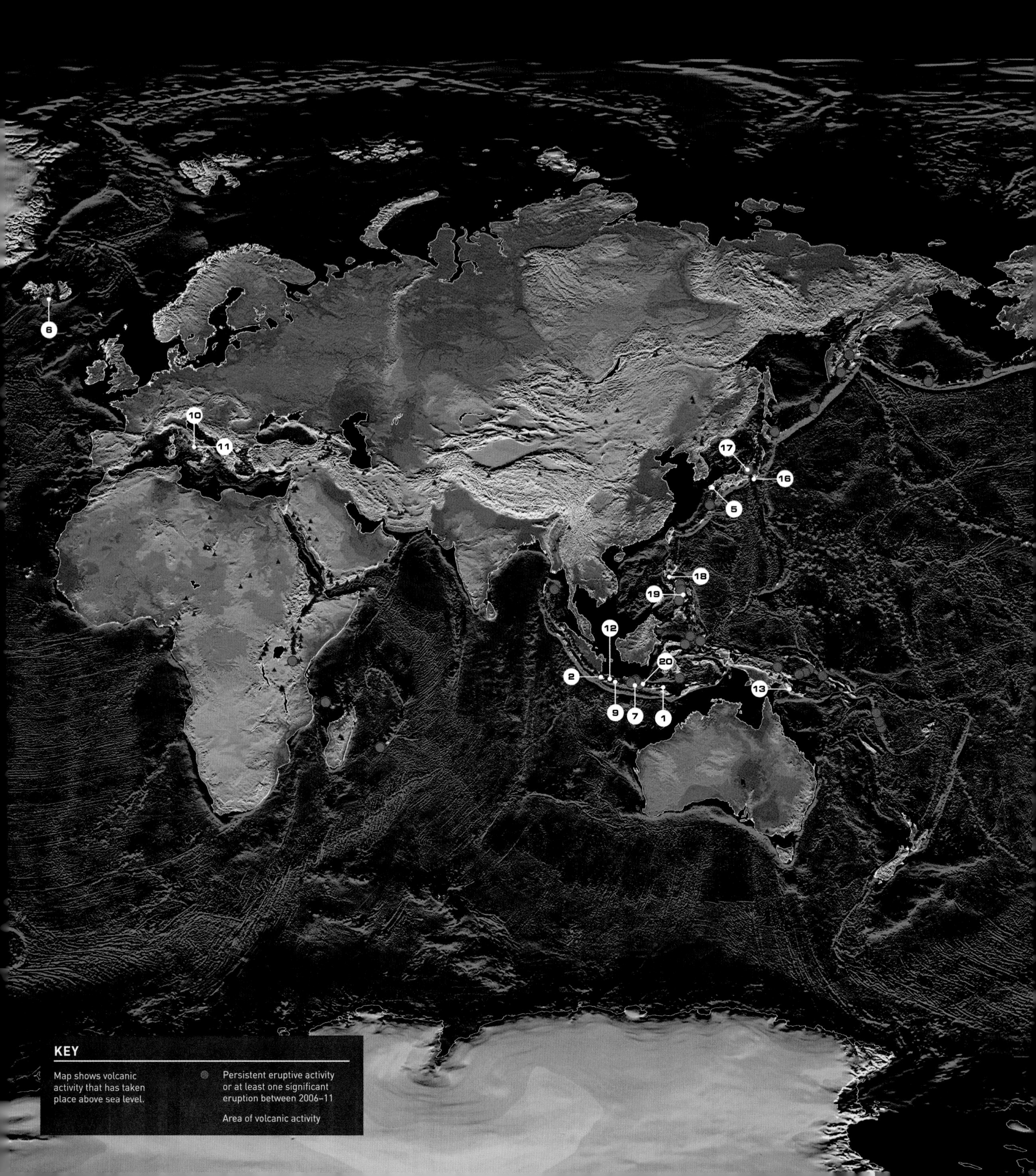

Volcanoes are heavily concentrated in a few areas of the world, mainly close to plate boundaries, particularly the "Ring of Fire" around the edges of the Pacific Ocean. Other concentrations are in Iceland, eastern Africa, the eastern Caribbean, at the "hotspots" of Hawaii in the central Pacific, and the Galápagos Islands in the eastern Pacific.

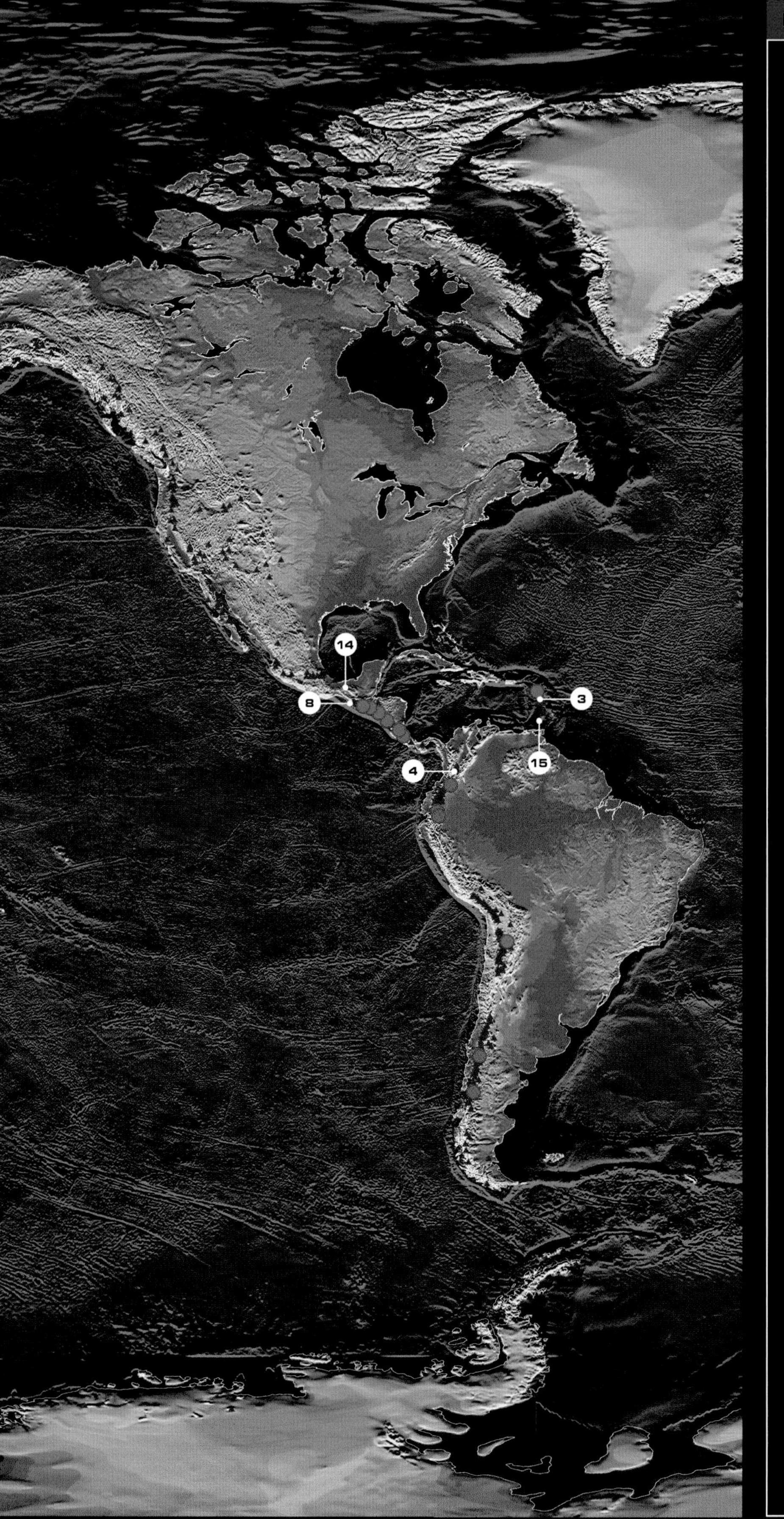

MOST LETHAL VOLCANIC ERUPTIONS IN HISTORY

#	Volcano	Country	Date	Deaths
1	MOUNT TAMBORA	Indonesia	1815	60,000
2	KRAKATAU	Indonesia	1883	36,417
3	MONT PELÉE	Martinique	1902	28,000
4	NEVADO DEL RUIZ	Colombia	1985	23,080
5	MOUNT UNZEN	Japan	1792	14,300
6	LAKI	Iceland	1783	9,350
7	KELUT	Indonesia	1919	5,110
8	SANTA MARÍA	Guatemala	1902	5,000
9	GALUNGGUNG	Indonesia	1882	4,011
10	MOUNT VESUVIUS	Italy	1631	4,000
11	MOUNT VESUVIUS	Italy	79CE	2,100
12	PAPANDAYAN	Indonesia	1772	2,957
13	MOUNT LAMINGTON	Papua New Guinea	1951	2,942
14	EL CHICHÓN	Mexico	1982	1,879
15	SOUFRIÈRE	St. Vincent	1902	1,680
16	OSHIMA-OSHIMA	Japan	1741	1,475
17	ASAMA	Japan	1783	1,491
18	TAAL	Philippines	1911	1,335
19	MAYON	Philippines	1814	1,200
20	AGUNG	Indonesia	1963	1,148

VOLCANIC ERUPTIONS

Volcanoes are of interest and potentially dangerous because of their eruptions. These fall into two broad categories, called effusive and explosive. In effusive eruptions there is a relatively quiet outpouring of lava. Explosive eruptions are characterized by explosions in which hot gas and magma are propelled into the air.

CAUSES AND TRIGGERS

Many factors affect when a volcano will erupt and what sort of an eruption it will be. These include the amount of magma (melted rock) in the volcano, its composition and viscosity (thickness), the amount of dissolved gases it contains, and the pressure in the magma chamber. In many cases, the trigger for an eruption is the upwelling of new magma. As the magma rises, the pressure inside it decreases and the dissolved gases form bubbles, which expand quickly, causing a further surge upward. If the magma contains little gas it may simply flow onto the surface, particularly if it is a nonviscous (runny) type. But in many volcanoes, the magma both contains a large amount of gas and is highly viscous, meaning that it can hold in this gas until the external pressure has fallen to almost nothing. As magma of this type approaches the surface, and the overlying pressure drops rapidly, the trapped gas suddenly escapes all at once. The result is a highly explosive eruption as pressure is released, and gases dissolved in the trapped magma turn into a mass of expanding bubbles. This causes the magma to fragment into ash particles and be expelled upward. If rising magma comes in contact with surface or groundwater, the result can be a violent explosion of steam together with ash formed from the sudden break-up of the magma. This is called a phreatomagmatic eruption.

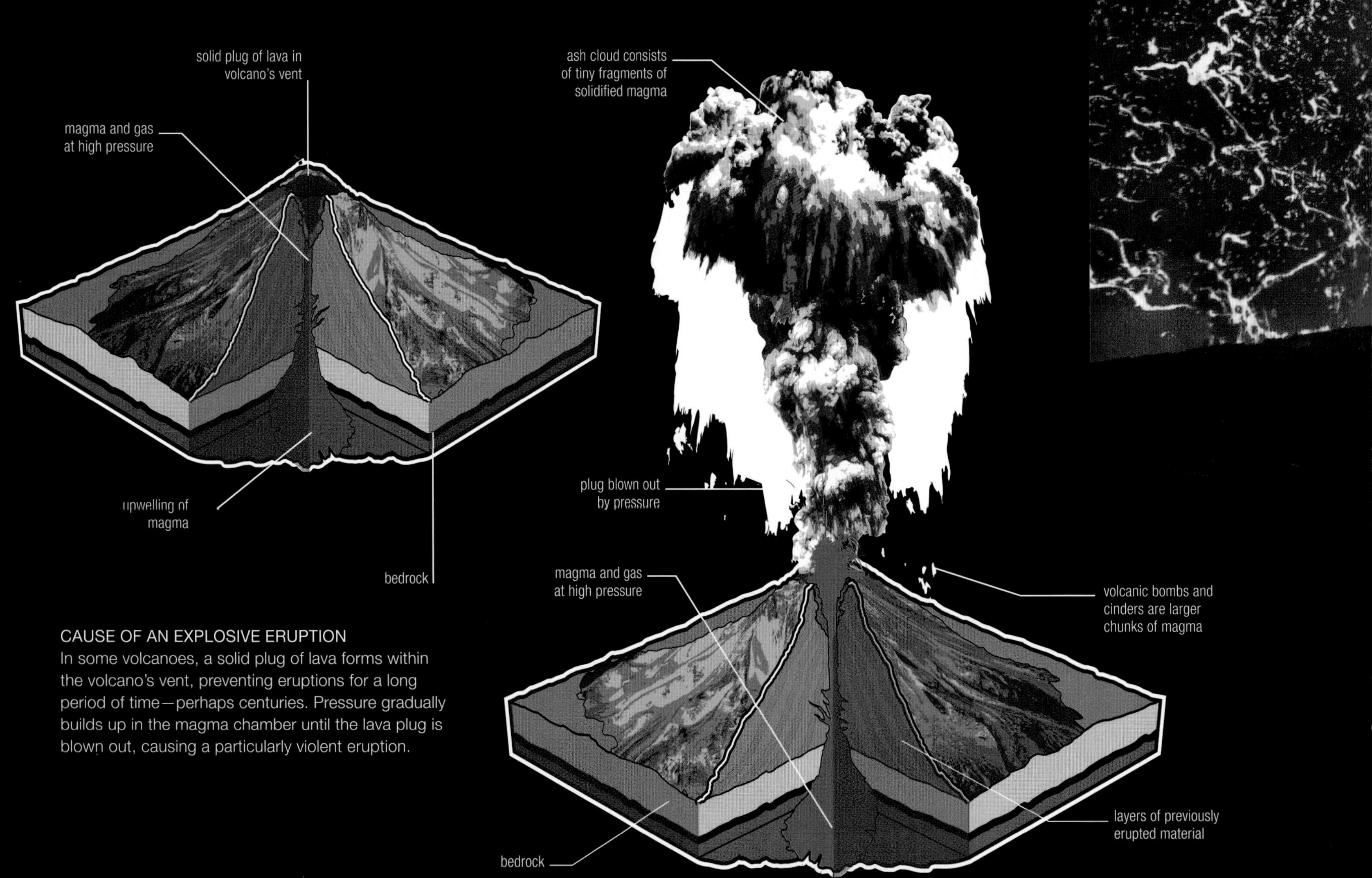

CAUSE OF AN EXPLOSIVE ERUPTION

In some volcanoes, a solid plug of lava forms within the volcano's vent, preventing eruptions for a long period of time—perhaps centuries. Pressure gradually builds up in the magma chamber until the lava plug is blown out, causing a particularly violent eruption.

> “THE REASON OF THESE FIRES IS THE ABUNDANCE OF SULPHUR AND BRIMSTONE... IN THE BOSOME OF THE HILL.”

SIR THOMAS POPE BLOUNT, IN A NATURAL HISTORY: CONTAINING MANY NOT COMMON OBSERVATIONS (1693).

VOLCANIC ACTIVITY

Volcanoes were once categorized as either active, dormant, or extinct, according to the frequency of their eruptions, but volcanologists no longer use this classification. Some volcanoes are still categorized as extinct if they clearly no longer have a magma supply. All other volcanoes are considered active, though a distinction is made between volcanoes that have erupted at least once in recorded history (called historically active), and those for which there is evidence only of an eruption in the past 10,000 years (Holocene active). There are about 1,550 holocene active volcanoes in the world of which 573 have historical eruptions.

ERUPTION MAGNITUDE

The strength of volcanic eruptions is measured on the Volcanic Explosivity Index or VEI scale (see below). The VEI value is based on the height of an eruption's ash plume or column and an estimate of the volume of material that is expelled. Particular eruption types (see pp.90–91) often have similar VEIs. For example, Strombolian eruptions usually have a VEI value of 1 or 2, while Plinian eruptions have a VEI of 5 or 6.

MAGMA EXPLOSION
The explosive eruption of magma at a volcanic vent—as here at the Kilauea volcano in Hawaii—is driven by gas expansion as pressure is suddenly released.

MEASURING VOLCANIC ERUPTIONS
The Volcanic Explosivity Index is the volcano equivalent of the Richter Scale. Eruptions with small VEI numbers are common, those with high numbers are rare. Thus, only five eruptions with a VEI of 6 or 7 have occurred since 1800, and eruptions with a VEI of 8 occur only about once in 100,000 years.

VOLCANIC EXPLOSIVITY INDEX (VEI)

VEI	DESCRIPTION	HEIGHT OF COLUMN	VOLUME OF MATERIAL	EXAMPLE	YEAR
0	**Nonexplosive**	Up to 330ft (100m)	Up to 350,000ft^3 (10,000m^3)	**Mauna Loa**	Various
1	**Gentle**	330–3,300ft (100–1,000m)	Over 350,000ft^3 (10,000m^3)	**Stromboli**	Various
2	**Explosive**	½–3 miles (1–5km)	Over 35 million ft^3 (1 million m^3)	**Tristan da Cunha**	1961
3	**Severe**	2–9 miles (3–15km)	Over 350 million ft^3 (10 million m^3)	**Etna**	2003
4	**Cataclysmic**	6–15 miles (10–25km)	Over 0.02 miles^3 (0.1km^3)	**Eyjafjallajökull**	2010
5	**Paroxysmal**	15 miles (Over 25km)	Over 0.2 miles^3 (1km^3)	**Mount St. Helens**	1980
6	**Colossal**	15 miles (Over 25km)	Over 2 miles^3 (10km^3)	**Krakatau**	1883
7	**Super-colossal**	15 miles (Over 25km)	Over 25 miles^3 (100km^3)	**Tambora**	1815
8	**Mega-colossal**	15 miles (Over 25km)	Over 240 miles^3 (1,000 km^3)	**Toba**	70,000 years ago

ERUPTION TYPES

Volcanic eruptions are usually thought of as sudden, cataclysmic explosions that produce large quantities of lava, ash, and other volcanic products. In practice, volcanoes can erupt in any of several different ways. A single volcano can erupt in different styles during separate eruptions or even during separate stages in the same eruption.

PICHINCHA PHREATIC ERUPTION
A large mushroom cloud of steam and ash rises from Guagua Pichincha, Ecuador, in 1999.

PLINIAN

These extremely violent eruptions produce colossal plumes of gas and ash. Rare, extra-large ones are termed ultraplinian. See also Plinian Eruptions, pp.154–55.

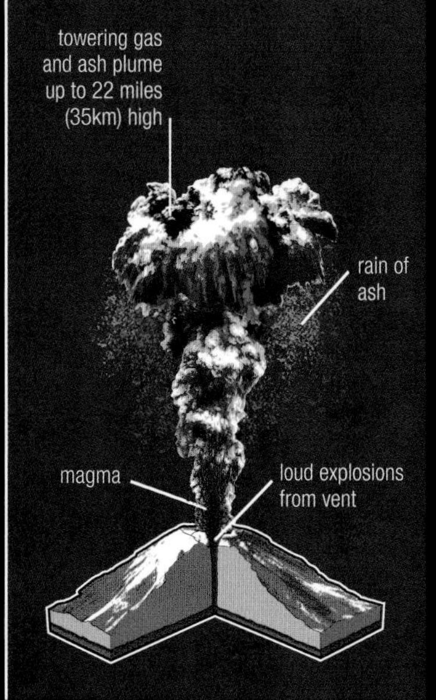

1991 ERUPTION OF PINATUBO
An eruption of Mount Pinatubo in the Philippines in June 1991 (above) was one of just a handful of Plinian eruptions in the 20th century. It produced a 21-mile (34-km) high ash column and killed more than 800 people.

VULCANIAN

Eruptions of this type start with a noisy explosion and feature volcanic bombs and an ash plume, often followed by a lava flow. See also Vulcanian Eruptions, pp.150–51.

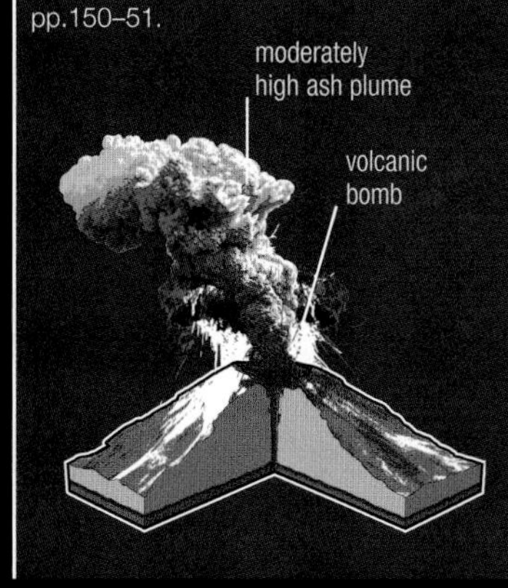

SURTSEYAN

Surtseyan eruptions result from the top of an underwater volcano reaching the surface and producing vigorous explosions. See also Surtsey, pp.256–57.

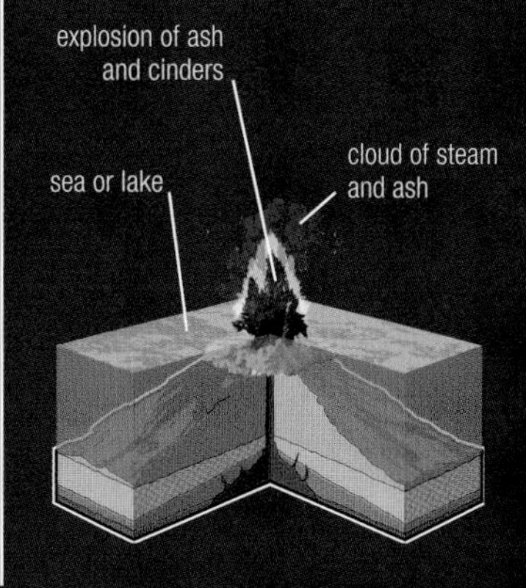

PHREATIC

Caused by volcanically heated rock coming in contact with cold groundwater or surface water, phreatic eruptions feature explosions of steam, ash, volcanic bombs, and rock. Related eruptions caused by interactions between magma and water are termed phreatomagmatic. In either case, no incandescent lava is produced, and there are no lava flows. See also Phreatic Eruptions, pp.164–65.

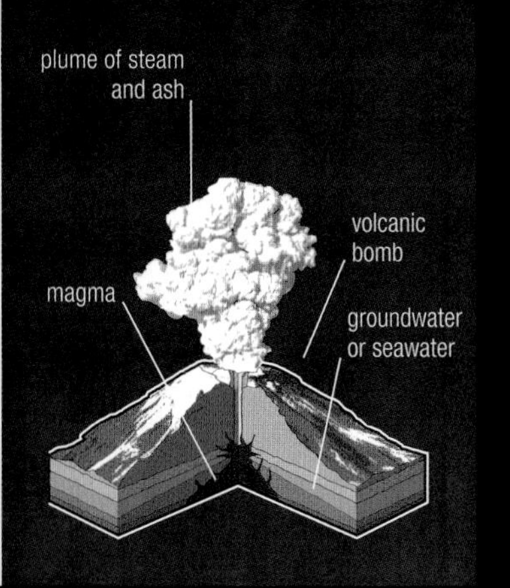

ASH AND GAS FLOW, MONTSERRAT
A pyroclastic flow—hot gas mixed with ash and rock fragments—flows down a slope on the Soufrière Hills volcano on the Caribbean island of Montserrat. Events like this are a hallmark of Peléan-type eruptions.

KILAUEA LAVA FOUNTAIN
Hawaiian-style eruptions are so-named because the large Hawaiian volcanoes Mauna Loa and Kilauea (right) usually erupt in this manner.

PELEAN

An important characteristic of this eruption type is a flow of pyroclastic material (mixture of hot gas and ash) at speeds up to 100mph (160kph) down the volcano's flank. See also Peléan Eruptions, pp.152–53.

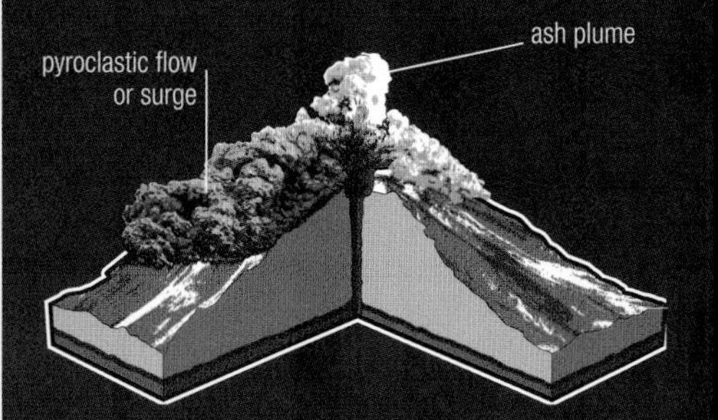

FISSURE AT KRAFLA
Fissure eruptions are also known as icelandic-style eruptions, since they occur commonly in Iceland, as in the eruption of the Krafla volcano (left).

STROMBOLI AT NIGHT
Strombolian eruptions are named after the small stratovolcano Stromboli (below), off the coast of Sicily, which almost continuously produces eruptions of this type.

FISSURE OR ICELANDIC

The main characteristic of this eruption type is the appearance of a long straight fissure in the ground. Large amounts of runny lava pour quietly out of the fissure, sometimes via a line of small lava fountains. See also Fissure Eruptions, pp.142–43.

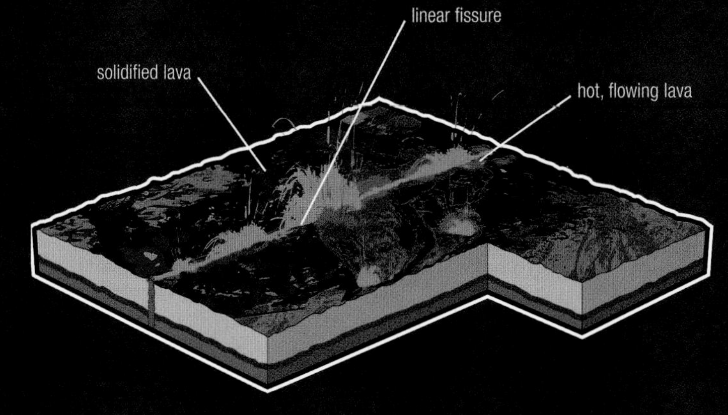

SUBGLACIAL

Where a volcano erupts under an ice cap or other type of glacier, it is called a subglacial eruption. See also Subglacial Volcanoes, pp.166–67.

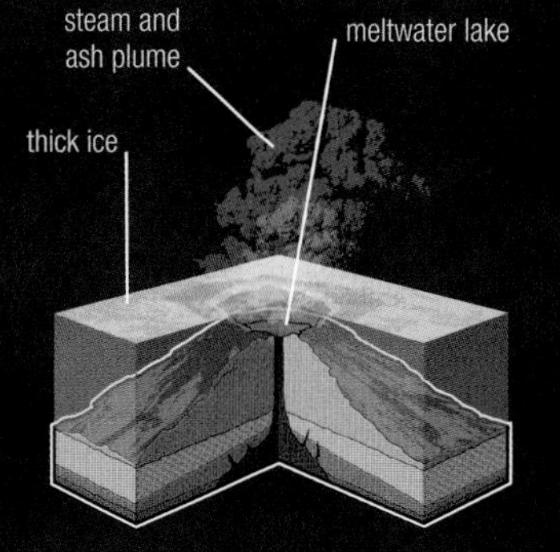

STROMBOLIAN

Eruptions of this type emit fountains and little bombs of lava at rhythmic intervals, sometimes interspersed with lava flows. See also Strombolian Eruptions, pp.148–149.

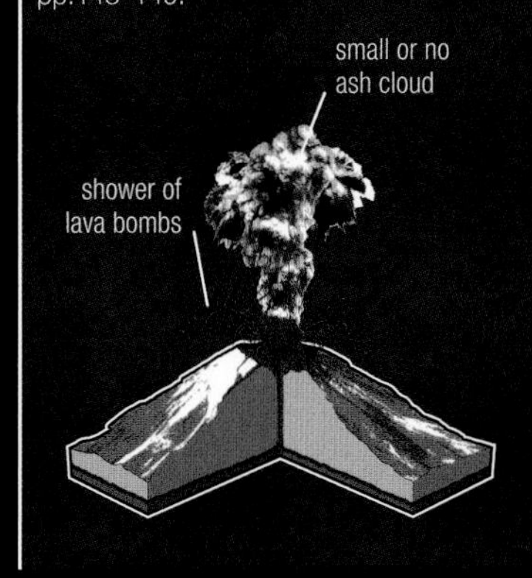

HAWAIIAN

In a Hawaiian-style eruption, there are relatively quiet outpourings of lava in the form of streams and fountains, usually from fissures on the volcano's flanks. Sometimes lava spills out of a lava lake in the summit crater. See also Hawaiian-Style Eruptions, pp.144–45.

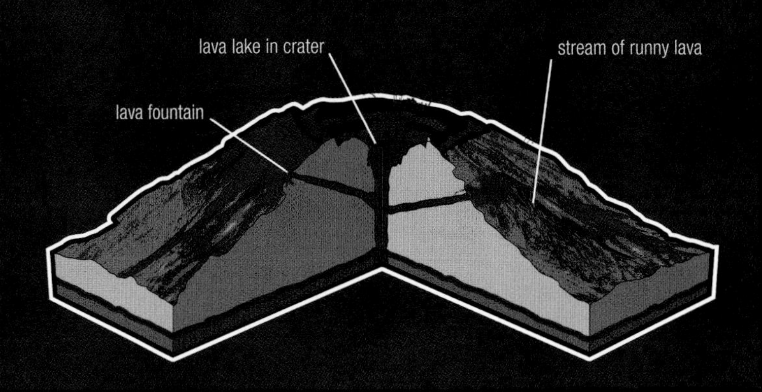

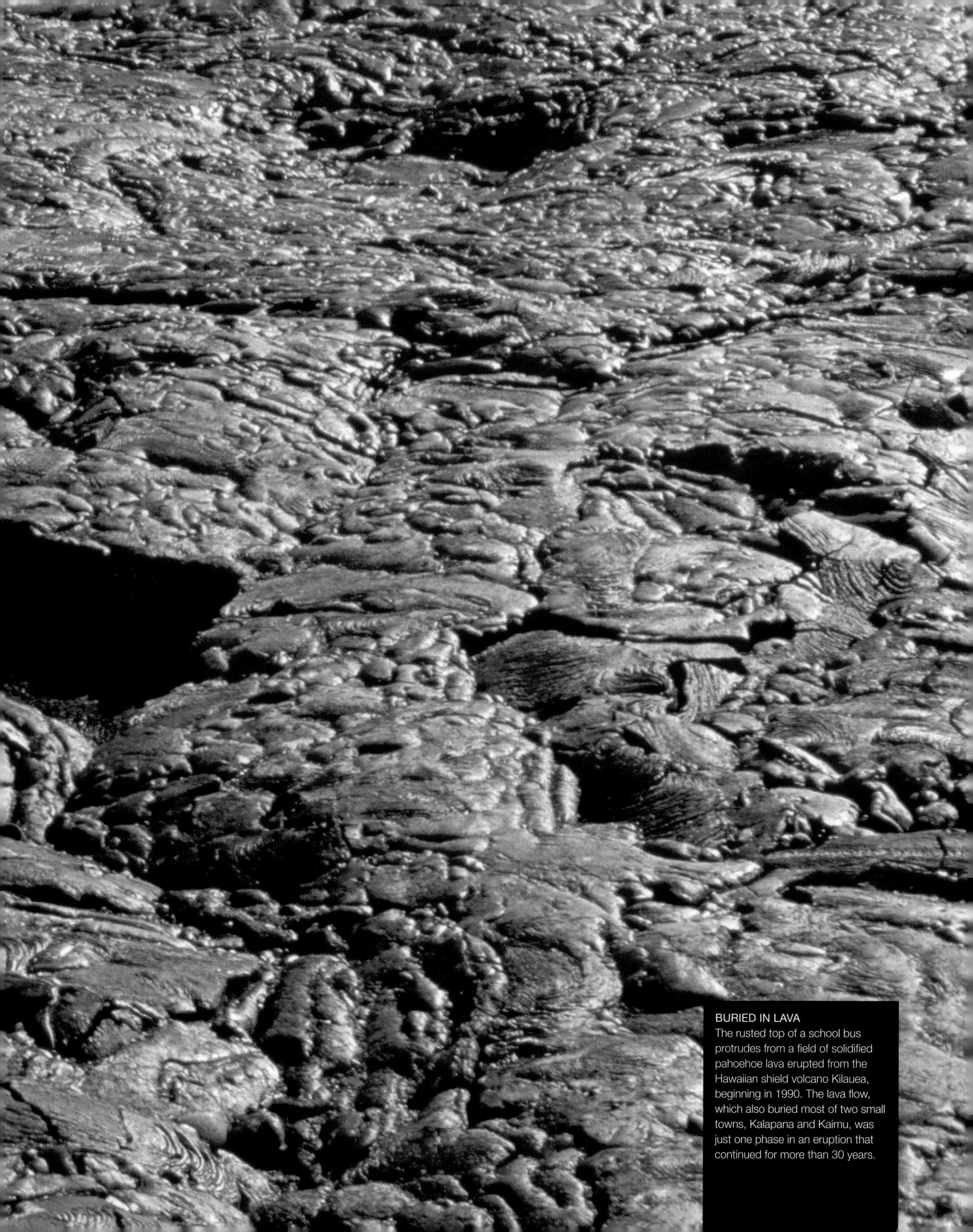

BURIED IN LAVA
The rusted top of a school bus protrudes from a field of solidified pahoehoe lava erupted from the Hawaiian shield volcano Kilauea, beginning in 1990. The lava flow, which also buried most of two small towns, Kalapana and Kaimu, was just one phase in an eruption that continued for more than 30 years.

VOLCANO TYPES

As well as erupting in a variety of ways, volcanoes take many different forms. The most familiar type is a large, steep-sided cone, but there are many other varieties, from much smaller cones to extensive, gently sloping areas of lava and water-filled depressions in the landscape. The form of volcano likely to be found in any part of the world depends on factors such as the type of magma (molten rock) formed within Earth's crust in that region.

SHIELD VOLCANOES

The giants of the volcano world are known as shield volcanoes. Shaped like broad, upturned shields, they are made of layer after layer of runny lava that flowed over the surface and then solidified. Shield volcanoes are typically formed above hotspots. See also pp.114–15.

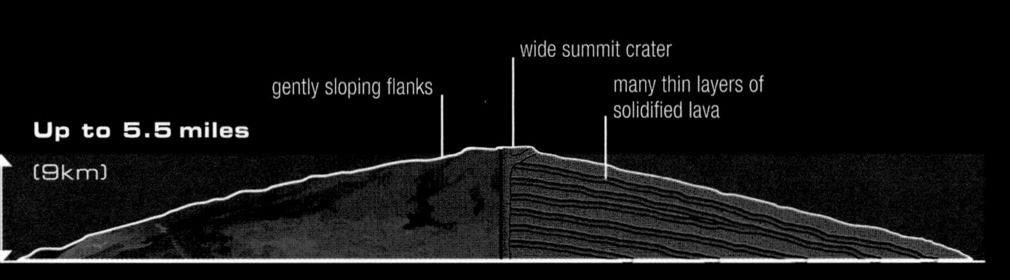

PITON DE LA FOURNAISE
Here, a small secondary cone is visible on the flank of the massive Piton de La Fournaise shield volcano, which forms part of the Indian Ocean island of Réunion.

REDOUBT VOLCANO
Situated on the coast of Alaska, Mount Redoubt is a large stratovolcano that erupted spectacularly in 1989–90 and again in 2009, producing enormous ash clouds.

STRATOVOLCANOES

Tall, steep-sided volcanoes, composed of successive layers of different types of volcanic product, are called stratovolcanoes or composite volcanoes. A common type of volcano, they form in parts of the world where viscous magma reaches Earth's surface. When they erupt, they often do so extremely violently. Many of the world's most famous volcanoes, such as Mount St. Helens and Etna, are of this type. See also pp.120–23.

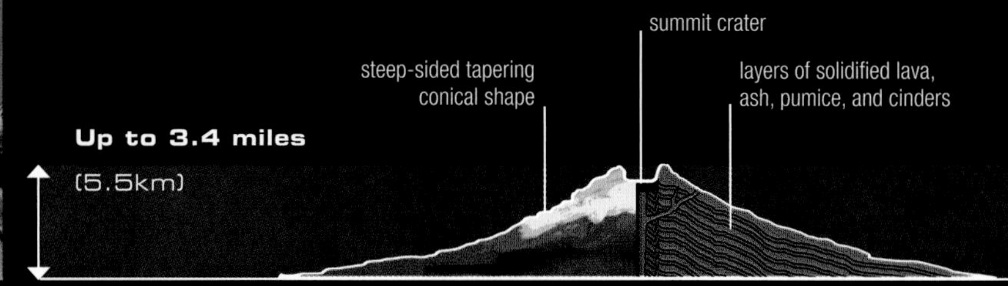

CALDERAS

If the upper part of a stratovolcano collapses following a cataclysmic eruption, the result is a caldera. This consists of a wide, deep crater, beneath which (in a still active volcano) lies a magma chamber. Many calderas are filled with water or even partly submerged. Some contain new volcanic cones or other volcanic features growing within them. See also pp.126–27.

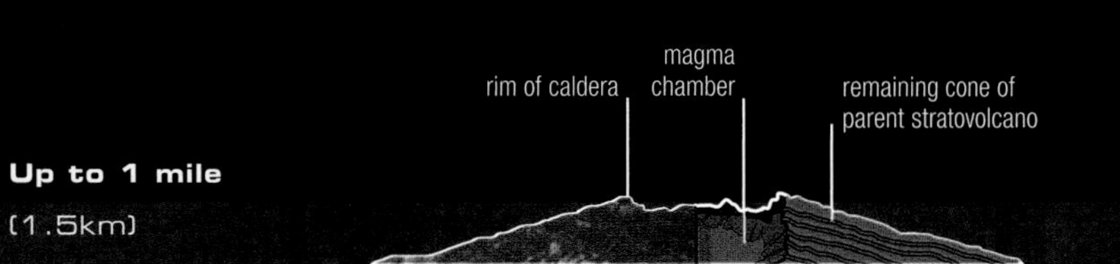

QUILOTOA CALDERA
This caldera in the Ecuadorian Andes formed about 800 years ago following the partial collapse of a stratovolcano. It now contains a deep lake, colored green by dissolved minerals.

ZUNI SALT LAKE
Located in New Mexico, Zuni Salt Lake occupies the lower part of a maar that is about 6,500ft (2,000m) across and 390ft (120m) deep. A sacred Native American site, the lake frequently dries out to leave salt flats. At its edge are two small cinder cone volcanoes.

MAARS

Volcanoes known as maars consist of relatively small, shallow, bowl-shaped craters sunk slightly into the ground. They are often filled with water to produce lakes, though there are also some dry maars in desert areas. They form as a result of explosions when magma comes into contact with groundwater or permafrost as it rises toward the surface. See also pp.130–31.

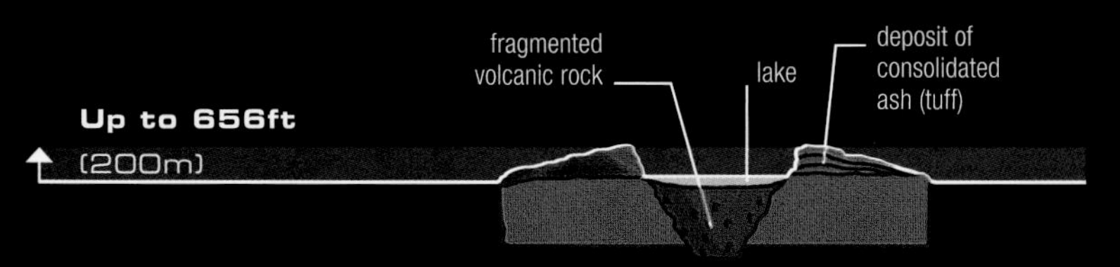

HAWAIIAN CONE
This tuff cone on O'ahu, Hawaii, is called Koko Crater. Like most tuff rings and cones, it formed in a single eruptive phase and is not expected to be active again.

TUFF CONES AND RINGS

These are small volcanoes with bowl-like central craters. They are formed in a similar way to maars—by rising magma coming in contact with groundwater or the sea. The resulting reaction produces a lot of volcanic ash, which accumulates in a cone or ring and consolidates to form a rock called tuff. See also pp.134–35.

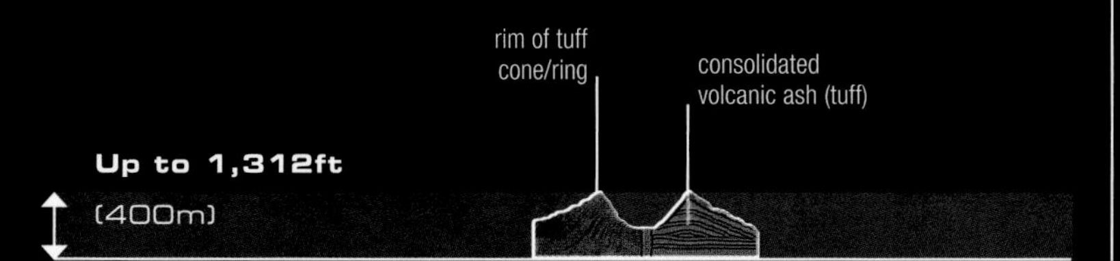

CINDER CONES

The relatively small cinder cones are composed mainly of loose volcanic cinders [glassy fragments of solidified lava] and ash. Also called scoria cones, they sometimes form on the sides of larger volcanoes. See also pp.116–17.

TWIN CONES
These two cinder cones form part of a volcanic landscape called the Mountains of Fire on Lanzarote, one of the Canary Islands.

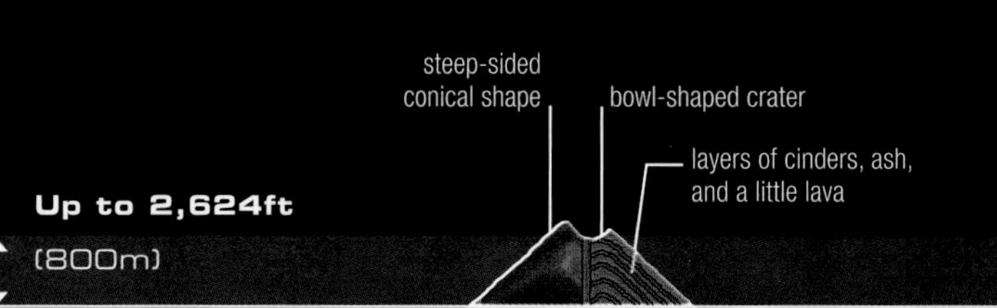

LAVA

Molten rock, or magma, that erupts onto Earth's surface from a volcano is called lava. As it spurts out it is red-hot, with a temperature between 1,290 and 2,190ºF (700 and 1,200ºC). Although much thicker and stickier than water, lava will flow over the ground, under the influence of gravity, as long as its temperature remains high enough. Eventually it cools to form a solid rock.

LAVA PROPERTIES

Different volcanoes produce different types of lava, which vary in temperature and composition, particularly their silica content. These properties affect how far the lava can flow. The hottest lavas, with the lowest silica content, are basaltic lavas and are quite runny. They can travel for tens of miles, even on gentle slopes, before solidifying to basalt. These lavas come in two main forms, called pahoehoe and ʻaʻa (see below). Andesitic lavas are cooler, more flow resistant, and can travel only short distances before solidifying. Dacitic and rhyolitic lavas make up a third group. They are the least fluid and form only slow-moving flows. They cool to form rocks called dacite and rhyolite.

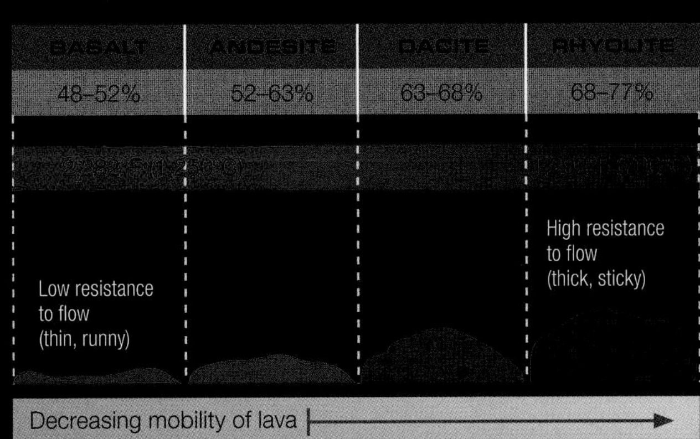

RANGE OF CHARACTERISTICS
From left to right, this chart shows lava with increasing silica content and viscosity (flow resistance), and decreasing temperature. Because it is relatively runny, the basaltic-type lava on the left can flow greater distances than the types on the right.

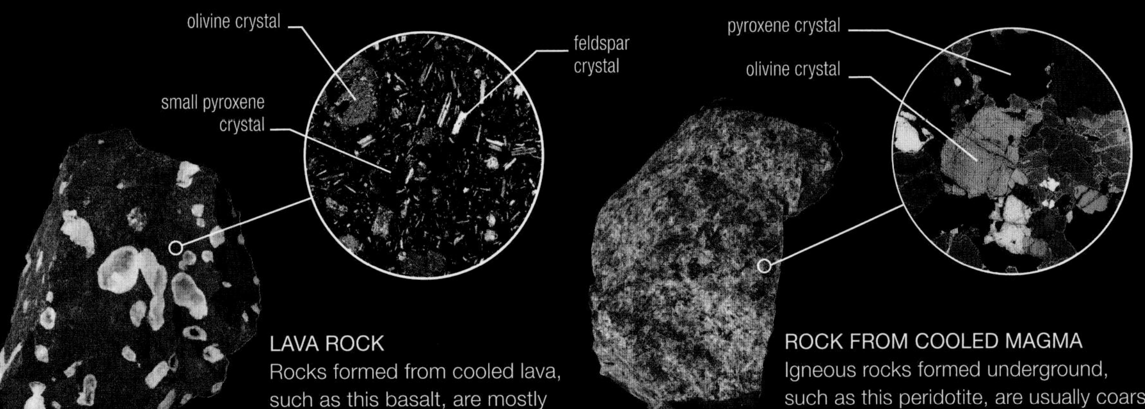

LAVA ROCK
Rocks formed from cooled lava, such as this basalt, are mostly fine grained, with small crystals, because the cooling was rapid.

ROCK FROM COOLED MAGMA
Igneous rocks formed underground, such as this peridotite, are usually coarse grained, with large crystals, because they come from magma that cooled slowly.

LAVA FOUNTAIN
Jets of lava sprayed forcefully, but not explosively, into the air are called lava fountains. They commonly occur in Hawaiian-style eruptions (see pp.144–45).

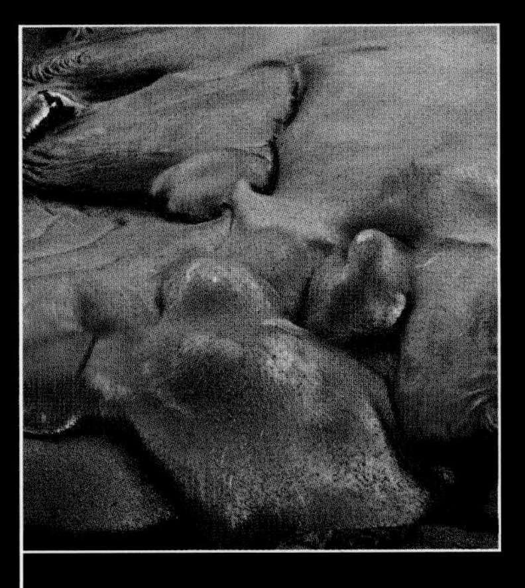

PAHOEHOE

Composition	Basaltic
Temperature	2,000–2,200°F (1,093–1,204°C)
Speed of advance	up to 6mph (10kph) (higher in channels)
Type	Shield volcanoes

Pahoehoe advances as a series of lobes, called toes. As its surface cools, it develops a thin, pliable skin, under which hot material streams. As the skin congeals it develops a ropelike texture.

ʻAʻA

Composition	Basaltic
Temperature	1,800–2,000°F (982–1,093°C)
Speed of advance	15–300ft/hour (5–100m/hour)
Type	Shield and stratovolcanoes

ʻAʻa is cooler and more flow-resistant than pahoehoe. Streams of ʻaʻa have extremely rough, fragmented surfaces. They can advance rapidly and are capable of pushing down houses and forests.

PILLOW LAVA

Composition	Basaltic, andesitic
Temperature	1,600–2,200°F (871–1,204°C) (interior)
Speed of advance	3–30ft/hour (1–9m/hour)
Type	Underwater volcanoes

Pillow lava occurs when a volcano erupts underwater. On contact with water, the lava solidifies as a pillow-shaped rock. Later, pillow lava may be exposed on dry land (as above) if the seabed is raised up.

BLOCK LAVA

Composition	Andesitic, rhyolitic
Temperature	1,400–1,700°F (760–927°C)
Speed of advance	3–15ft/day (1–5m/hour)
Type	Stratovolcanoes, lava domes

Relatively cool and flow-resistant, block lava advances slowly and produces short and stubby flows. When solidified (as above), it forms roughly cube-shaped lumps of rock with relatively smooth surfaces.

FRESHLY ERUPTED LAVA
Flows of freshly erupted molten lava, some of it surface-solidified (black), advance across a crater on the Hawaiian volcano Kilauea. Underneath is an older lava crust (gray), part of which, in the foreground, has just collapsed downward.

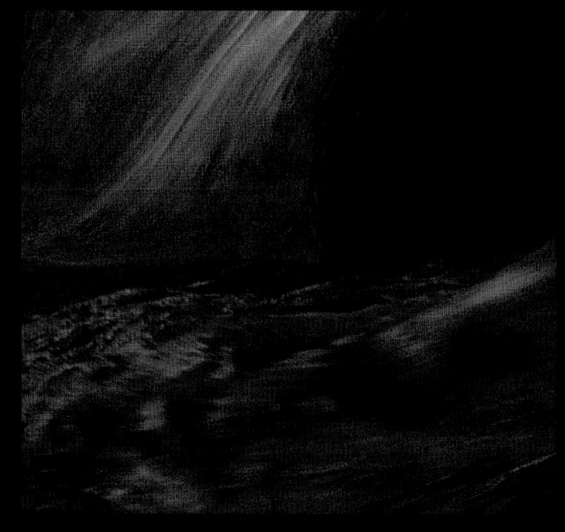

FAST-MOVING FLOW
This long-exposure photograph shows fast-flowing pahoehoe lava. On a moderate slope, as here, pahoehoe can attain speeds of 30mph (50kph) or faster.

MEETING THE SEA
Steam rises as the lava reaches the coast of Hawaii's Big Island. All the land here is made of black, solidified lava, the flows of which continually extend the island.

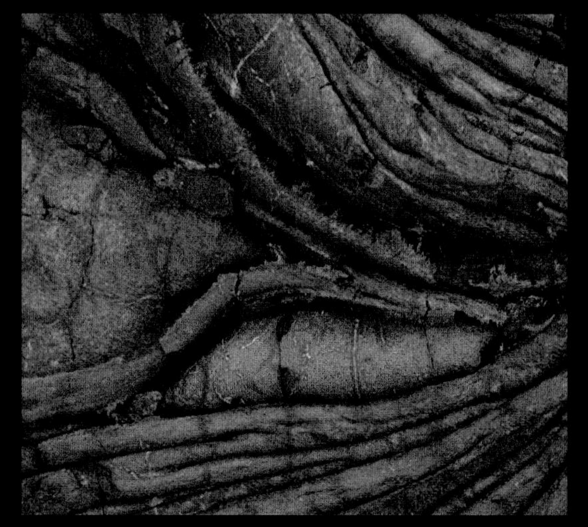

SOLIDIFIED PAHOEHOE
This pahoehoe lava produced by the Piton de la Fournaise volcano on Réunion Island, Indian Ocean, took on a typical ropelike texture as it solidified.

AERIAL PRODUCTS

Other than molten lava flowing out over the ground, the main products of eruptions are gases—some of them poisonous—and solid particles produced from magma and rock that has been blasted into the air. Each can pose hazards to people near and far.

SOLID PRODUCTS

The solid products that originate from materials hurled into the air by a volcano are referred to as tephra. They come in a range of sizes. The larger pieces are volcanic bombs—blobs of magma (molten rock) that solidified as they fell—and chunks of a rock called pumice, which forms from magma blown into the air as a froth containing gas bubbles. Medium-sized fragments are called cinders, while the smallest make up volcanic ash. Although much of this smaller-sized material comes from finely fragmented and solidified magma, in violent eruptions some of it comes from preexisting hardened lava around the volcano's vent, sometimes from deep within the volcano that is blasted skyward with the hot magma.

Tephra pose a varying degree of danger to people in eruption zones. Volcanic bombs are a hazard because of their size. Very heavy ash falls can cause death by suffocation, and combined with torrential rain can result in dangerous mudflows (see pp.106–07), while wet ash on house roofs is heavy and can lead to collapse. Inhaling even tiny amounts can cause problems for people with respiratory illnesses. Cinders are less of a danger, although if you are in the cinder zone (see below) you are at risk of being hit by falling cinders, which can burn or cause head injuries.

VOLCANIC GASES

Gases given off during volcanic eruptions include water vapor, nitrogen, and various asphyxiating, poisonous, or irritant gases such as carbon dioxide, carbon monoxide, and sulphur dioxide. The harmful gases pose the greatest hazard close to the volcanic vent, where concentrations are greatest. Rarely they can be of imminent danger to people living downwind of the eruption. Eruptions of carbon dioxide from volcanoes have, occasionally, caused mass human deaths from asphyxiation. In addition, volcanic gases cause air pollution. Sulphur dioxide reacts with moisture in the air to form acid rain, which is damaging to vegetation and a health hazard for the elderly and infirm.

FALLOUT DISTANCES

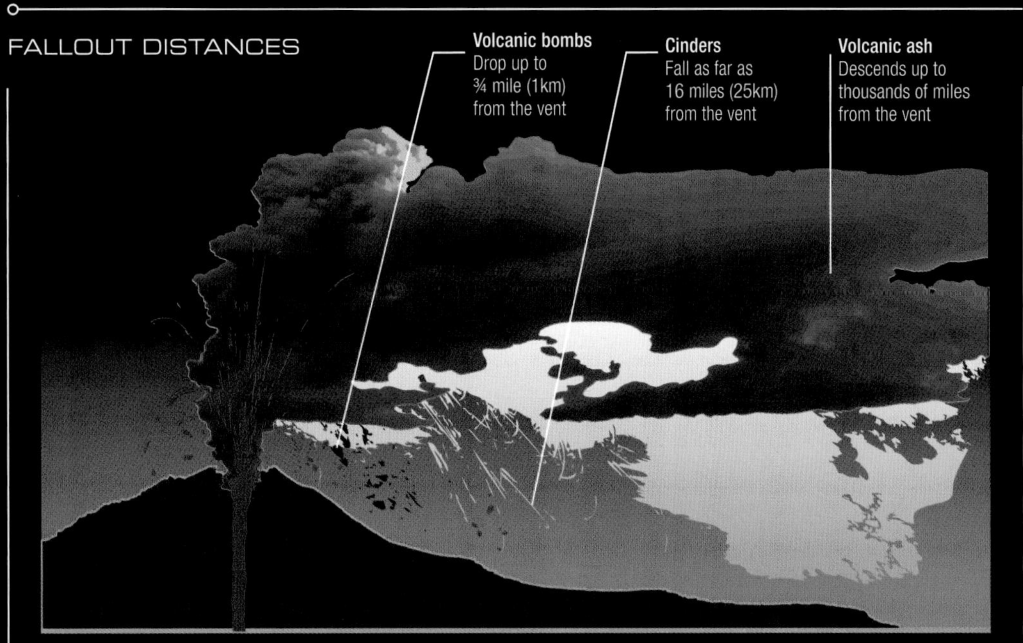

Different types of tephra tend to fall to the ground at varying distances from the site of the eruption, due to their different size ranges and susceptibility to being transported in the wind. The heaviest, volcanic bombs, drop closest to the volcanic vent, usually within a mile. The next heaviest, cinders, can fall further away. The lightest, volcanic ash, may be carried for tens, hundreds, or thousands of miles, depending on wind strength. In 1883, ash from the eruption of Krakatau, in what is now Indonesia, eventually fell to the ground all around the world.

HOT ASH
The eruption of Iceland's Eyjafjallajökull volcano in 2010 led to the cancellation of aircraft flights due to the large concentration of tiny glassy fragments it threw into the atmosphere.

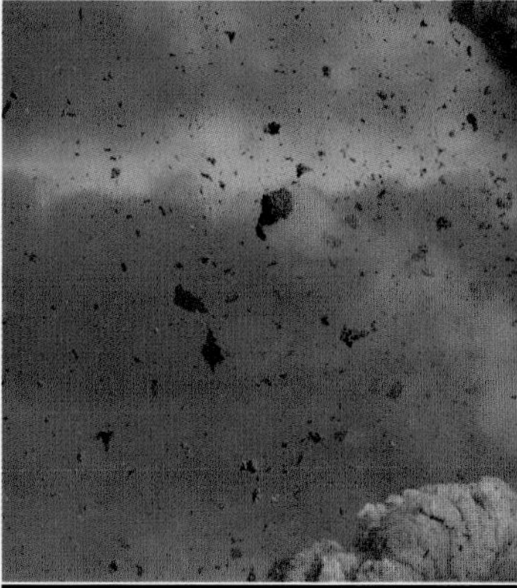

CINDERS

Also called lapilli, cinders are solid particles between 1/10in (2mm) and 2½in (6.4cm) in diameter. Often teardrop- or button-shaped, they fall in showers, sometimes welding together as they hit the ground.

LAVA BOMBS

These blocks and bombs of lava can be anything from 2½in (6.4cm) in diameter to boulder sized. The larger ones are solid on the outside but still molten in the center as they hit the ground.

VOLCANIC ASH

The smallest particles—less than 1/10in (2mm) in diameter—are carried into the atmosphere as a plume that can interfere with aviation. Where the ash eventually falls, it forms a dustlike layer.

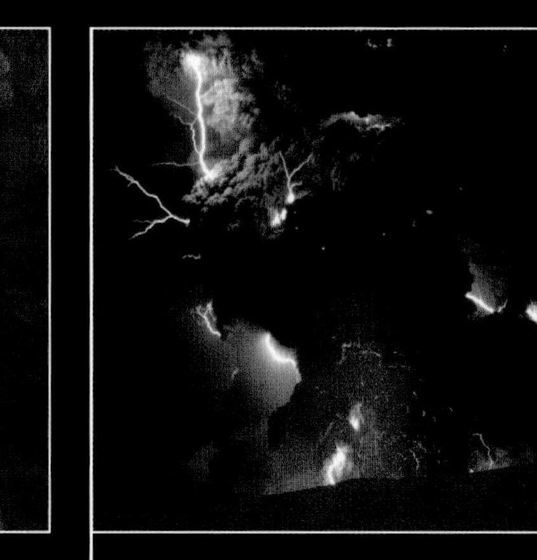

LIGHTNING

Lightning flashes occur commonly in volcanic ash plumes, due to friction between ash-laden cloud and the normal atmosphere. The combination of ash and lightning is called a "dirty thunderstorm."

VOLCANIC GASES

During eruptions, gases that were dissolved in magma are released. These gases typically have a temperature above 752ºF (400ºC). Gases are also released from magma that remains below ground.

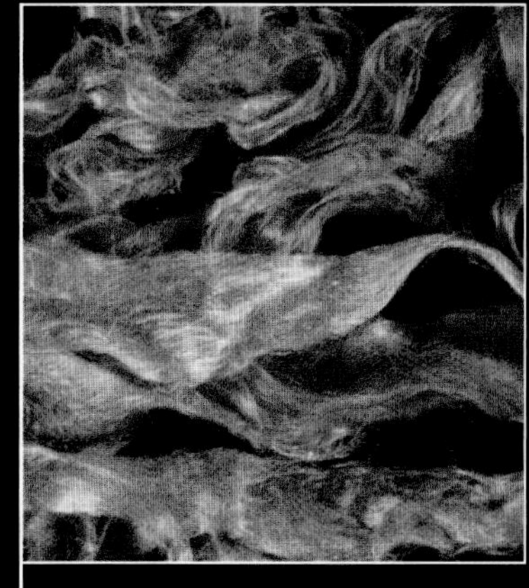

PELÉ'S HAIR

This curious, yellow-colored material is formed when airborne particles of magma are spun by the wind into glassy, hairlike strands. It is named after Pelé, the Hawaiian goddess of volcanoes.

PYROCLASTIC FLOWS AND SURGES

One of the most dangerous phenomena associated with a volcanic eruption are those that produce pyroclastic flows—hot, fast-moving, ground-hugging mixtures of ash, rock, and hot gas. Just as devastating are related events called pyroclastic surges.

FLOWS

Also known as *nuées ardentes* (glowing clouds), pyroclastic flows flatten, burn, and bury everything they encounter. Most travel for around 3–6 miles (5–10km). They are most often caused by the collapse of an ash column following a large eruption. Normally, the erupted ash heats the surrounding air, and the ash-gas mixture rises by convection. However, if the air hasn't heated up sufficiently, both ash and gas fall back down the flanks of the volcano. Other causes include eruptions in which a large lateral blast comes from the side of a volcano, a large lava dome at the volcanic vent collapses, or a thick, slowly moving lava flow front collapses.

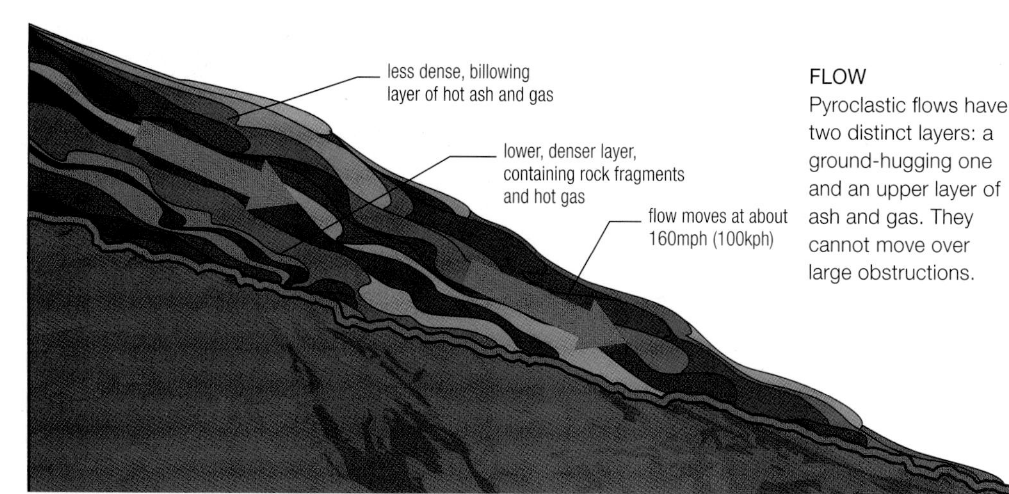

FLOW
Pyroclastic flows have two distinct layers: a ground-hugging one and an upper layer of ash and gas. They cannot move over large obstructions.

SURGES

Surges contain a higher proportion of gas to rock and are faster than pyroclastic flows. Hot surges contain gas and steam at temperatures of 212–1,472°F (100–800°C). Cold ones usually have a temperature below 212°F (100°C) and are produced when magma comes into contact with a large quantity of water known as a phreatomagmatic eruption. They often contain poisonous gases.

SURGE
A pyroclastic surge consists mainly of gas with some ash and small rock fragments. It is more turbulent than a flow with no distinct layers.

COLLAPSING ASH CLOUD
A searing combination of hot gas and ash roars downhill during an eruption of the Soufrière Hills volcano on the Caribbean island of Montserrat.

AFTER THE PYROCLASTIC FLOW
These houses on Montserrat were blasted and buried by a pyroclastic flow following the eruption of the Soufrière Hills volcano in June 1997.

ENVIRONMENTAL IMPACT

Pyroclastic flows and surges incinerate all vegetation they encounter, devastating huge areas. A pyroclastic flow from Mount St. Helens in 1980, which impacted an area of 230 sq miles (600 sq km) previously covered by dense forest, left not a single plant alive. Thousands of mammals and millions of fish and birds were also wiped out. Areas affected by these events can take decades to recover, though encouragingly, they eventually do so.

HUMAN IMPACT

Pyroclastic flows from Mount Lamington in Papua New Guinea in 1951, killed almost 3,000 people, while those from El Chichón volcano in Mexico in 1982, caused 2,000 fatalities. As well as destroying buildings, pyroclastic events kill people through a combination of burning, asphyxiation, and poisoning. They have been implicated in some of the deadliest volcanic eruptions in recorded history, for example, the 4,000 killed by an eruption from Vesuvius in 1631 and 30,000 killed in a matter of minutes in Martinique in 1902 (see p.153). As well as causing high numbers of fatalities, pyroclastic flows and surges can also result in a huge burden of injury and illness, ranging from severe burns to respiratory problems from ash inhalation.

PYROCLASTIC FLOW
Photographers who had been observing the 1991 eruption of Mount Pinatubo in the Philippines fled from a huge pyroclastic flow bearing down on them. The eruption killed several hundred people, but the occupants of this truck—and the person who was photographing them—escaped.

VOLCANIC MUDFLOWS

Also called "lahars," volcanic mudflows are violent, fast-moving slurries of water, ash, rocks, and other debris, resembling flowing wet concrete, that—in response to various triggering factors—may surge down the flanks of a volcano. Large flows can rip trees out of the ground, transport boulders the size of houses, sweep away buildings and people, and bury extensive areas of land in thick mud.

OCCURRENCE

Lahars can be set off by anything that causes large quantities of water to become mixed with ash and other debris on a volcano. Scenarios include eruptions from water-filled volcanic craters, torrential storms, rainfall on fresh ash deposits, and eruptions or earthquakes causing glaciers around a volcano's summit to disintegrate and melt. Lahars can also result from pyroclastic flows (see pp.104–05) running into mountain lakes or melting glacier ice. Large lahars move at up to 60 miles (100km) per hour. They can surge for miles along river valleys at the base of a volcano, leaving mud deposits several feet thick in their wake. These deposits quickly congeal, making it difficult for anybody trapped in them to escape.

Deadly lahars have affected many parts of the world historically. In May 1919, for example, lahars from the Kelut volcano in Indonesia killed more than 1,500 people. Other examples include a lahar in Nicaragua that killed more than 1,500 people in 1998, triggered by hours of intense rainfall onto the Casita volcano.

MUDFLOW ON SOUFRIÈRE HILLS
This mud deposit was left on the western flank of the Soufrière Hills volcano, on the Caribbean island of Montserrat, following an eruption in 2006. Soufrière Hills has been erupting on and off since 1995, often producing pyroclastic flows and lahars.

> "WE WERE SWEPT ALONG, IN OUR CAR, FOR ABOUT FIVE MINUTES IN THICK WARM MUD."
>
> **EYEWITNESS** CAUGHT UP IN A LAHAR TRIGGERED BY THE 1980 MOUNT ST. HELENS ERUPTION

EMERGENCY AID
Apart from any fatalities caused, volcanic mudflows often leave huge numbers of people injured or homeless—24,000 in the case of the Nevado del Ruiz disaster—and fast provision of aid is essential.

NEVADO DEL RUIZ

The most deadly lahars in history followed an eruption of the Nevado del Ruiz volcano in Colombia in November 1985. Pyroclastic flows interacted with extensive glaciers covering the volcano's summit, triggering massive slurries of mud that surged down the side of the volcano. Heavy rain had recently fallen over the whole region, so the lahars took in enormous amounts of additional water and rock debris from river channels as they descended. This increased their volume and momentum. Worst hit was the town of Armero, where more than 23,000 people were either drowned or asphyxiated by burial in mud. Armero became the focus of the disaster because it was located at a depression in the landscape just downstream from where two lahars flowed into each other. The economic cost of the disaster was estimated at $7.7 billion. Today, a lahar warning system is in place in the region.

ARMERO DESTROYED
As a result of lahars from Nevado del Ruiz in 1985, 5,000 homes were destroyed in this town in central Colombia. Thousands were killed by a thick wall of mud that arrived in the dead of night.

THE MOUNT RAINIER THREAT

Some of the world's volcanoes have been identified as the source of large historic lahars, and a potential source of future ones. Mount Rainier, a large stratovolcano in Washington State, is one example of this. Although it hasn't erupted since 1894, Mount Rainier has several large glaciers on its upper slopes, which, if they disintegrated and melted, could cause catastrophic lahars. About 5,600 years ago, the largest known lahar at Mount Rainier, the Osceola Mudflow, surged more than 60 miles (100km) across Washington to reach areas on which parts of the city of Tacoma, and other communities to the south of Seattle, are now built. Deposits from this lahar extended over more than 212 sq miles (550 sq km) of northwestern Washington and produced mud deposits that were 262ft (80m) deep in places. Other smaller mudflows have flowed from Mount Rainier since then, and around 150,000 people live in communities that are built on old lahar deposits. If, as is possible, another lahar of comparable size to the Osceola Mudflow were to occur, it would bury a number of these towns and might even reach the center of Seattle.

To protect the population who live around Mount Rainier, a lahar warning system was installed in 1998. Seismometers are placed at various locations around the volcano to sense tremors that might indicate the start of a large mudflow. Alarms caution people to move to higher ground until the danger has passed.

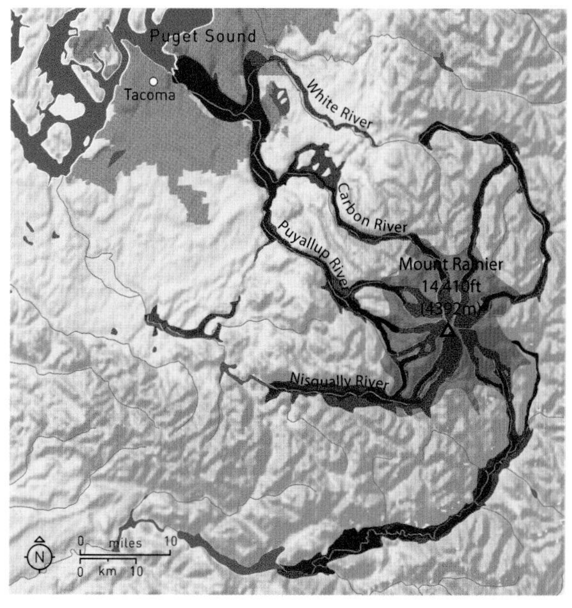

KEY

- Small lahars with recurrence interval of less than 100 years
- Moderate lahars with recurrence interval of 100–500 years
- Large lahars with recurrence interval of 500–1,000 years
- Area most likely to be affected by lava flows and pyroclastic flows
- Urban areas

LAHAR PATHS
This map of Mount Rainier and the surrounding region shows the paths of some historic lahars, which are also believed to be the most likely routes for future possible large mudflows.

CONTINENTAL VOLCANIC ARCS

All around the eastern and northern Pacific Ocean, where the edges of oceanic lithospheric plates push down beneath continents, some impressive arrays of volcanoes have formed inland on the continental sides of the plate boundaries. These are known as continental volcanic arcs.

CENTRAL AMERICAN ARC

The Pacific coastline of Central America is marked by a chain of more than 50 active volcanoes. This chain extends for 900 miles (1,500km), from Guatemala to western Panama. As with all continental volcanic arcs, these volcanoes have formed tens of miles inland from the coast and run parallel to it. They were created when the Cocos Plate on the west subducted beneath the Caribbean Plate on the east. This volcanic chain has produced some colossal eruptions in the past, including an eruption of Guatemala's Santa María volcano in 1902, which was one of the four largest of the 20th century.

ANDEAN VOLCANIC BELT

In western South America, intermingled with many nonvolcanic mountains, lies an interrupted chain of almost 200 volcanoes that were created by the Nazca and Antarctic Plates pushing beneath the South American Plate. Diverse in forms and activity, these volcanoes fall into four arcs, or zones. The first stretches across Colombia and Ecuador (northern arc), the second between southern Peru, southwestern Bolivia, and northern Chile (central arc), the third in Chile (southern arc), and the fourth in southern Chile and Argentina (austral arc). Many of these volcanoes pose a major hazard as they lie in densely populated regions.

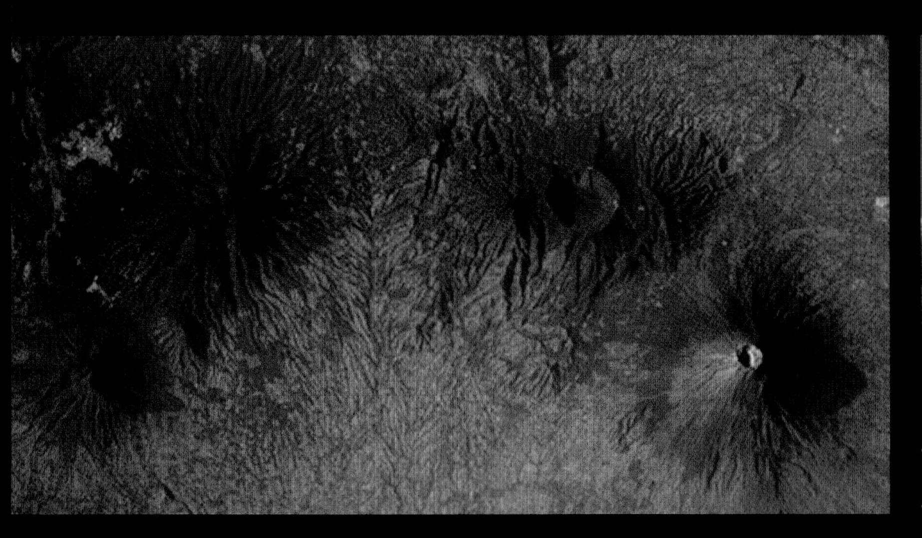

FOUR STRATOVOLCANOES
Four volcanoes in El Salvador, called Usulután, El Tigre, Chinameca, and San Miguel (left to right), lie near the middle of the Central American Arc. San Miguel last erupted in 2002.

ON THE BORDER
These two stratovolcanoes, called Licancabur and Juriques, are close to the southern end of the Chile–Bolivia border, within the central zone of the Andean volcanic belt.

KAMCHATKA VOLCANIC ARC

About 30 volcanoes are situated along the Kamchatka Peninsula in eastern Russia. They occur above a region where the Pacific Plate descends beneath the Okhotsk Plate. Like all continental volcanic arcs, the Kamchatka consists mainly of stratovolcanoes, many of which can erupt violently. However, because of the sparse population of the region, they are not a major threat to human life. The peninsula contains Kliuchevskoi, the tallest active volcano in Europe and Asia, which regularly emits ash plumes to 19,000 ft (6.000m) or higher. Its most active volcano is Karymsky, which has been erupting continuously since 1996.

KLIUCHEVSKOI AND NEIGHBORS
The symmetrical volcano in the foreground is Kliuchevskoi, with neighbors Kamen on the left and Ushovsky on the right.

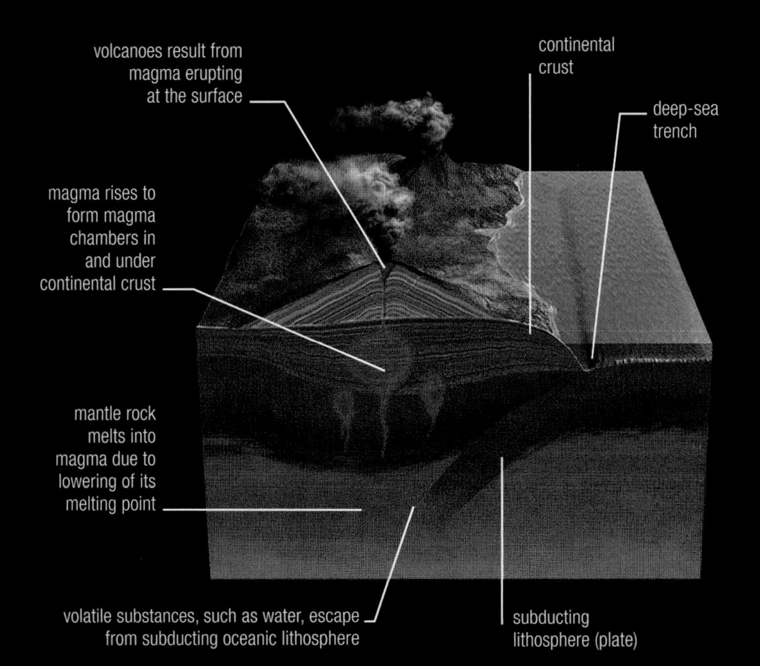

VOLCANO FORMATION

At plate boundaries, where the oceanic lithosphere is drawn down, or subducted, beneath the continental lithosphere, magma forms and leads to the creation of volcanoes. This process is thought to always occur in the same way. At depth, water and other volatile substances escape from the subducting oceanic lithosphere into the mantle region beneath the neighboring continent. There, the presence of these volatile substances acts as a flux and lowers the melting point of the mantle rocks. As a result, the mantle rocks melt to form magma—the hot, molten combination of melted rock and gases. This rises up to form magma chambers within the overlying continental crust. From these chambers, the magma erupts onto the surface to form volcanoes.

MAGMA FORMATION
The key process in the formation of continental volcanic arcs is the melting of mantle rocks at depth, which occurs when volatile substances escape from the subducting plate.

CASCADE VOLCANIC ARC

One of North America's arcs is known as the Cascade Volcanic Arc, or the Cascade Volcanoes. This chain of about 20 snow-capped volcanoes extends north for more than 680 miles (1,100km), from northern California, through Oregon and Washington, and into British Columbia (Canada). It includes well-known peaks such as Mount St. Helens and Mount Rainier in the US, and Mount Garibaldi in Canada. Many of these volcanoes are potentially dangerous, as they lie close to highly populated areas, such as Portland, Seattle, and Vancouver. They occur above a region where the Juan de Fuca Plate, a small tectonic plate in the northeastern Pacific, was subducted beneath the North American Plate. Of these volcanoes, Mount St. Helens has experienced the most eruptions, most recently in 2008 (see pp.160–61).

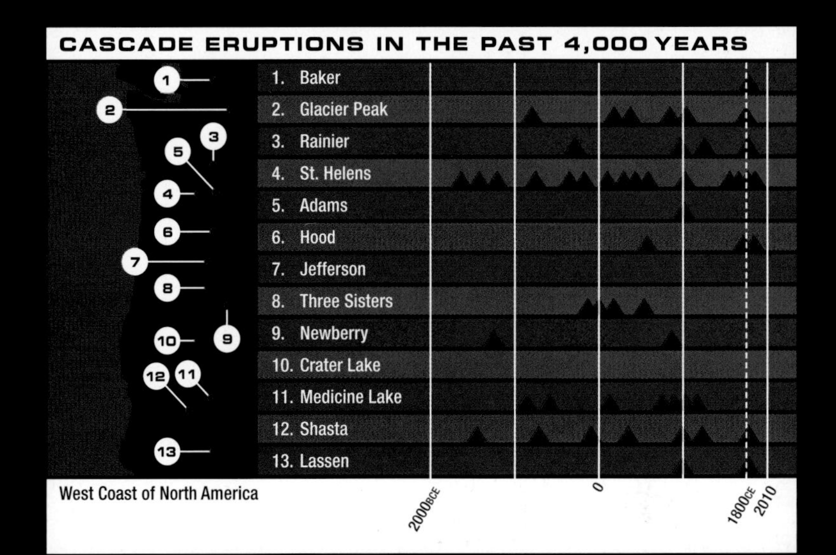

CASCADE ERUPTIONS
This chart of Cascade Volcanoes within the US shows that although only Mount St. Helens and Mount Lassen have erupted in very recent times, many others have done so repeatedly in the past 4,000 years.

VOLCANIC ISLAND ARCS

At numerous plate boundaries around the world, slabs of oceanic lithosphere on separate plates come together, or converge, with the edge of one plate subducting beneath its neighbor. The result is both a deep trench on the sea floor, and a gently curving line, or arc, of volcanic islands formed about 125 miles (200km) from the trench and parallel to it. These lines of islands are called volcanic island arcs.

ARC FORMATION

As in continental volcanic arcs, volcanic island arcs occur when magma forms at depth, close to the descending plate (see p.109), and then rises up to erupt at the surface. In an all-oceanic setting, the only difference is that the magma erupts onto the sea floor, eventually forming an arc of volcanic islands. A classic example is the Lesser Antilles Arc, which formed when the North and South American Plates subducted under the Caribbean Plate. It consists of a dozen small islands stretching in a perfect curve across the eastern Caribbean. Some of these have produced lethal eruptions over the past 200 years. Another, the Sunda Arc, was caused by the Australian Plate pushing beneath the Eurasian Plate. This arc includes many of the main islands of Indonesia. In this case, more than 70 distinct volcanoes have joined together to form the cores of two very large islands, Java and Sumatra, and many smaller ones. The Sunda Arc includes the notorious volcanoes Tambora and Krakatau, responsible for two of the most violent, lethal eruptions in history.

STRING OF ISLANDS
This view from space shows some of the Kuril Islands along with the Japanese Island of Hokkaido in the background. Both are parts of volcanic island arcs.

> "VOLCANOES FORM NO EXCEPTION TO THE PRINCIPLES OF UNIVERSAL ORDER."
>
> **JOHN KENNEDY**, *VOLCANOES: THEIR HISTORY, PHENOMENA, AND CAUSES*, 1852

ISLAND ARC FORMATION
About 60 miles (100km) under the sea, volatile substances escape from the oceanic lithosphere and act as a flux, lowering the melting point of the mantle rock above. Magma forms and rises to the surface, causing volcanoes to erupt on the sea floor, eventually forming islands.

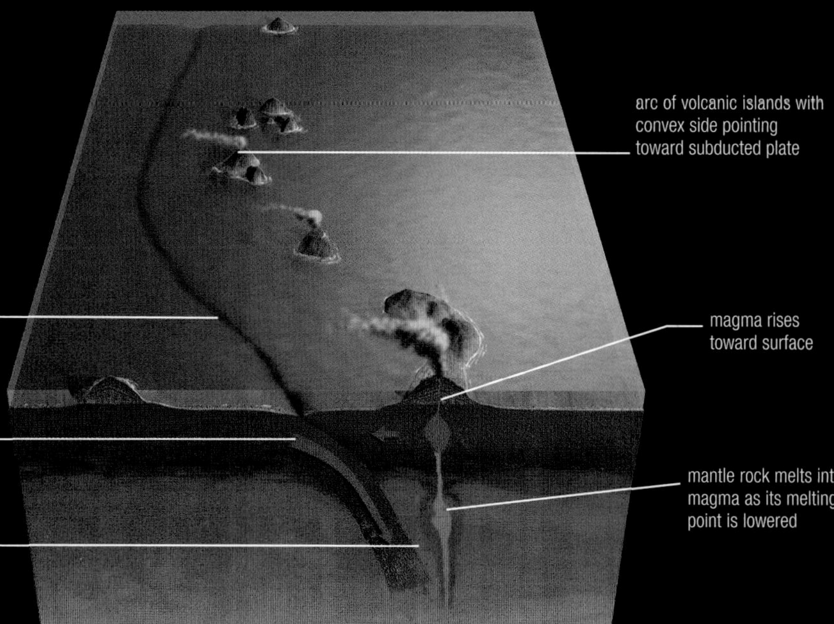

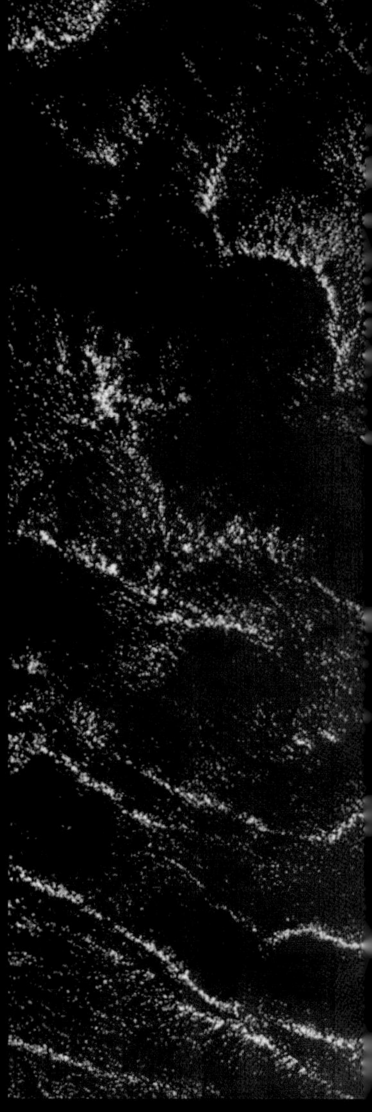

LESSER ANTILLES ARC
This Caribbean arc is about 530 miles (850km) long. Nearly every island has a volcano, with Dominica having nine.

ONEKOTAN ISLAND
Part of the Kuril Island Arc, which stretches between northern Japan and Russia's Kamchatka Peninsula, Onekotan Island consists of two connected volcanoes. The larger, Tao-Rusyr, is a caldera with a central crater lake, within which a small stratovolcano is growing. The other is Nemo Peak, which last erupted in 1938.

PACIFIC ARCS

Other than the Lesser Antilles and Sunda Arcs, most volcanic island arcs are situated around the edges of the Pacific—making up a large part of the Pacific Ring of Fire (see pp.204–05). One of the longest, the Aleutian Island Arc, lies in the Northern Pacific, where the Pacific Plate subducts beneath the North American Plate. To its southwest lies the Kuril Island Arc, and continuing to the southwest, the islands of Japan, which are themselves a volcanic arc. Further south is the 470 miles (750km) long Mariana Arc, lying approximately 110 miles (180km) to the west of the Mariana Trench, the deepest of all deep-sea trenches. Numerous other Pacific island arcs include the Japanese Izu and Ryukyu Islands, the Philippines, the Solomon Islands, and Vanuatu. Another example is the Bismarck Volcanic Arc, off the northeast coast of Papua New Guinea, which contains the dangerous volcanoes Ulawun and Rabaul.

RABAUL CALDERA
A flooded caldera, Rabaul is the easternmost volcano in the Bismarck Arc. It has two stratovolcanoes at its edge, one of which, Tavurvur, is visible here. A combined eruption from the two stratovolcanoes in 1937 killed more than 500 people.

VOLCANIC ISLAND CHAINS

If a plate consisting of oceanic lithosphere gradually moves over a hotspot at the top of the mantle (see pp.32–33), it can create a chain of islands. It is usual to find a volcano—or volcanoes—energetically building an island above the hotspot location, while the rest of the chain displays evidence of the past volcanic activity of the same hotspot.

HAWAIIAN CHAIN

The Hawaiian chain in the central Pacific is a classic example of a volcanic island chain. The Big Island of Hawaii—in particular its two large active volcanoes, Mauna Loa and Kilauea, and a young submarine volcano off Hawaii's south shore called Loihi—stands over a strong, active, and persistent hotspot under the Pacific Plate. Stretching away in a line to the northwest from the Big Island are other substantial islands. Beyond these are smaller islands, atolls, reefs, and submerged seamounts—mostly extinct submarine volcanoes—stretching out for 1,500 miles (2,500km) to a location called Kure Atoll. The whole chain is thought to have been created during the course of the past 30 million years as the Pacific Plate moved over a hotspot. A number of other volcanic chains in the central and southern Pacific, such as the Tuamoto Archipelago, show a similar pattern.

> "...WE ARE LED TO BELIEVE THAT WITHIN A PERIOD GEOLOGICALLY RECENT THE UNBROKEN OCEAN WAS HERE SPREAD OUT."
>
> **CHARLES DARWIN**, IN *THE VOYAGE OF THE BEAGLE*, 1845, REFERRING TO THE RELATIVELY RECENT ORIGIN OF THE GALÁPAGOS ISLANDS

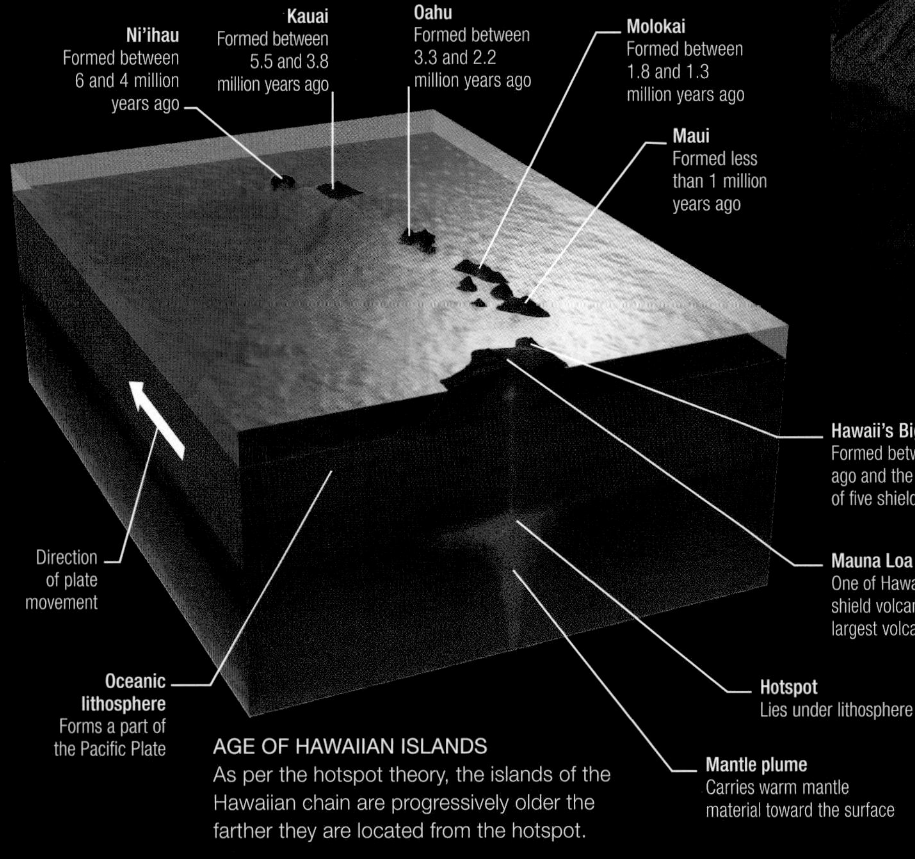

AGE OF HAWAIIAN ISLANDS
As per the hotspot theory, the islands of the Hawaiian chain are progressively older the farther they are located from the hotspot.

GALÁPAGOS ISLANDS

The Galápagos Archipelago is a group of volcanic islands that lie on the Nazca Plate in the eastern Pacific. They are thought to have formed because of the eastward movement of the Nazca Plate over a persistent hotspot. Unlike the simple pattern of the Hawaiian chain, the Galápagos hotspot has produced several lines of volcanoes that have formed over the past 5–10 million years. This complex formation has been attributed to the location of the Galápagos—it is close to a mid-ocean spreading ridge, where new plate is created. Varying activity at this ridge may have led to the unusual grouping of these islands.

GALÁPAGOS CRATERS
Signs of recent or past volcanic activity, caused by the presence of the Galápagos hotspot, can be seen all over the Galápagos Islands. The central volcanic crater here is named Beagle, after the ship that carried Charles Darwin to the islands in 1835.

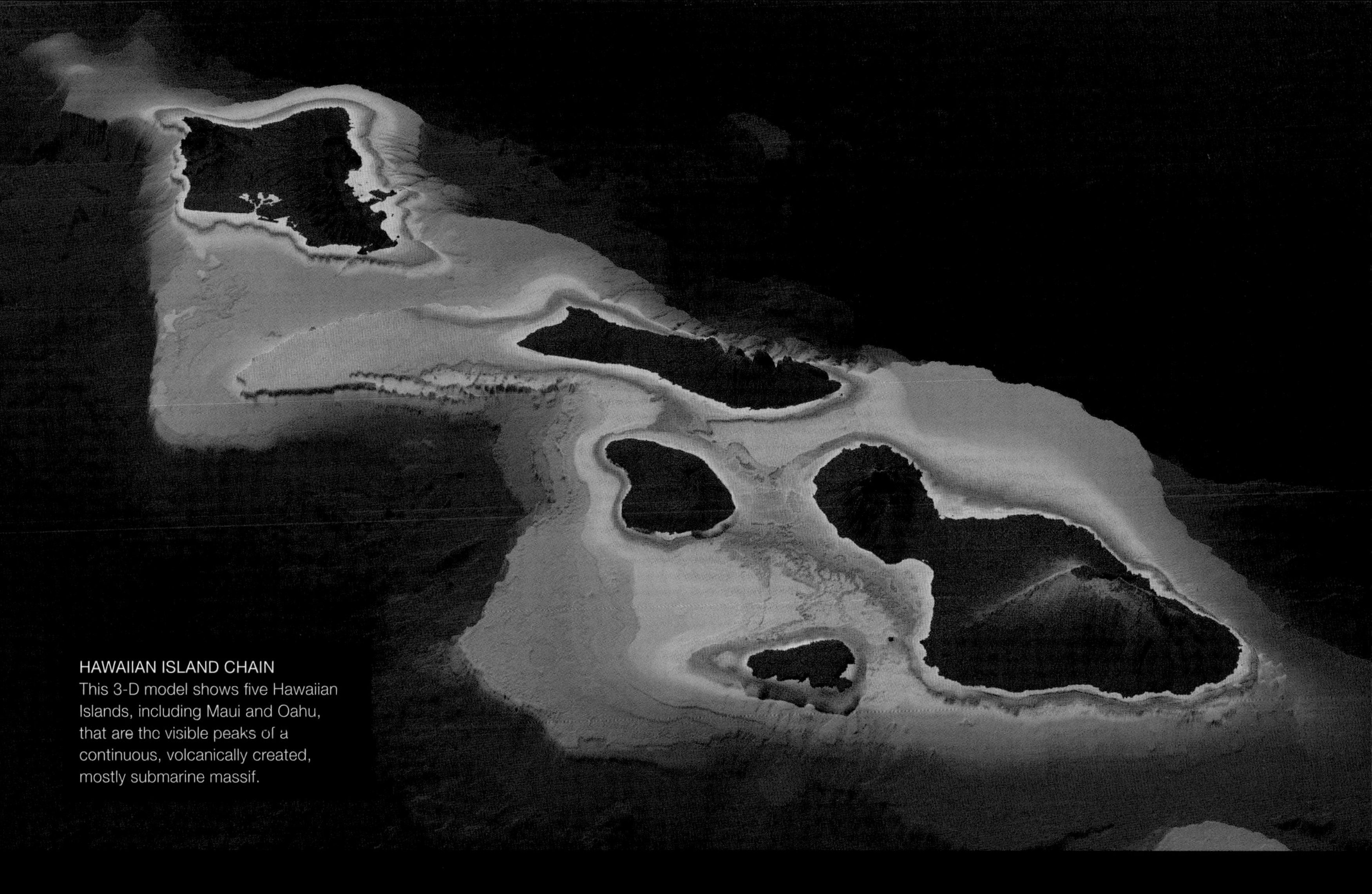

HAWAIIAN ISLAND CHAIN
This 3-D model shows five Hawaiian Islands, including Maui and Oahu, that are the visible peaks of a continuous, volcanically created, mostly submarine massif.

INDIAN OCEAN VOLCANIC CHAIN

One of the most remarkable volcanic chains believed to have been caused by a hotspot lies in the Indian Ocean. About 67 million years ago, plate movements are thought to have led India to pass over the hotspot, resulting in the formation of thick lava flows that make up the Deccan Traps. Later, as the Indian Plate moved in a northeasterly direction, more volcanic islands were created in the Indian Ocean. About 30 million years ago, a mid-ocean ridge passed over the same hotspot. Since then, the hotspot has been located under the African Plate, which has moved relative to the hotspot in a roughly easterly direction. After a period of quiet, the hotspot created more islands, including Mauritius and, most recently, Réunion.

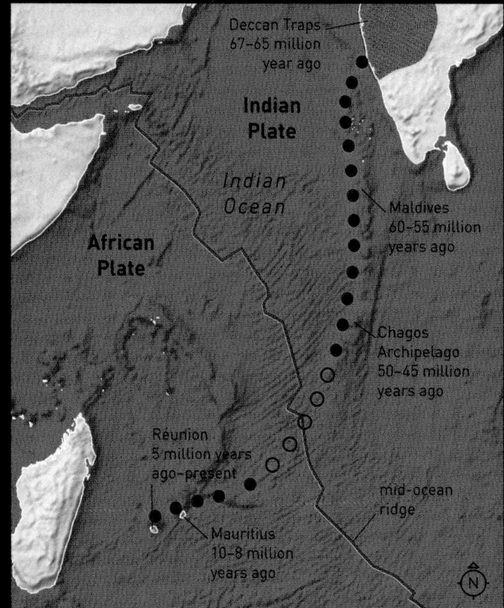

HOTSPOT TRACK
Over the past 67 million years, plate movements over the hotspot currently under Réunion are thought to have created the Deccan Traps, the Maldives, the Chagos Archipelago, Mauritius, and Réunion itself.

MORNE BRABANT, MAURITIUS
Mauritius shows many signs of past volcanism, including this large basaltic rock called Le Morne Brabant, probably caused by the Réunion hotspot.

SHIELD VOLCANOES

Shield volcanoes are broad, shield-shaped volcanoes formed from multiple layers of runny lava that erupted from the volcano, flowed over its flanks, and solidified. Shield volcanoes are limited in number and are found in most areas of the world but are most common in locations such as Hawaii, Iceland, the Galápagos Islands, and the East African Rift Zone.

STRUCTURE AND FORMATION

Shield volcanoes form in places where magma with a basaltic composition rises and erupts as lava at the surface. This typically occurs where there is a hotspot beneath the crust (see pp.32–33)—such a hotspot may be under oceanic crust, under a mid-ocean ridge, or under a continental rift zone. Basaltic lava is very fluid and can flow a long distance before it solidifies, which accounts for the broad shape of these volcanoes. Although some shield volcanoes are extinct, others undergo almost nonstop Hawaiian-style eruptions. In these eruptions, copious amounts of lava are quietly spouted out onto the surface, but in rare cases there have been huge explosions or ash columns.

SHIELD VOLCANO ERUPTION
During an eruption, runny lava is spewed onto the volcano's surface, often in the form of lava fountains, from fissures and parasitic cones on its flanks. Some lava may also spill out of the summit crater. Channels of lava develop and spread the erupted material over a wide area.

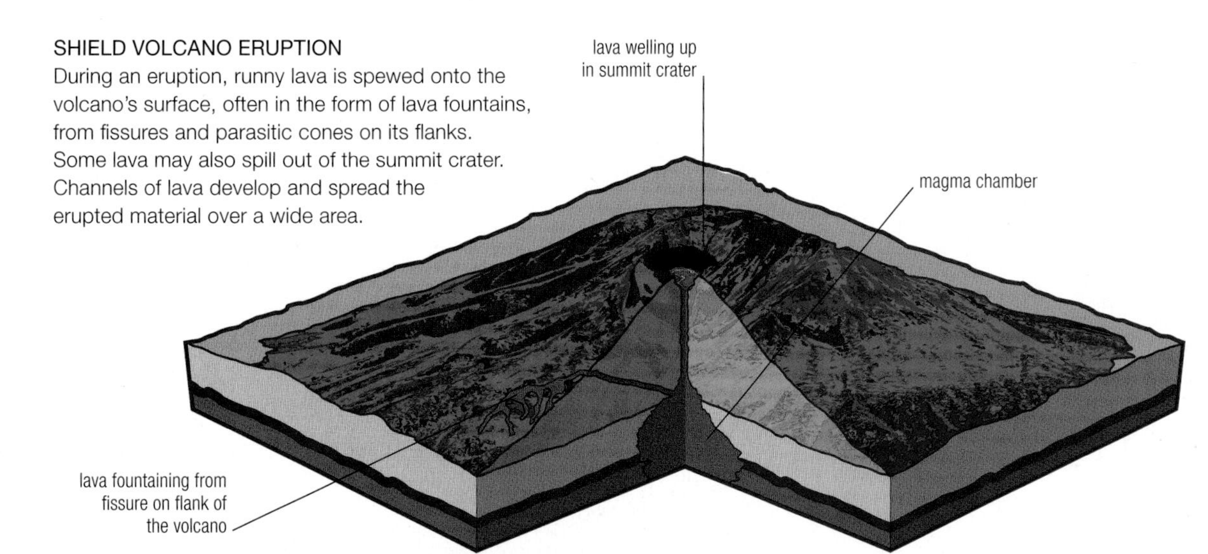

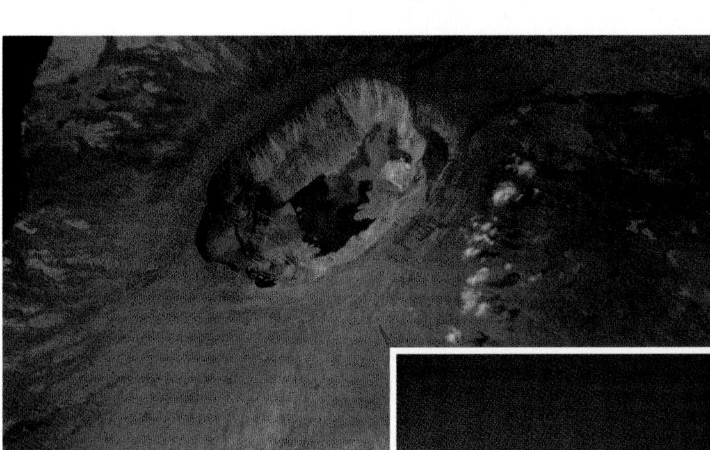

FERNANDINA
This massive shield volcano occupies the whole of the island of Fernandina, the youngest and most volcanically active of the Galápagos Islands. At its summit is a partially collapsed caldera about 4 miles (6km) wide and several hundred feet deep. Fernandina's last major eruption was in 2009.

ERUPTING CONE ON KILAUEA
Lava is seen here fountaining out of a secondary, or parasitic, cone on the flank of Kilauea, a shield volcano in Hawaii. Kilauea is one of the world's most active volcanoes—it has been erupting continually since 1983.

LAVA LAKE
The summit of Erta Ale in Ethiopia contains pits that are often filled with lava lakes. The lake surface seen here has a dark skin of solid lava, but splits have appeared in it, revealing the searingly hot, bright, molten lava underneath.

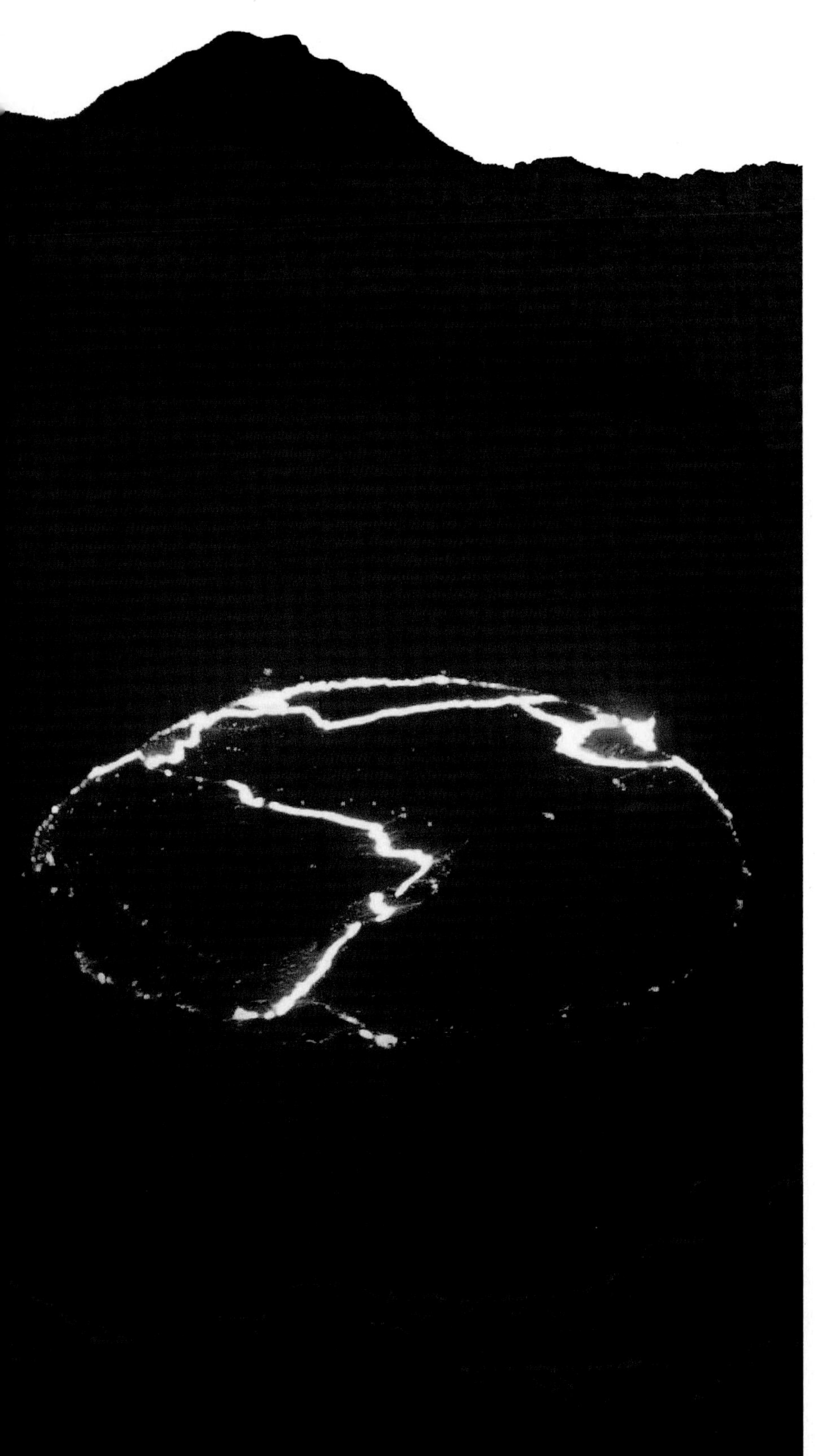

DISTINCTIVE FEATURES

At the summit of a shield volcano is a wide crater sometimes in the form of a caldera, which may be partially collapsed. In a few, the summit crater contains one or two pits that contain lakes of red-hot lava. Sometimes, where the volcano has not recently erupted, the crater may be partly filled with water. The flanks of the volcano usually slope gently and are covered in dark, solidified flows of lava. Also visible are fissures and parasitic cones, the sites of past or ongoing eruptions of lava. On some shield volcanoes, channels of flowing lava become enclosed in conduits called lava tubes. Once the lava has drained, these leave long, cavelike tunnels under the surface.

SPATTER CONES
These erupting cones, photographed on the shield volcano Piton de la Fournaise on Réunion Island in 2010, are called spatter cones. They are built from chunks of lava that were blown into the air and congealed in a heap once they hit the ground.

LARGEST ACTIVE SHIELD VOLCANOES

LOCATION	PROFILE	SUMMIT HEIGHT	MAXIMUM WIDTH OF BASE
Mauna Loa, Hawaii		13,677ft (4,169m)	59 miles (95m)
Erta Ale, Ethiopia		2,011ft (613m)	50 miles (80km)
Sierra Negra, Galápagos		4,921ft (1,500m)	31 miles (50km)
Nyamuragira, Democratic Republic of the Congo		10,033ft (3,058m)	28 miles (45km)
Kilauea, Hawaii		4,091ft (1,247m)	31 miles (50 km)

CINDER CONES

Also called scoria cones or pyroclastic cones, cinder cones are relatively small volcanoes built mainly of loose volcanic cinders (glassy fragments of solidified lava). Some contain appreciable quantities of volcanic ash or lava. They often form on the flanks of larger volcanoes, sometimes as single cones or in groups.

FORMATION AND ERUPTION

Cinder cones often start as small fissures that suddenly appear in the ground and start spouting cinders and lava bombs. They then grow rapidly for a few months or years, producing Strombolian and Vulcanian-type eruptions (see pp.148–49 and pp.150–51), with showers of cinders, bombs, and some lava flows. After a period of intense activity they may then go quiet. A typical example is a cone known today as Parícutin in Mexico, which began in 1943 as a fissure in a cornfield. Within a year it had grown to 984ft (300m) high, but in 1952, after reaching a maximum height of 1,391ft (424m), its eruptions stopped.

simple conical shape
bowl-shaped crater
single conduit
cinders with some layers of ash and lava

CINDER CONE STRUCTURE
A cinder cone has steep sides and is built from volcanic cinders, sometimes with layers of lava and ash. At its summit is a crater from which cinders and ash are spewed out. Where lava is erupted, it tends to flow from a breach on the side of the crater.

ACTIVITY LEVEL

Most cinder cones appear, erupt and grow for a few years, and then go quiet. The many volcanoes around the world like this, which have just one main eruptive phase, are known as monogenetic. Others are polygenetic, meaning that they have more than one eruptive phase. A large cinder cone in Nicaragua called Cerro Negro, for example, has erupted more than 23 times since 1850, and poses a hazard to people living close to it today.

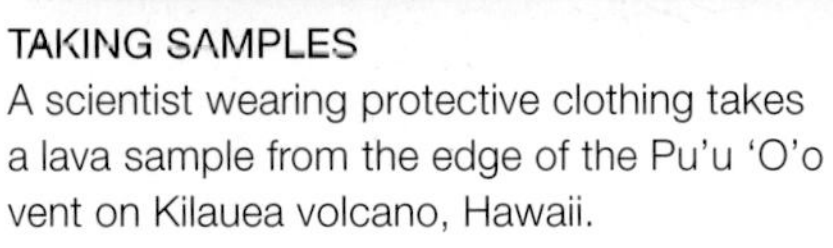

TAKING SAMPLES
A scientist wearing protective clothing takes a lava sample from the edge of the Pu'u 'O'o vent on Kilauea volcano, Hawaii.

SYMMETRICAL CONE
Eve cone is an almost perfectly symmetrical cinder cone, one of a group of 30 that sits on the flank of a shield volcano in British Columbia, Canada. It formed around 1,300 years ago, which in geologic terms is recent. It is 564ft (172m) high and about 1,476ft (450m) wide.

GROUP OF CONES
This group of cones lies in part of a volcanically active region of northwestern Saudi Arabia, called Harrat Lunayyir. The area is close to a divergent plate boundary, marking a rift, where Arabia is moving away from Africa, thus explaining its volcanicity. More eruptions may occur in the future.

PU'U 'O'O CINDER CONE
This is a parasitic cone sitting on the flank of a much larger parent volcano, the shield volcano Kilauea in Hawaii. Pu'u 'O'o has been erupting continually since 1983, producing cinders, lava fountains and flows, and volcanic gases.

MADAGASCAN CINDER CONE
This extinct, heavily cultivated, and partially eroded cinder cone is part of the Itasy Volcanic Field. It lies near Lake Itasy in Madagascar within a volcanic region that also contains many hot springs. The last eruptions in the area took place about 8,000 years ago.

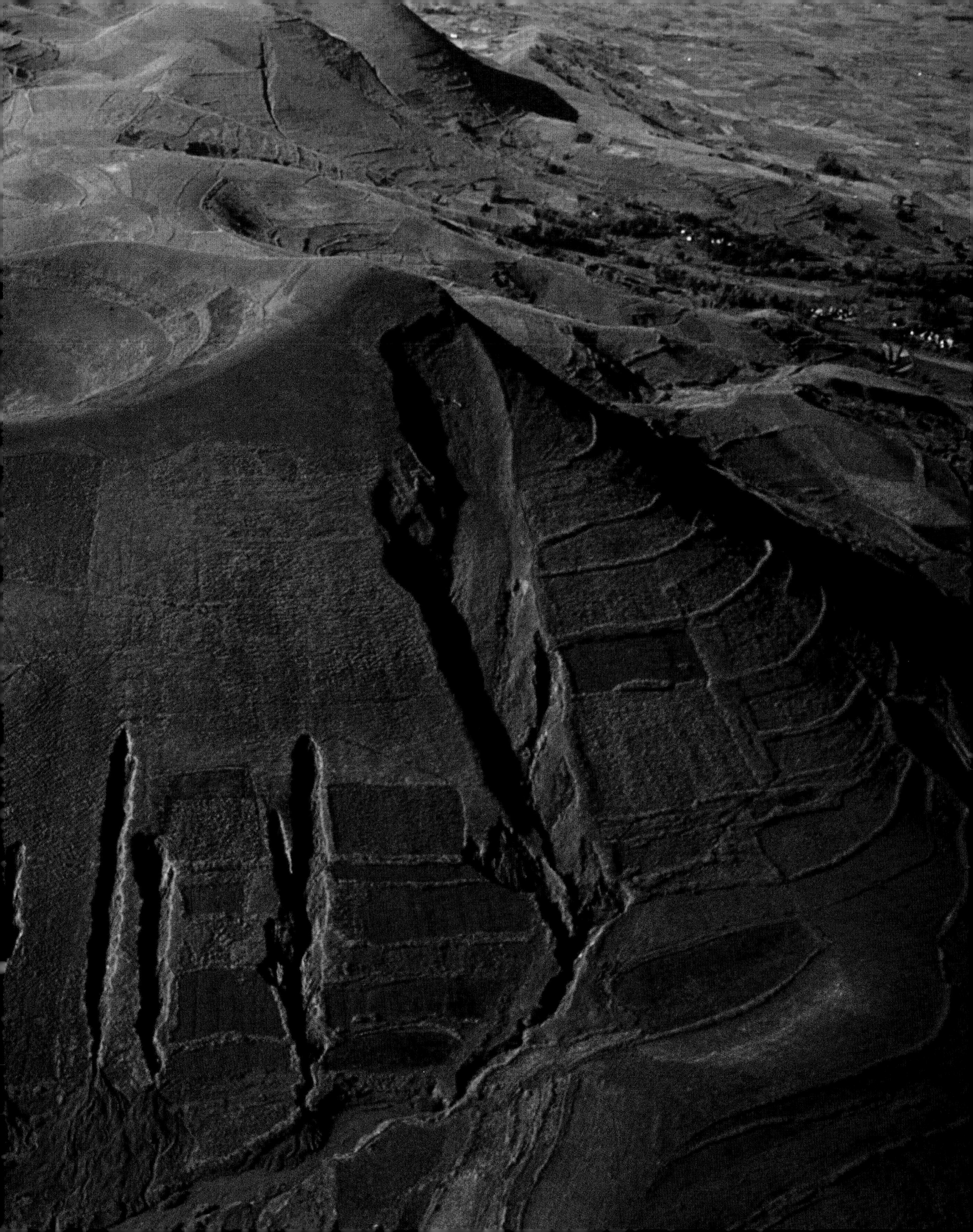

STRATOVOLCANOES

Stratovolcanoes, or composite volcanoes, are large, conical volcanoes built from layers of hardened lava and materials such as ash and cinders that are produced when magma (molten rock) is blasted into the air. They include some of the best known and most spectacular-looking—but also some of the most dangerous—volcanoes in the world.

FORMATION AND ACTIVITY

Although some stratovolcanoes develop at hotspots, most grow near ocean-continent and ocean-ocean convergent plate boundaries. In contrast to the runny lava that builds shield volcanoes (the other main type of large volcano), the lava erupted by a stratovolcano usually does not flow far. Instead it has a tendency to solidify around and within, and from time to time to block up the volcano's main vent. As a result, much of the material that a stratovolcano erupts is not lava flowing over the surface, but tephra (cinders, ash, pumice, and volcanic bombs) produced in explosive eruptions as the volcano clears its main vent. This helps explain not only the structure of stratovolcanoes but also their long-term behavior, with phases of violent activity interspersed with quiet periods, which can last for anything from a few years to several thousand years.

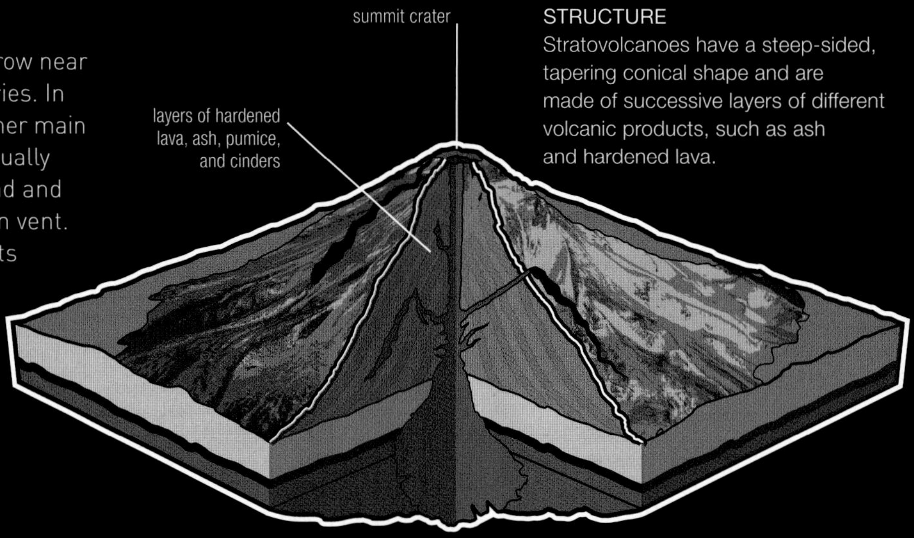

STRUCTURE
Stratovolcanoes have a steep-sided, tapering conical shape and are made of successive layers of different volcanic products, such as ash and hardened lava.

COTOPAXI
This almost perfectly symmetrical stratovolcano in the Ecuadorean Andes rises to a height of 19,393ft (5,911m). It has erupted more than 50 times since 1738, most recently in 1940. The main risk of large eruptions at Cotopaxi comes from melting of the ice and snow at its summit, which can lead to devastating lahars.

KRAKATAU: A FAMOUSLY DANGEROUS STRATOVOLCANO

One of the deadliest volcanic eruptions in history occurred in August 1883 from a stratovolcano on the island of Krakatau in Indonesia. An explosion at the end of the eruption literally blew Krakatau apart. Vast amounts of rock and ash were spewed into the atmosphere, or despatched toward nearby islands in the form of pyroclastic flows, and a series of powerful tsunamis followed. According to official records, 165 towns and villages were destroyed and some 36,000 people died, mainly as a result of the tsunamis. Krakatau's final explosion is famous for being the loudest sound ever reported—it was heard distinctly in Perth, Australia, some 1,925 miles (3,100km) away.

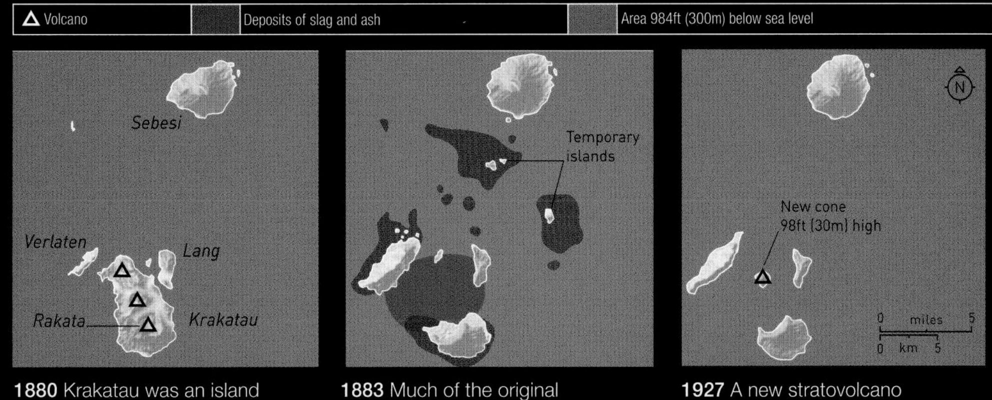

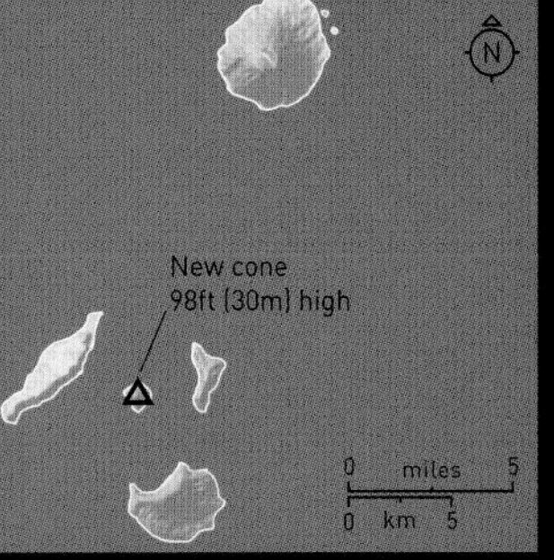

1880 Krakatau was an island containing three coalescing volcanic cones (triangles) of which at least one, Rakata, was a stratovolcano.

1883 Much of the original island had gone, through being blown apart. New islands and underwater deposits had formed from the debris.

1927 A new stratovolcano had appeared where some of Krakatau's original cones stood. This was named Anak Krakatau or "Child of Krakatau."

CHARACTERISTICS

Stratovolcanoes can erupt in a variety of "styles" from relatively mild Strombolian eruptions to more dangerous vulcanian, Peléan, or Plinian eruptions. They can produce a range of effects from showers of lava bombs and cinders, accompanied by loud explosions, to huge ash clouds and pyroclastic flows. Due to the high altitude of their summits, many stratovolcanoes carry large glaciers or snowfields on their upper slopes, which pose the risk of dangerous lahars (mudflows) when an eruption occurs. Eruptions can last for anything from a few hours or days to many tens of years. Many of the most destructive eruptions of the past have come from stratovolcanoes, including Krakatau.

ARENAL ERUPTION
This young stratovolcano in Costa Rica has been erupting regularly since 1968, after centuries of inactivity. In recent years, glowing fountains and streams of lava have been visible almost every night.

MOUNT RUAPEHU ERUPTING
The largest active volcano in New Zealand, Ruapehu erupted spectacularly in 1996, producing a tall dark ash plume. A later euption in 2007 triggered a powerful lahar.

ETNA

Europe's largest volcano, Etna covers 460 sq miles (1,190 sq km) of eastern Sicily. Standing at 10,922ft (3,329m) high, it is a stratovolcano with a complex structure that includes four separate summit craters and more than 300 smaller parasitic vents and cones on its flanks. Etna started forming 500,000 years ago on the floor of the Mediterranean and emerged from the sea around 100,000 years ago. Over the past several thousand years, it has been almost continuously active, producing eruptions of two main types. Spectacular explosive eruptions, from one or more of its summit craters, produce volcanic bombs, cinder showers, and large ash clouds. Etna also has Hawaiian and Strombolian-type eruptions (see pp.144–49) from vents and fissures on its flanks. These eruptions typically feature lava fountains and extensive flows of runny basaltic lava, which can be either pahoehoe or 'a'a (see p.97).

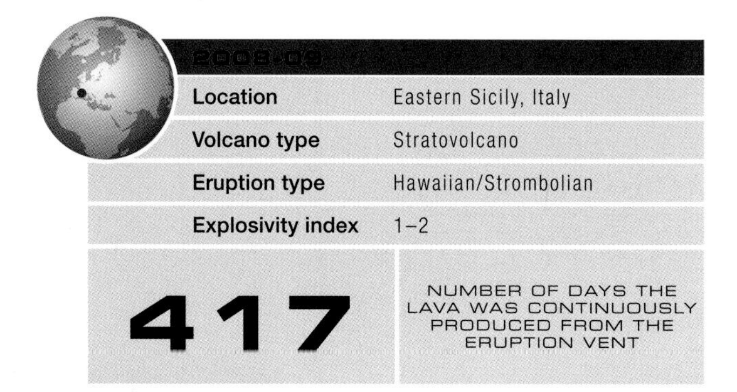

2008–09	
Location	Eastern Sicily, Italy
Volcano type	Stratovolcano
Eruption type	Hawaiian/Strombolian
Explosivity index	1–2
417	NUMBER OF DAYS THE LAVA WAS CONTINUOUSLY PRODUCED FROM THE ERUPTION VENT

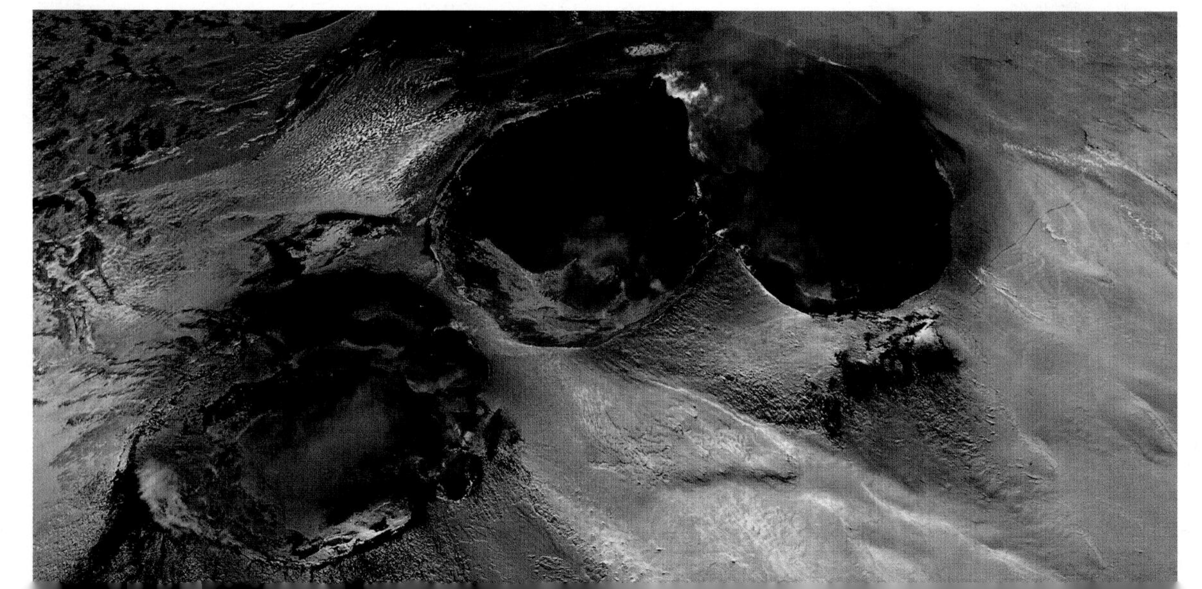

SUMMIT CRATERS
Etna is unusual in having four separate summit craters, of which three are visible here. From the left, they are the Northeast Crater, La Voragine or The Chasm, and Bocca Nuova or New Mouth. These craters are 985–1,310ft (300–400m) across. In addition, there are hundreds of smaller cones and craters on Etna's slopes.

ETNA ERUPTS
In this night-time photograph, a bright lava fountain and lava flows are visible on Etna's northern flank. In the background lies the city of Catania, which in the past has been invaded by large lava flows from Etna.

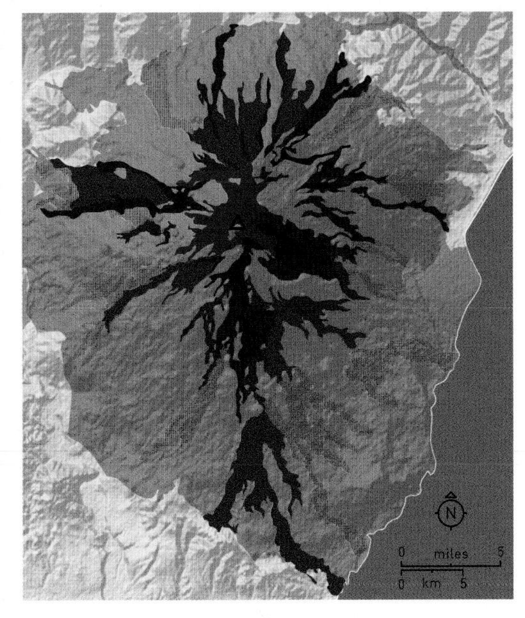

LAVA FLOWS
This map shows the various times lava streams have reached the base of Etna on every side, and also beyond it.

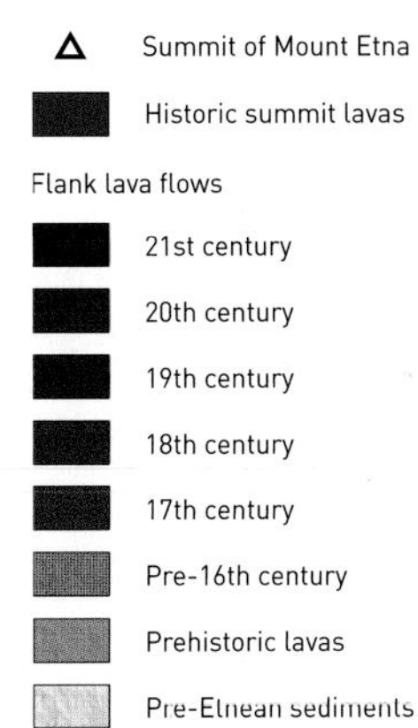

> **...THIS TERRIBLE MOUNT UPON WHOSE CHARR'D AND QUAKING CRUST I STAND—THOU, TOO, BRIMMEST WITH LIFE!**
>
> **MATTHEW ARNOLD**, BRITISH POET, IN HIS POEM *EMPEDOCLES ON ETNA*, 1852

ERUPTION TIMELINE

1673 EXTENDED ERUPTION

An eruption that lasted from 1634 to 1638 is estimated to have produced 5,300 million cu ft (150 million cu m) of lava. This engraving of the eruption is from a book by the German scholar Athanasius Kircher, who witnessed it in 1637.

1766 ETNA RESHAPED

This eruption produced an estimated 4,000 million cu ft (115 million cu m) of lava that partially reshaped Etna and threatened the town of Nicolosi on its southern flank. This depiction is by the French engraver Jean-Baptiste Chapuy.

SMOKE RINGS

Etna's summit craters occasionally emit "smoke rings." These are actually rings made of steam. In February 2002, Swiss volcano-watchers witnessed a series of such rings, which are extremely rare, coming out of the volcano's Bocca Nuova crater. Some lasted for up to 10 minutes, as they slowly drifted upward and away.

SMOKING ETNA
This is one of the steam rings venting from Etna's summit in February 2002. The diameter of the ring was estimated at 655ft (200m).

2002 DEVASTATION ON ALL SIDES

Lava spewed from vents on two sides of the volcano, earthquakes shook its eastern flank, while a 2.5-mile- (4-km-) high ash column rose from a crater on its southern side. Lava flows destroyed part of a forest as well as a tourist complex and skiing station.

2008 MASSIVE LAVA FLOWS

In May, a powerful eruption occurred from a fissure near Etna's summit, and several lava streams surged toward the city of Riposto. Days later, more than 200 earthquakes led to a new eruption from a flank fissure that lasted for 14 months.

MERAPI ERUPTION 2010

The eruption of Mount Merapi, a large stratovolcano on the Indonesian island of Java, in late 2010 ranks among the most serious volcanic eruptions of the 21st century. This dangerous stratovolcano (see pp.120–21) produced a series of earthquakes, explosions, ash plumes, incandescent lava avalanches, fireballs, lahars (mudflows), and pyroclastic flows that killed more than 300 people. Several hundred million cubic feet of ash and other volcanic material were blasted over the surrounding area. Indonesia's most active volcano, Gunung Merapi (which translates as "Fire Mountain") lies in one of the world's most densely populated areas. With a height of 9,737ft (2,968m), it dominates the region immediately north of the major city of Yogyakarta. A particular problem with Merapi is that it harbors a steep-sided active lava dome at its summit that is prone to partial collapse. When a dome collapse happens, the resulting pyroclastic flows and lahars are a huge danger to people living on the slopes of the volcano, where they cultivate the fertile soil. Merapi's 2010 eruption lasted for three months, from early warning signs in September to the quietening down of the volcano in early December. The most lethal phase of the eruption, however, was from October 25 onward.

OCTOBER–NOVEMBER 2010

Location	Central Java, Indonesia
Type	Stratovolcano
Fatalities	353

350,000

NUMBER OF PEOPLE DISPLACED

ERUPTION
At the height of the eruption, in November 2010, an incandescent glow pervaded Merapi's summit as massive quantities of hot ash and lava surged toward the volcano's lower slopes.

THE UNFOLDING DEVASTATION

1 BEGINNING OF ERUPTION
Incandescent lava avalanches were noticed coming from the lava dome at Merapi's summit on October 25, 2010.

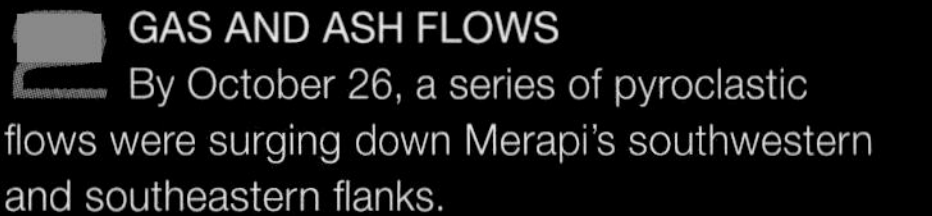

2 GAS AND ASH FLOWS
By October 26, a series of pyroclastic flows were surging down Merapi's southwestern and southeastern flanks.

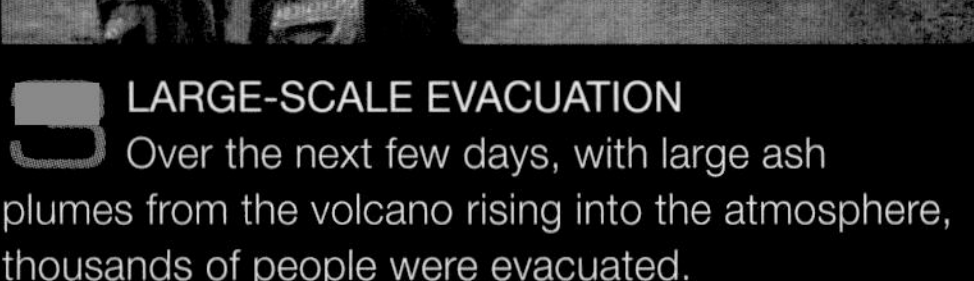

3 LARGE-SCALE EVACUATION
Over the next few days, with large ash plumes from the volcano rising into the atmosphere, thousands of people were evacuated.

KEEPING TRACK
A government scientist monitors activity at Merapi from the Volcano Monitoring Centre in Yogyakarta on October 25, 2010. That same day, volcanologists recommended evacuation of a region near the volcano and raised the alert to its highest possible level.

IMPACT AND CASUALTIES

Toward late October when a sharp increase in minor earthquakes near the volcano and a swelling of Merapi's lava dome was detected, villagers living within 6 miles (10km) of Merapi were advised to evacuate the area and seek refuge in emergency shelters. By November 5, the recommended evacuation zone had been expanded to 12 miles (20km). Unfortunately, many villagers did not comply with the advice, either remaining behind or returning to their homes while eruptions continued. By the end of the eruption, the death toll had risen to 353, with most deaths due to suffocation and burns. The eruptions had blanketed a large area of forest, farms, and plantations in volcanic ash, and ash plumes from the volcano caused major disruption to air traffic across Java.

> "VOLCANIC ASH IS RAINING DOWN, IT'S DARK HERE, AND THE VISIBILITY IS ONLY TWO METERS."
>
> **BAGYO SUGITO**, A 39-YEAR-OLD DRIVER LIVING IN YOGYAKARTA

4 SEARCH AND RESCUE EFFORTS
By early November, scores of people were injured or killed as the eruption reached

5 THE DESTRUCTION
Some of the worst damage was caused by lahars, which flowed for up to 10 miles (16km)

6 THE AFTERMATH
Relatives in Umbulharjo village, Sleman, pray during a funeral ceremony for the victims

ANIAKCHAK CALDERA
This Crater-Lake type caldera is in the Aleutian Range of Alaska. Formed in a gigantic eruption 3,400 years ago, it is 6 miles (10km) across.

CALDERAS

The word "caldera" means cauldron and is a 0.6-mile- (1-km-) wide to 60-mile- (100-km-) wide usually circular depression caused by the eruption of vast quantities of magma. It is commonly used to refer to either of two different types of structure. One is a type of volcano in itself, the other a feature of a stratovolcano or large shield volcano. Calderas can be gigantic—for example the Aniakchak volcano in the Aleutian Range of Alaska has a 6-mile- (10-km-) wide caldera that at its deepest is 1,341ft (408m) from the caldera rim.

TYPES OF CALDERAS

The type of landform to which the description "caldera" is most commonly applied is the remnant of a large stratovolcano (see pp.120–21) that underwent a catastrophic Plinian-style eruption and collapse, usually many thousands of years ago. These structures are volcanoes in themselves. Although there is no universally agreed name for them, they are sometimes referred to as Crater-Lake type calderas, after a caldera of this type in Oregon. The second type, sometimes called subsidence calderas, result from the gradual recent subsidence of the summit of a shield volcano (see pp.114-15). Some authorities also define a third class of calderas that are too huge to have been caused by collapse of a single stratovolcano. There are only a few of these structures in the world and they are sometimes referred to as "supervolcanoes" (see pp. 128-29). Each has produced some cataclysmic eruptions in the past

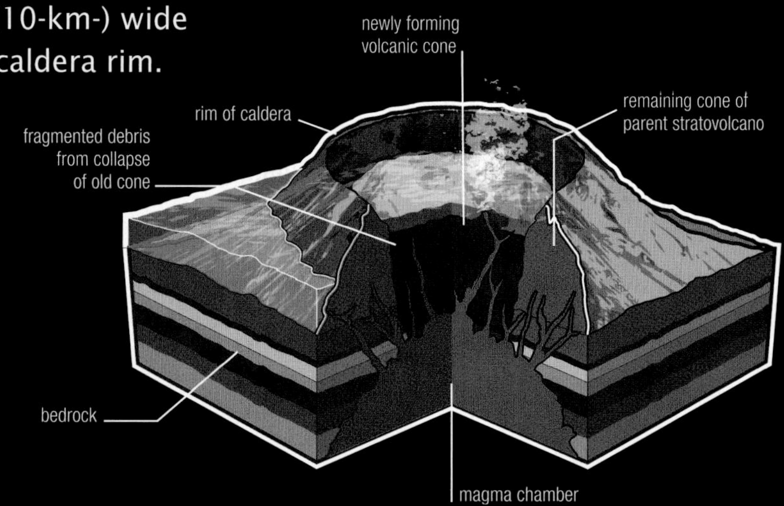

CALDERA STRUCTURE
A Crater-Lake type caldera is a wide deep crater that may contain a lake. Its floor contains debris from the collapse of a stratovolcano some time in the past

CRATER-LAKE TYPE CALDERAS

These calderas take the form of wide, often near-circular, craters, typically anything from about 3 to 12 miles (5 to 20km) across, with rims that are usually raised hundreds of feet above the surrounding land. The archetypal example, Crater Lake itself, is filled with water to a depth of 1,968ft (600m), making it the deepest freshwater lake in North America, but not all calderas of this type contain lakes. A few Crater-Lake type calderas, such as Ngorongoro Crater in Tanzania, are definitely extinct, but many still have large chambers full of magma beneath them and may well erupt again in future. Many have new stratovolcanoes or cinder cone volcanoes growing within them.

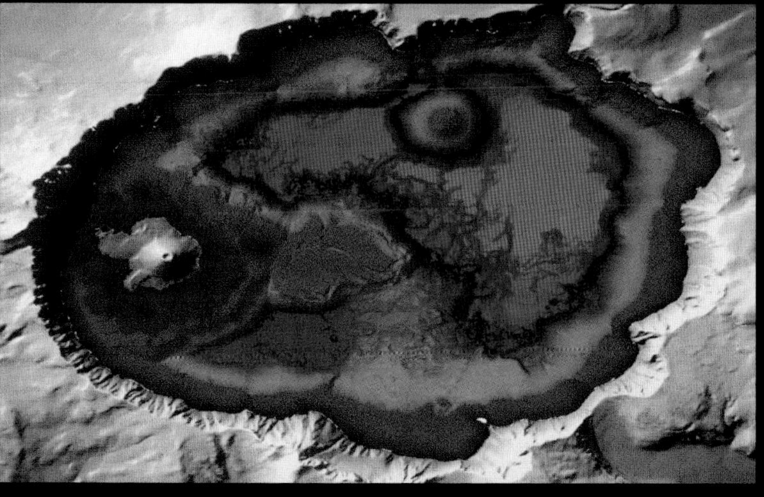

CRATER LAKE, OREGON
Depicted here in colored 3-D relief is Crater Lake. The depth of the crater-lake floor below water level is color coded, from red (shallowest) to purple (deepest). Crater Lake formed from the violent eruption and collapse of a gigantic stratovolcano about 6,850 years ago.

FORMATION OF A CRATER-LAKE TYPE CALDERA

Calderas of this type usually form from the collapse of a stratovolcano. This may occur as the result of a single cataclysmic Plinian-style eruption, or in stages as the result of a series of eruptions. The total area that collapses may be hundreds of square miles.

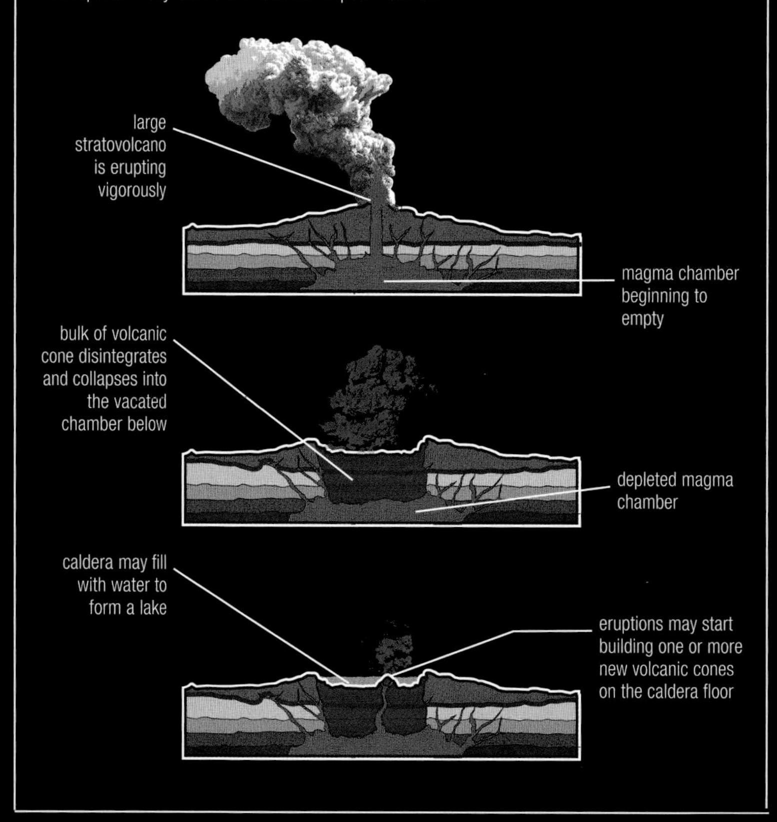

SHIELD VOLCANO CALDERAS

This type of caldera is not a volcano in itself, merely the summit area of a shield volcano that has subsided over time. An example can be seen at the volcano Kilauea, in Hawaii. Various collapses of its summit area have formed a roughly circular depression that is about 540ft (165m) deep and 3.1 miles (5km) across, with a floor that is a fairly flat but roughly surfaced bed of lava. Nested within it is a much smaller circular crater named Halema`uma`u. This occasionally fills with a lava lake or explosively emits gases, lava, and ash.

KILAUEA CALDERA
This view of the caldera at Kilauea's summit looks across from the caldera rim to the vertical wall on the opposite side.

THE SANTORINI CALDERA

Santorini in the southern Aegean Sea, southeast of Greece, is a series of overlapping shield volcanoes cut by at least four overlapping calderas. It has erupted many times over the past several hundred thousand years. The last major eruption, some 3,600 years ago, was one of the biggest volcanic events in recorded history. More than 14 cu miles (60 cu km) of material was blasted into the atmosphere, and a destructive tsunami was generated that may have contributed to the collapse of the Minoan civilization on the nearby island of Crete.

AN AEGEAN CALDERA
The Santorini caldera, as seen here from space, measures about 4 miles (7km) by 7 miles (12km). The island that forms about two thirds of its rim is called Thera.

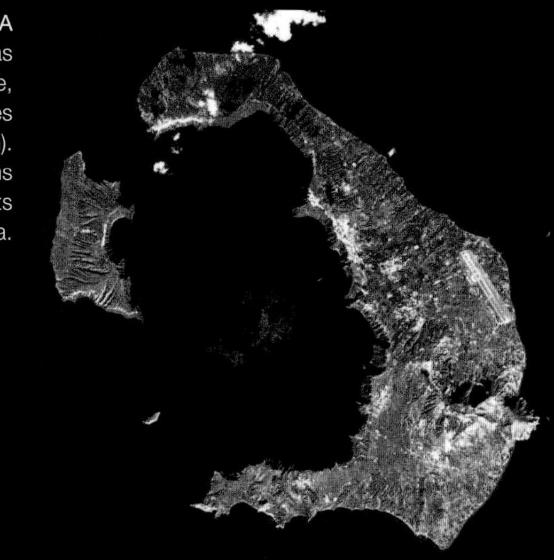

SUPERVOLCANOES

A small number of volcanoes across the world, popularly known as "supervolcanoes," have produced truly cataclysmic eruptions in the past—and are capable of future eruptions that could radically alter landscapes and severely impact the world's climate.

CHARACTERISTICS

A supervolcano is a volcanic site that has seen at least one eruption rating an 8 on the Volcanic Explosivity Index (see p.89) and is considered capable of future, similar eruptions. Eruptions of this size are about a thousand times bigger than anything witnessed in recent centuries, such as the 1980 Mount St. Helens eruption. Only a handful of volcanic sites worldwide qualify for the supervolcano description, and all are large calderas (see pp.126–27) with underlying, active magma chambers. A prime example is Yellowstone caldera, which makes up a large part of Yellowstone National Park in Wyoming. Yellowstone's last big eruption, about 640,000 years ago, blasted an estimated 230 cu miles (1,000 cu km) of rock and magma into the air, and covered a large part of the western United States in volcanic ash. Further large eruptions from this caldera, with catastrophic consequences both locally and for the world's climate, are considered likely, though geologists do not believe one is at all imminent. Two other supervolcanoes are calderas underlying large, scenic lakes—Lake Taupo in New Zealand's North Island, and Lake Toba in Sumatra. The largest-known eruption at Taupo, which happened about 22,600 years ago, blasted an estimated 280 cu miles (1,170 cu km) of material into the air and caused the collapse of several hundred square miles of land, but the largest eruption of Toba was even bigger (see above right).

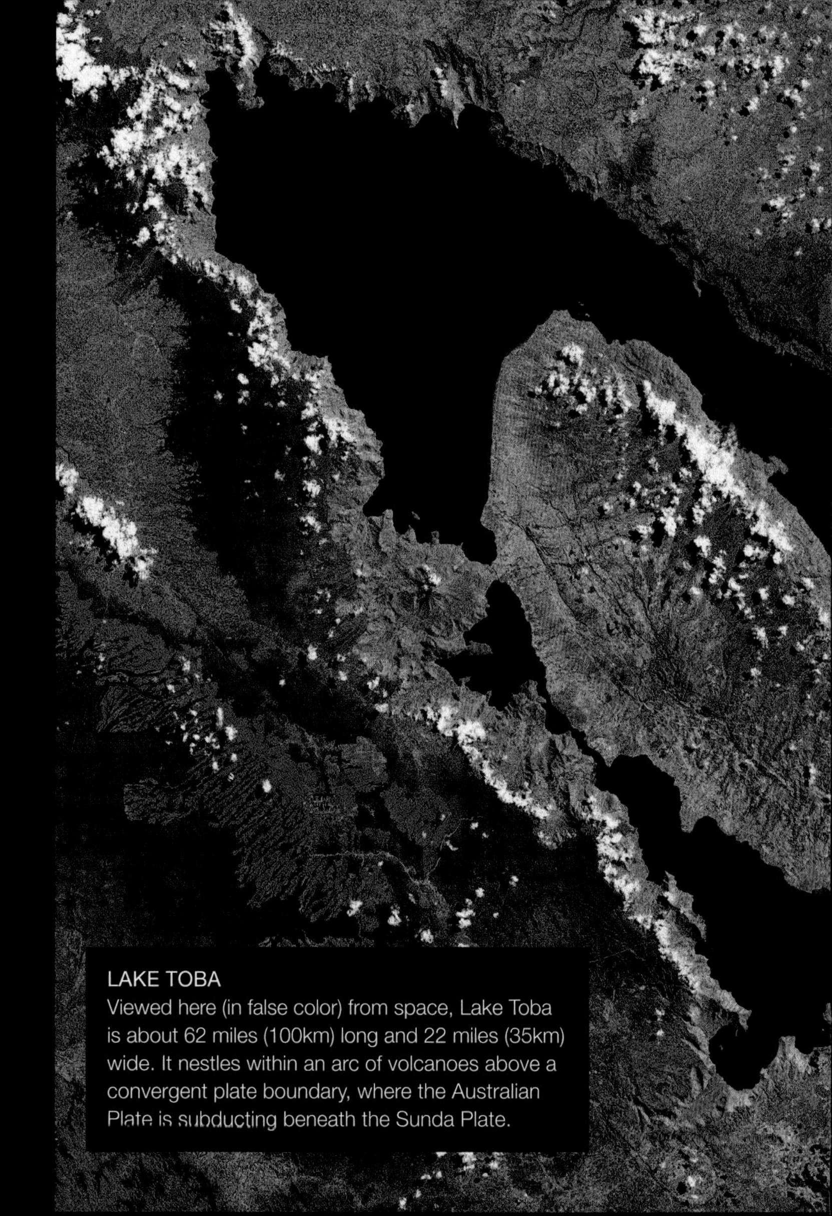

LAKE TOBA
Viewed here (in false color) from space, Lake Toba is about 62 miles (100km) long and 22 miles (35km) wide. It nestles within an arc of volcanoes above a convergent plate boundary, where the Australian Plate is subducting beneath the Sunda Plate.

STRUCTURE OF YELLOWSTONE CALDERA
A large magma chamber lies about 5 miles (8km) below the caldera. Uplifting of the rock dome above the magma chamber (called a resurgent dome) or a big increase in earthquake activity could herald a new eruption.

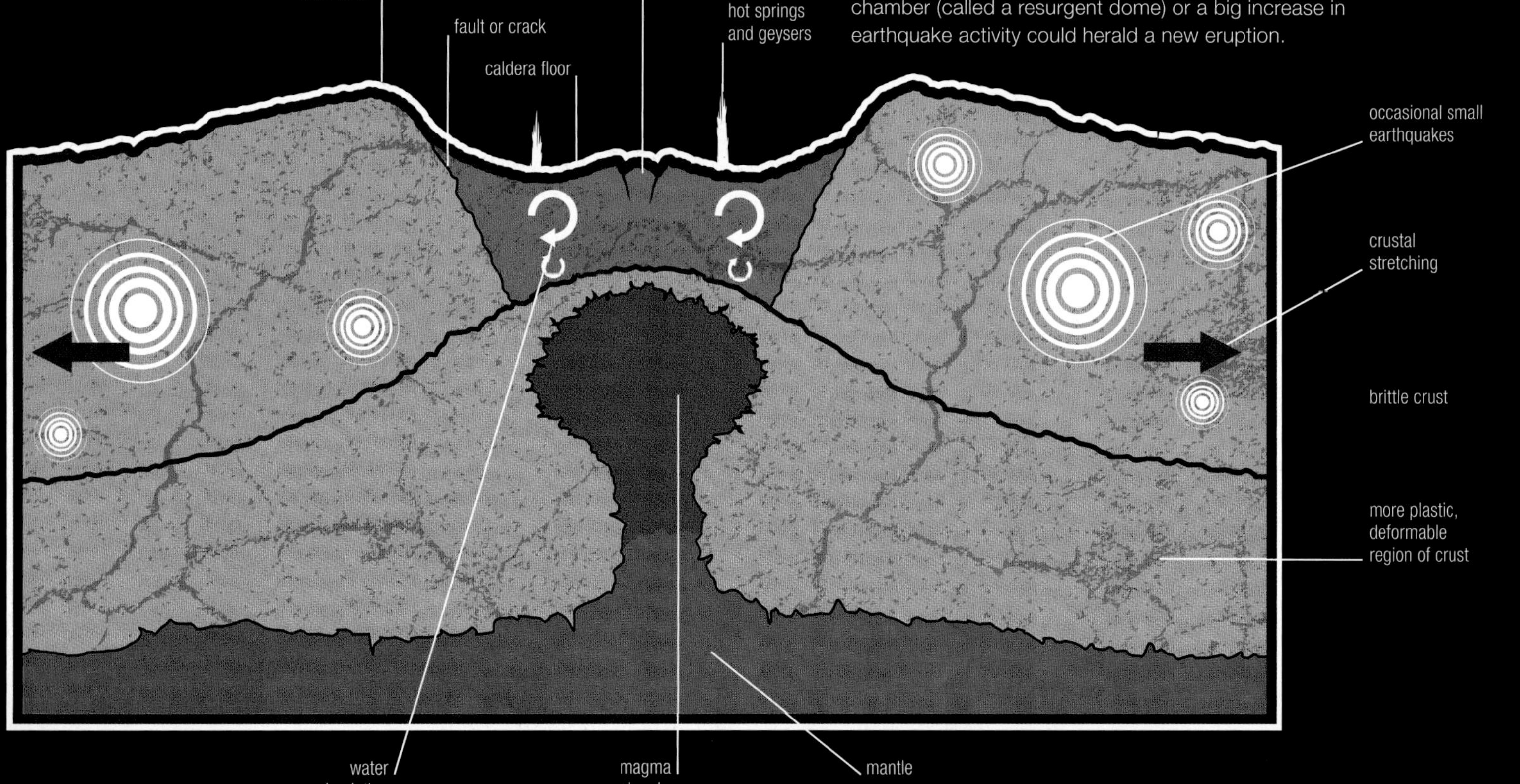

THE TOBA ERUPTION

About 74,000 years ago, Lake Toba in Sumatra was the site of the largest volcanic eruption of the past two million years. The fact that the eruption occurred has been worked out from the thick volcanic ash deposits it left over a vast region of south Asia—it is estimated to have blasted 670 cu miles (2,800 cu km) of pulverized rock into the air. As the ash cloud traveled around the world and blocked out sunlight, it probably caused temperatures to cool by about 5 to 9°F (3 to 5°C) . There is some evidence (from genetic studies) that the effects drastically reduced the world's human population at the time to about 10,000 individuals. A large magma chamber still exists under Lake Toba, and many earthquakes have occurred in its vicinity over the past century. It will probably erupt spectacularly again in the future.

BEAUTY SPOT
From its tranquil-looking setting, few would suspect Toba's catastrophic history or its potential for future cataclysm.

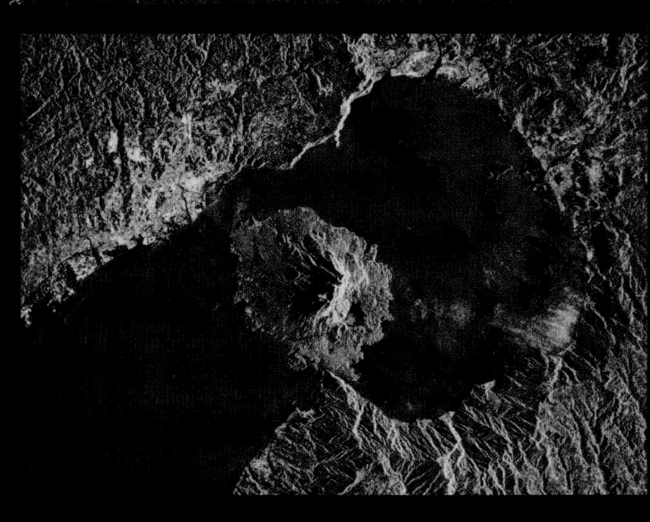

AIRA CALDERA, JAPAN
Some 22,000 years ago, 14 cubic miles (40 cubic km) of material was blasted out of the ground to form this caldera, a candidate for supervolcano status.

DISTRIBUTION OF SUPERVOLCANOES

The three volcanic sites with the best supervolcano credentials are Yellowstone Caldera, Lake Toba, and Lake Taupo. Of these, Yellowstone sits over a continental hotspot (see pp.32–33), while Toba and Taupo are close to plate boundaries. In addition, a few other volcanic sites are close to supervolcano status, with their largest-known past eruptions scoring a 7, rather than a maximum 8, on the Volcanic Explosivity Index. These include Long Valley Caldera in California and Aira Caldera in Japan. Many other sites worldwide were once supervolcanoes but are now probably extinct.

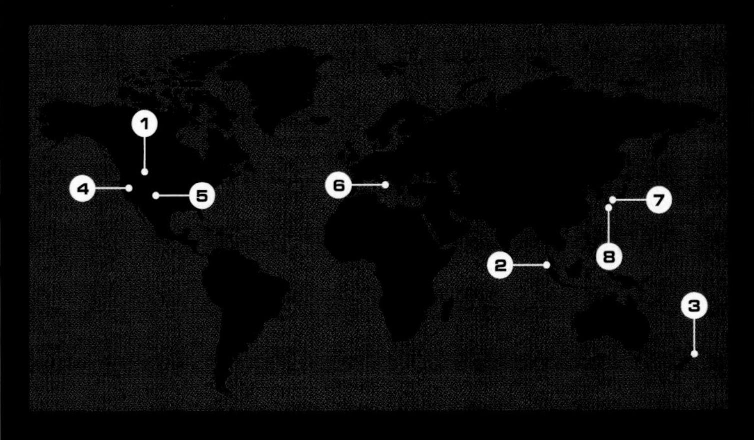

1. Yellowstone Caldera, Wyoming
2. Lake Toba, Sumatra
3. Lake Taupo, New Zealand
4. Long Vallley, California
5. Valles Caldera, New Mexico
6. Phlegraean Fields, Italy
7. Aira Caldera, Japan
8. Kikai Caldera, Ryukyu Islands, Japan

YELLOWSTONE CALDERA
Visible on the right here is the rim of the caldera, and in the middle-distance, its floor. Geologists constantly monitor the floor for any upward bulging that could indicate an imminent eruption.

MAARS

Also known as volcanic explosion craters, maars are shallow bowl-shaped volcanic craters, usually sunk slightly into the ground. These relatively small volcanic features are often filled with water to form circular lakes—the word "maar" is a German word derived originally from the Latin word "mare" (sea).

FORMATION AND SIZE

Maars are produced when magma (melted rock) reaches Earth's surface and comes into contact with groundwater or, in polar regions, ice-laden permafrost (frozen soil). The resulting steam-driven explosion excavates a shallow crater that can be anything from 200ft (60m) in diameter to considerably larger. The widest known, found on the Seward Peninsula in western Alaska, are up to 1.2 miles (2km) in diameter. They were caused by magma encountering permafrost, which results in particularly large explosions. Although there are a few dry maars in desert regions of the world, the majority are filled with lakes, which can be anything from 30 to 650ft (10 to 200m) deep.

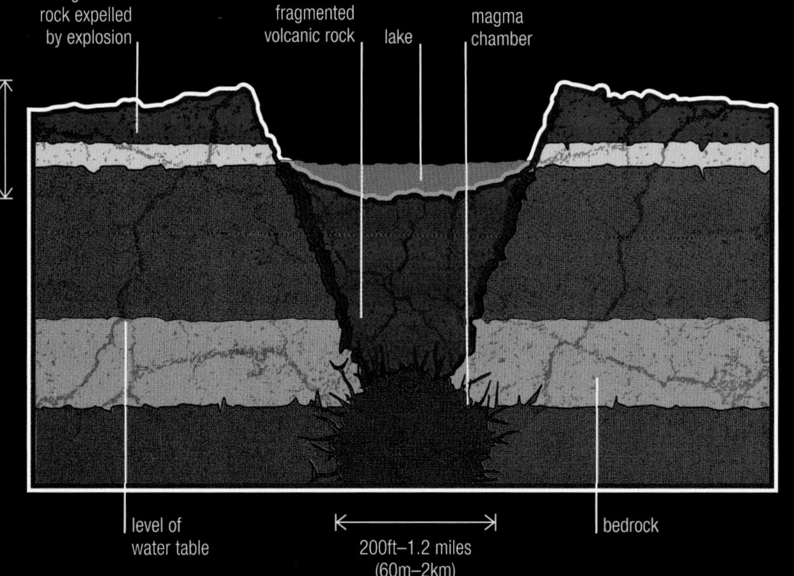

MAAR STRUCTURE
A maar contains a mass of fragmented rock beneath it in an inverted conical structure. Below that is an extinct, or sometimes active, magma chamber. The crater is surrounded by a low rim composed of ash and loose fragments of rocks torn from the ground when the explosion happened.

DISTRIBUTION AND ACTIVITY

Maars can occur anywhere in the world where magma has come in contact with groundwater or permafrost. Some are in volcanic regions near to plate boundaries, others in locations that have experienced past or recent hotspot activity (see pp.32–33). A group of maars in the Eifel region of Germany, for example, was caused by the Eifel hotspot. Many maars are extinct, but with others, future volcanic activity is not ruled out. Two particularly dangerous maars, because of their potential for gas eruptions, are found in Cameroon and have been referred to as "exploding lakes" (see pp.132–33).

"WE LIKEWISE FORGET, IN THESE COOL DISTRICTS OF THE EARTH, WE ARE NOT QUITE BEYOND THE HAZARD OF SUBTERRANEAN FIRES."

PUBLISHED IN THE NATIONAL MAGAZINE, "A POSSIBLE EVENT – DANGERS OF OUR PLANET", 1854

LAC D'EN HAUT, FRANCE
This maar is part of the Chaîne-des-Puys group of extinct volcanoes in central Fance, created 70,000 to 7,000 years ago. Its name reflects its relatively high altitude, 4,064ft (1,239m) up in the Massif Central.

PULVERMAAR, GERMANY
Its rim completely covered by a forest, the Pulvermaar is one of several in the Eifel volcanic field, Germany. Though formed in eruptions ending 11,000 years ago, the field is regarded as still possibly active.

VITI MAAR, ICELAND
About 500ft (150m) in diameter, this Icelandic maar was formed in 1875, during an eruption of a nearby stratovolcano, Askja. Its lake waters are 115ft (35m) deep and always warm, suggesting an underlying magma chamber

DALLOL CRATER, ETHIOPIA
This colorful maar, in the Afar Depression, formed in a 1926 eruption. Lying 150ft (45m) below sea level, the crater and others nearby are the world's lowest volcanic vents on land

EXPLODING LAKES

In the highlands of Cameroon in west Africa are two unusual lakes, Nyos and Monoun, that sit in the craters of volcanic maars (see pp.130–31). During the 1980s, they became known as "exploding lakes" due to disasters caused by the lakes suddenly releasing large clouds of carbon dioxide.

LAKE NYOS DISASTER

The more lethal of the two disasters, known as limnic eruptions, occurred on August 21, 1986 at the larger of the two lakes, Lake Nyos. On an otherwise normal day, more than 1,750 people living in villages close to the lake suddenly died, apparently asphyxiated by something suspected to have come from the lake. Following scientific investigation, a picture of what had happened emerged. Lying in a volcanic region, Nyos has pockets of magma deep beneath it that release carbon dioxide (CO_2) and other gases. These dissolve in groundwater and feed into the lakes, so the water near the lake floor holds high concentrations of CO_2 under pressure. Eventually, when the overlying water pressure at the bottom of the lake could no longer hold the CO_2, bubbles began to form, and the low density bubble-water mixture caused the lake water to "overturn," leading to a sudden escape of CO_2 via the lake surface. At Lake Nyos, it is thought that about 0.25 cu miles (1 cu km) of the gas, which is heavier than air and tends to hug the ground, was released. It overspilled the rim of the maar and channeled down into nearby valleys, asphyxiating all people and animals in its path.

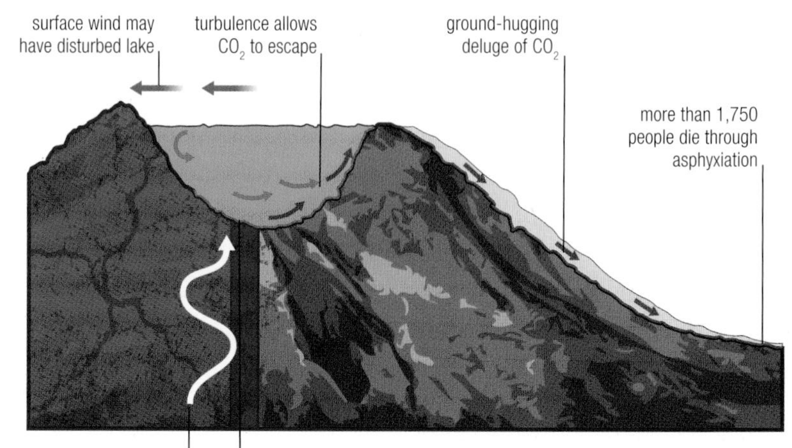

DEVELOPMENT OF LIMNIC ERUPTION
Lake Nyos contains high levels of CO_2 at depth. In the 1986 disaster, something disturbed its waters, causing the release of a vast amount of the gas.

LAKE NYOS, CAMEROON
What makes the lake particularly dangerous is its setting in a crater elevated above the surrounding land—so if a cloud of CO_2 is produced it spills down into surrounding valleys.

DEGASSING LAKE NYOS

The magma pocket underlying Nyos continues to recharge the lake with CO_2, so disasters are likely to recur unless excess gas is prevented from building up. In 1995, initial tests began on a degassing method. This involves setting up a strong plastic pipe vertically between the lake bottom and the surface, and initially using a pump to draw water out of and up the pipe. This triggers a self-sustaining process in which the gas-saturated bottom water is continuously sucked upward, driven by the expansion of bubbles in the rising water. At the surface, the gas is released. Degassing has been proceeding at Lake Nyos since 2001 and by the end of the decade the CO_2 levels had been greatly reduced. However, there are new concerns that an earthquake could cause partial disintegration of the Nyos maar, resulting in a disastrous flood as well as a dangerous release of the remaining gas.

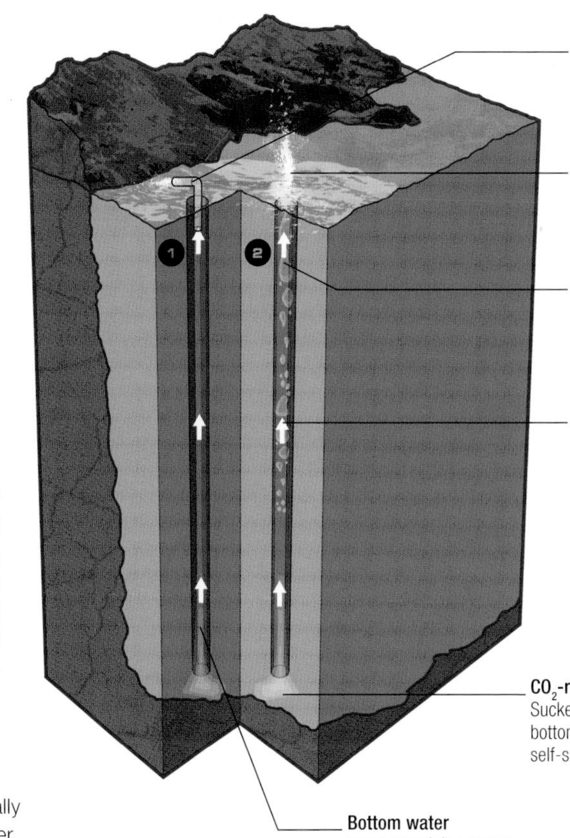

EXTRACTION TECHNOLOGY
A plastic pipe is set up (1) and water pumped out at the top, causing deep water to start rising up it. The upward flow becomes self-sustaining (2) due to gas bubble formation and expansion in the rising water.

SCIENTISTS STUDY LAKE NYOS
Investigations by an international team of scientists into the geology around Nyos and its chemical make-up eventually unravelled the cause of the 1986 disaster.

LAKE MONOUN AND LAKE KIVU

When the cause of the disaster at Lake Nyos was clarified, it was realized that a similar but smaller catastrophe, with a death toll of 37, had preceded it, at Lake Monoun in 1984. Degassing work has been in operation at Monoun since 2003, and by 2008 it was largely degassed. Scientists have also sought out other lakes that might pose similar dangers. The main one identified is Lake Kivu. This doesn't lie in a maar, but in a volcanic region within the western arm of the East African Rift Valley (see pp.174–75). There is some evidence that Kivu may have undergone large limnic eruptions in the past, and that 2 million people living close to it are in danger. As yet, no degassing system has been installed, but a plan initiated in 2010 to extract methane from the lake has led to some degree of CO_2 degassing, too.

DEGASSING LAKE MONOUN
Japanese scientists conduct final tests from a newly assembled raft prior to the installation of one of three degassing pipes now working successfully at Lake Monoun.

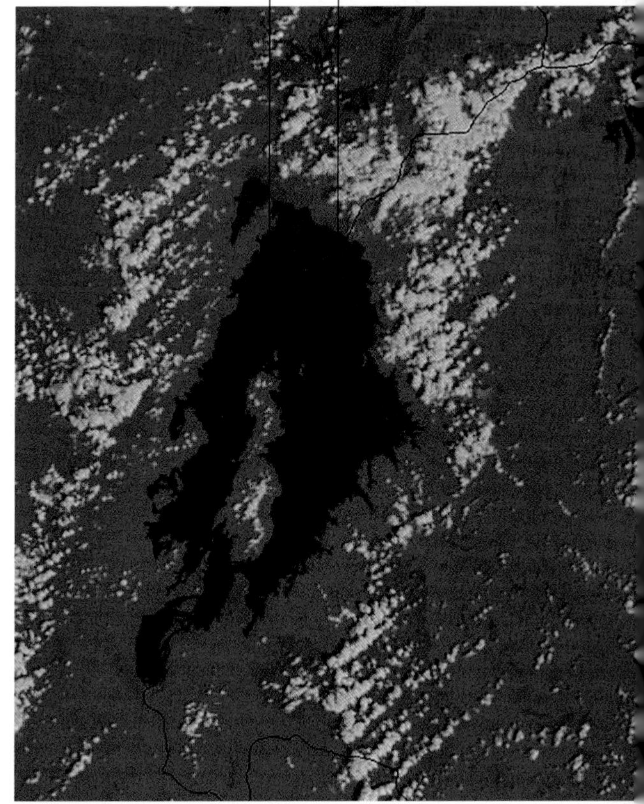

LAKE KIVU
This large lake lies in part of the East African Rift Valley—a region where Earth's crust is slowly stretching with intrusion of magma into the crust.

TUFF RINGS AND CONES

Two relatively small and simple types of volcanically created landform are tuff rings and tuff cones. Like other volcanic features called maars, they are created by violent interactions between magma (molten rock) and water. Of the two, tuff cones are more compact but have higher crater rims.

FORMATION AND STRUCTURE

Tuff rings and cones are found in areas of the world where volcanic activity is generally high (or has been in the past)—that is, near plate boundaries and hotspots—and where upwelling magma has come in contact with groundwater or surface water in the form of a lake, marsh, or shallow sea. The contact leads to a vigorous eruption with the production of ash, which falls in a ring around the eruption vent and later consolidates into a type of rock called tuff. The water involved in the creation of the cone or ring may later disappear. Tuff rings have wide, low-rimmed craters, and are often 0.5–1 mile (1–2km) in diameter. Tuff cones are smaller and more conical, with higher crater rims. Although formed in a similar way to maars (see pp.130–31), tuff rings and cones are usually not sunk into the ground, nor filled with water.

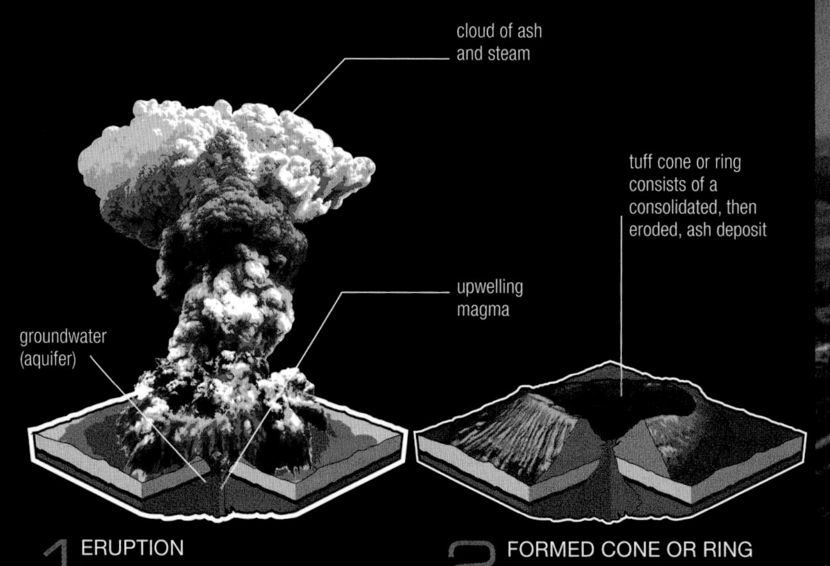

1 ERUPTION
Where magma contacts groundwater, a cloud of steam and ash is blown into the air. The ash rains down to form a ring or cone around the vent.

2 FORMED CONE OR RING
Over time, the ash deposit consolidates to form a ring (here) or cone of tuff. Later, this is weathered and eroded by the action of wind and water.

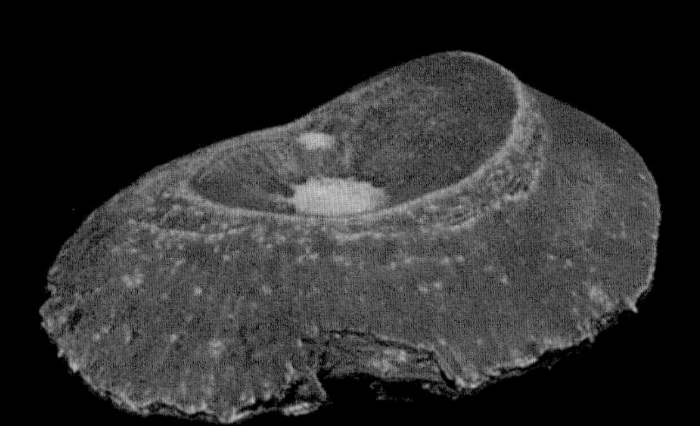

DAPHNE MAJOR
This small island, one of the Galápagos Islands in the eastern Pacific, is a heavily eroded tuff cone whose rim currently rises 390ft (120m) above sea level. It is thought to have formed about 1.8 million years ago.

FORT ROCK, OREGON
This tuff ring was created tens of thousands of years ago when rising magma encountered wet mud at the bottom of an ancient lake. Once the ring had formed, waves from the lake eroded its outside walls to form terraced cliffs.

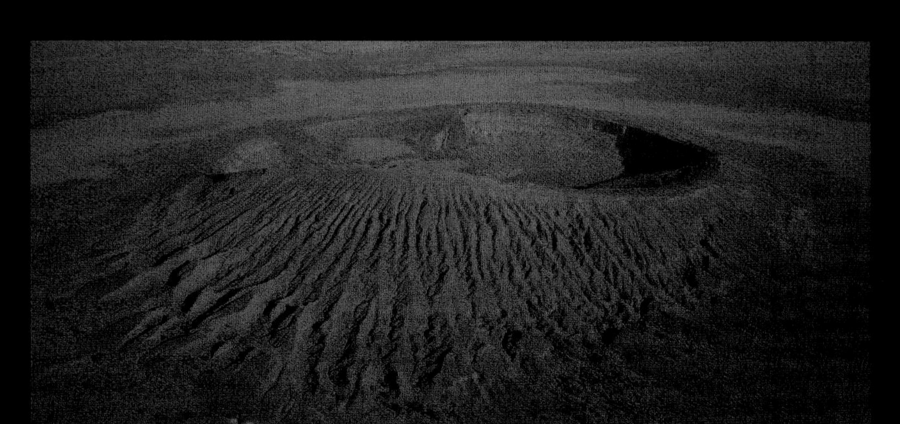

DESERT CONE
This tuff cone in the Sonoran Desert of northwest Mexico has a 3,280-ft- (1-km-) wide crater. Called Cerro Colorado, it is part of a volcanic field (group of small volcanoes).

KOKO CRATER, HAWAII
This tuff cone on the island of O'ahu, in Hawaii, is 1,207ft (368m) high and now houses a botanical garden in its crater. No eruption has occurred here or elsewhere in O'ahu in more than 10,000 years, although future eruptions are possible.

LAVA DOMES AND SPINES

A lava dome is a craggy, bulbous, slow-growing, and potentially dangerous mass of congealed lava that develops at a volcanic vent when viscous lava oozes out. A lava spine is a bizarre-looking object that sometimes grows up vertically up out of a lava dome.

DOME FORMATION

Lava domes develop at volcanic vents that extrude highly viscous lava, such as rhyolitic or dacitic lava (see p.97). Lava of this type cannot flow far from the vent from which it is being squeezed out. Instead, it piles up to form a slow-growing mound that blocks the vent. Most lava domes have developed in the main vent of a larger volcano, and sit within its summit crater, although a dome can also grow on a side vent or can even be a large separate volcano. The time it takes a dome to form and grow to its maximum size can be anything from a few weeks to several thousand years.

lava dome
crater of parent volcano
vent
rising viscous magma

STRUCTURE
A lava dome has a rough surface and a characteristic mound shape. Although its surface is solid, an actively growing dome contains large amounts of hot, viscous, molten lava.

NEWLY FORMED DOME
In 2008, this new lava dome, some 394ft (120m) high, appeared in a volcanic caldera called Chaitén in Chile as the volcano came to life after 9,500 years of inactivity.

DANGEROUS DOMES

Some lava domes across the world are extinct volcanic remnants, but others are active, evolving structures undergoing processes such as growth, erosion, occasional collapse, and regrowth. As a dome enlarges, its edges creep outward, and if one of these edges oversteepens, the dome may partially collapse to produce a dangerous landslide of hot rubble called a pyroclastic flow. Following an earthquake in 1792, the partial collapse of a lava dome on Mount Unzen in Japan created a huge landslide. This triggered a tsunami that killed about 15,000 people—Japan's worst-ever volcano-related disaster.

FIERY DOME
This glowing lava dome appeared in the middle of a crater lake during a 2007 eruption of the Kelut volcano in Indonesia. It grew to a height of 394ft (120m) before cracking open. Hot lava oozed into the lake, producing huge plumes of steam.

LAVA SPINES

Lava spines are dramatic-looking spires or fingers of solidified lava, usually cylinder shaped, that are sometimes pushed up out of lava domes on large stratovolcanoes (see pp.120–21). They are caused by highly viscous, pasty lava partially solidifying inside a volcanic vent and then being squeezed upward—rather like hardened toothpaste emerging from a toothpaste tube. Lava spines can reach a considerable height. One that grew at the summit of Mont Pelée in Martinique (see pp.152–53), after its eruption in 1908, attained a height of 984ft (300m) and a volume comparable to that of the Great Pyramid of Giza (Egypt's largest pyramid).

Over the past 30 years, notable lava spines have developed on volcanoes such as Mount St. Helens in the US, Mount Unzen in Japan, and Mount Pinatubo in the Philippines. After growing for a few weeks or several months, a lava spine becomes unstable and starts to collapse under its own weight, eventually disintegrating into a pile of rubble.

SPINE ON MOUNT ST. HELENS
A series of lava spines, some of them reaching 295ft (90m) in height, grew out of a lava dome on Mount St. Helens in Washington State between 2004 and 2008.

FINGERLIKE SPINE
This short-lived lava spine was pushed out of a lava dome on Mount Unzen, a group of stratovolcanoes in Japan, in 1994.

PUY DE DÔME
A large extinct lava dome in central France, Puy de Dôme (literally "dome-shaped hill") has a summit height of 2,759ft (841m). The last eruption here occurred about 10,700 years ago. In 1875 a physics laboratory was built at the summit, and in 1956 a television antenna was added.

COLIMA LAVA DOME
This dome at the summit of the Colima Volcano in Mexico has been growing for nearly a century. In early 2010, it had almost filled the volcano's summit crater. Occasionally, explosive eruptions of magma occur from the dome, causing pyroclastic flows and ash plumes.

VOLCANIC FIELDS

Volcanic fields are areas of past or present volcanic activity containing clusters of small volcanoes. From air or space, they look almost like a rash on the landscape. Like volcanoes in general, volcanic fields are usually found either on, or near, plate boundaries or above hotspots that exist, or once existed, beneath Earth's crust.

FORMATION

Volcanic fields develop where magma (molten rock) rises up beneath Earth's crust, but the channels through which the magma reaches the surface are spread over too broad an area, or the supply of magma is too little, for a single large volcano to form. Instead, many smaller volcanic features develop, although not necessarily all at the same time. Depending on the type of magma and many other factors (such as whether there is much groundwater or surface water at the eruption site), the individual volcanoes may be cinder cones (see pp.116–17), maars (pp.130–31), lava domes (pp.136–37), small stratovolcanoes (pp.120–21), or a mixture of several of these different volcano types.

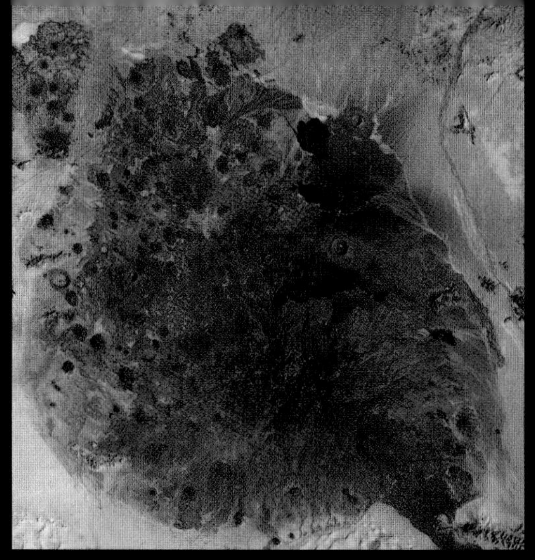

PINACATE VOLCANIC FIELD
The Pinacate National Park in Mexico, seen here in a satellite photograph, contains more than 300 small volcanoes, including cinder cones, maars, tuff rings, and small lava flows.

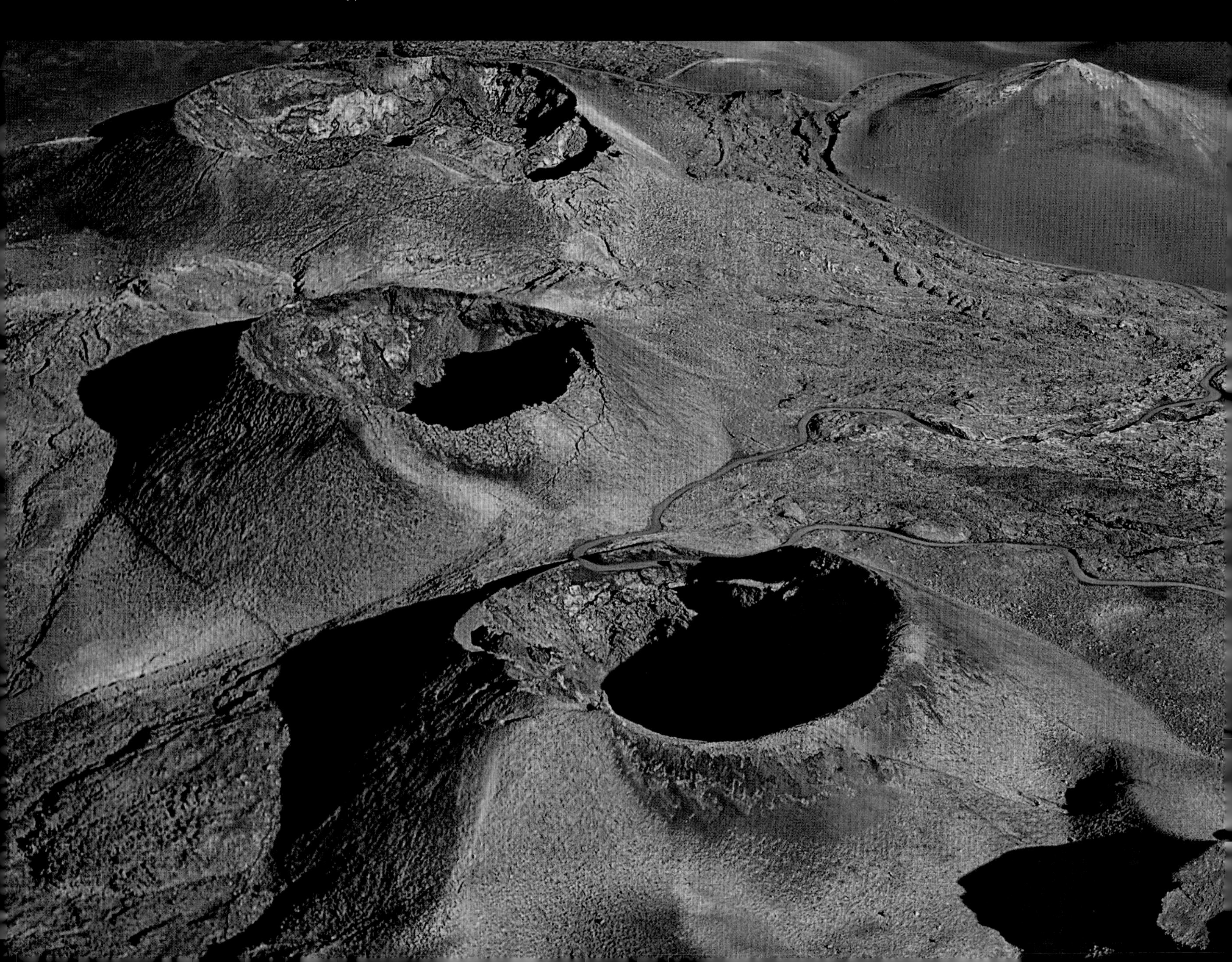

SOME NOTABLE VOLCANIC FIELDS

1. Hopi Buttes, Arizona
2. Timanfaya National Park, Lanzarote
3. Crater fields at Marsabit, Kenya
4. Chaîne des Puys, France
5. Harrat Khaybar, Saudi Arabia
6. Michoacán-Guanajuato, Mexico
7. Pinacate Biosphere Reserve, Mexico
8. West Eifel volcanic field, Germany

LANZAROTE CRATERS
Timanfaya National Park contains more than 100 volcanoes, mainly cinder cones, that erupted between 1730 and 1736.

ACTIVITY LEVELS

In most volcanic fields, each volcano is monogenetic—that is, it only erupts once, for a relatively short period of time (anything from a few days to a few years), and then goes quiet. If a further eruption occurs later in the same region, it produces new volcanic cones and other features. Many monogenetic volcano fields are extinct, with no further eruptions expected to happen, such as the Chaîne des Puys volcanic field in France. Others are still volcanically active, with more eruptions considered possible, such as the Timanfaya National Park in Lanzarote. In some parts of the park, the temperature of the ground just 43ft (13m) below the surface can reach as high as 1,112°F (600°C). This indicates the continuing presence of reserves of hot magma.

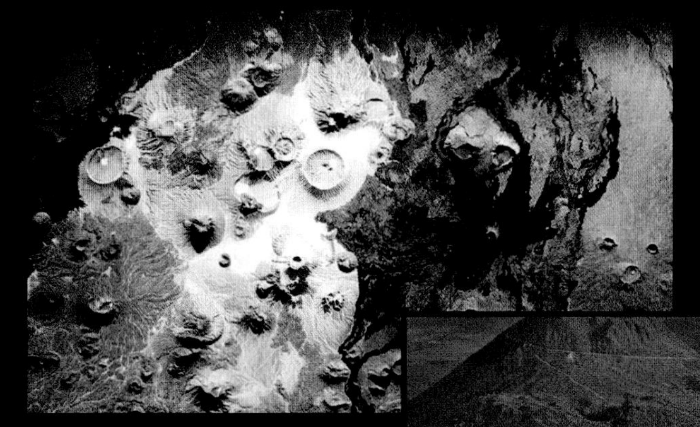

HARRAT KHAYBAR
Viewed here from the International Space Station, the Harrat Khaybar volcanic field in Saudi Arabia contains a small stratovolcano as well as tuff cones and lava domes. Eruptions last occurred here about 1,400 years ago.

CHAÎNE DES PUYS
This vegetated cinder cone is one of a group of more than 70 small, extinct volcanoes in Chaîne des Puys ("chain of volcanic hills"), located in the Massif Central region of France.

ANCIENT VOLCANIC REMNANTS

The Hopi Buttes volcanic field consists of some 300 erosion-resistant volcanic remnants spread over an area of 960 sq miles (2,500 sq km) of northeastern Arizona. Formed between 8 and 4 million years ago, its most prominent features are scattered dark volcanic plugs. These are mixed in with numerous maars (shallow craters formed by explosive interactions between magma and groundwater).

Dark volcanic plugs
These mounds were originally hardened plugs of lava or cemented fragments of volcanic rock that formed below the land surface. They were exposed by erosion of the surrounding softer rock.

VOLCANIC COMPLEXES

Although many volcanoes have a simple form, with a single main vent and crater at the top of a single cone, others have more complicated structures. They include volcanoes with overlapping cones and multiple craters, and others where new cones develop within the remains of older stratovolcanoes.

COMPOUND VOLCANOES

Sometimes called complex volcanoes, these structures consist of two or more volcanoes, usually stratovolcanoes (see pp.120–21) that have formed close to each other, with separate main conduits and vents, and partially overlapping cones. They can form as a result of small shifts in the spot where rising magma reaches Earth's surface. Usually, the separate cones have formed at different times. Compound volcanoes typically have multiple summit craters. One of the strangest is Kelimutu, on the island of Flores in Indonesia. This has three summit craters, each containing a lake of a different color.

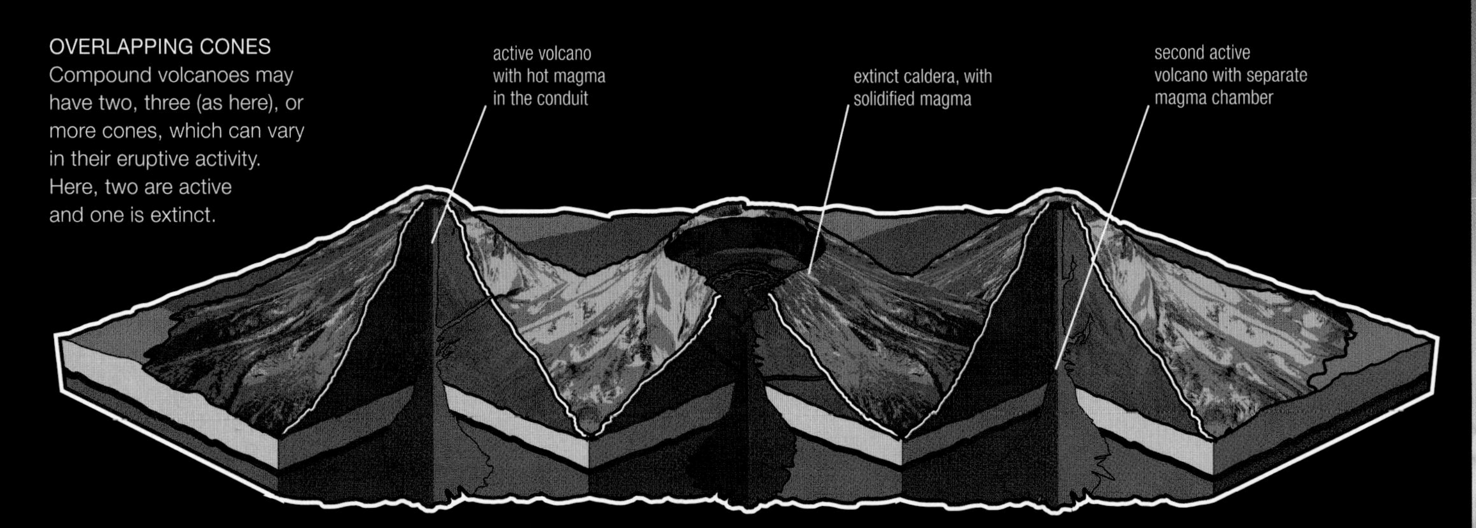

OVERLAPPING CONES
Compound volcanoes may have two, three (as here), or more cones, which can vary in their eruptive activity. Here, two are active and one is extinct.

SOMMA VOLCANOES

A somma volcano consists of one or more stratovolcanoes that have grown to occupy a large part of the caldera that survives from an older, collapsed, stratovolcano. The best-known somma volcano in the world is the Vesuvius/Somma volcanic complex in southern Italy (see pp. 156–57). This consists of a stratovolcano, Vesuvius that has grown up within the caldera of a larger, more ancient volcano, called Monte Somma—hence the name for this type of volcanic complex. Somma volcanoes are not common. Most identified examples are in the remote Kuril Islands of the northwestern Pacific or in the Kamchatka Peninsula of eastern Russia. Another classic example is the Teide/Pico Viejo/Las Cañadas complex on Tenerife in the Canary Islands. In this somma volcano, two stratovolcanoes, Teide and Pico Viejo, have developed over the past 150,000 years within the ancient Las Cañadas caldera, which originally formed at least 3.5 million years ago.

TEIDE, TENERIFE
In this view of the island of Tenerife based on satellite imagery, the Teide and Pico Viejo volcanoes sit at the center, within the surrounding steep walls of the roughly elliptical Las Cañadas caldera. Tenerife is an entirely volcanic island—the third largest in the world by volume.

CALDERA COMPLEXES

Caldera complexes consist of one or more large volcanic calderas that contain several additional, more recently formed volcanic features growing within them. Where there is more than one caldera, they may overlap, and the newer volcanoes—which may be stratovolcanoes, cinder cones (see pp.116–17), or lava domes (see pp.136–37)—may be coalesced together, as for example in the Taal caldera in the Philippines. One of the largest caldera complexes in the world is the Masaya caldera complex in Nicaragua. This consists of several partially overlapping pit-like craters at the summit of twin volcanic cones, which lie within the Masaya caldera.

CINDER CONE IN TAAL CALDERA
This cone is part of the Taal caldera complex, a group of volcanoes within a large lake-filled caldera. The cone is joined with others to form a volcanic island called Volcano Island—the site of many eruptions.

TENGGER COMPLEX
This large volcanic complex in Java, Indonesia, consists of five stratovolcanoes within an older caldera. On the edge of the complex, and seen in the background here, is an active stratovolcano, Semeru.

FISSURE ERUPTIONS

Fissure or Icelandic-type eruptions are voluminous outpourings of lava and poisonous gases from long linear cracks, called fissure vents, that appear in the ground. These eruptions usually occur fairly quietly, without loud explosions, but their effects can be dramatic—in the past they may have caused climate change and mass extinctions.

PRODUCTION OF FLOOD BASALTS

Fissure eruptions occur mainly in parts of the world where Earth's crust is rifting or being stretched, usually at a divergent plate boundary (see pp.28–29), or where the crust lies over a mantle plume at a hotspot (see pp.32–33). The crust lava that emerges is usually runny basaltic lava, which flows a considerable distance—up to tens of miles—before it solidifies. At various times over the past 300 million years, fissure eruptions have built up extensive thick deposits or plateaus of hardened basaltic lava in various parts of the world.

The largest of these flood basalts, as they are called, are the Siberian Traps, which formed 250 million years ago and cover 0.77 million sq miles (2 million sq km) of northern Russia. Other flood basalts include the 6,562-ft- (2,000-m-) thick Deccan Traps of west-central India (formed 68–60 million years ago), the Columbia River Basalt Group of the northwestern US (formed around 15 million years ago), the Chilcotin Plateau Basalts in British Columbia, Canada, and the Antrim Plateau of Northern Ireland.

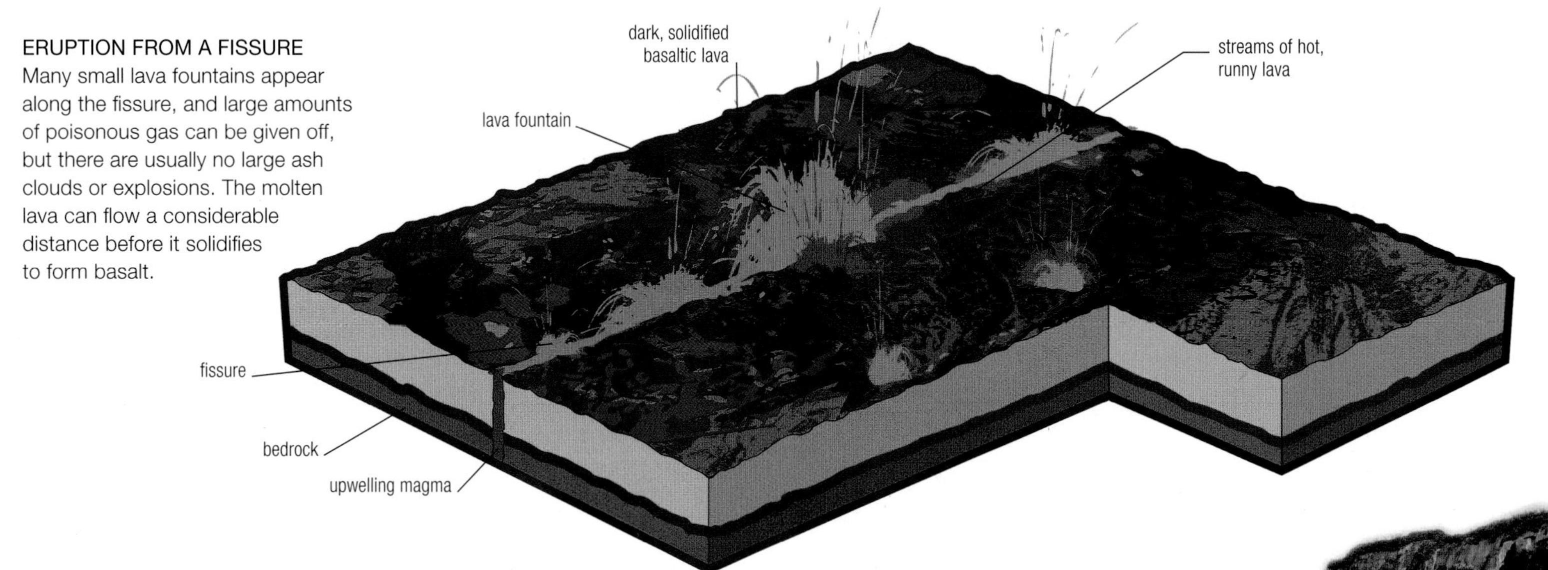

ERUPTION FROM A FISSURE
Many small lava fountains appear along the fissure, and large amounts of poisonous gas can be given off, but there are usually no large ash clouds or explosions. The molten lava can flow a considerable distance before it solidifies to form basalt.

RATHLIN ISLAND
This small island off the coast of Antrim, Northern Ireland, is part of a group of flood basalts created 60–50 million years ago during a rifting process that led to the opening of the North Atlantic.

CHILCOTIN BASALTS
Over the past 10 million years, fissure eruptions created vast flood basalts in British Columbia, forming the Chilcotin Plateau. This canyon was formed by a river eroding the plateau.

SVARTIFOSS WATERFALL
This waterfall in Iceland is famous for the dark hexagonal basalt columns that hang off the cliffs like organ pipes at each side. These formed as lava cooled 15 million years ago.

FIMMVÖRÐUHÁLS ERUPTION
This lava flow is from a 984-ft- (300-m-) long fissure that appeared in Fimmvörðuháls in southern Iceland in March 2010. The fissure eruption was merely a prelude to the Eyjafjallajökull eruption in April.

THE LAKI ERUPTION

An eruption that occurred in 1783–84 from a fissure called Laki in Iceland is reckoned to be the most deadly volcanic catastrophe in recent history, causing more than six million human deaths worldwide. Clouds of poisonous gases from the fissure killed more than half of Iceland's livestock, leading to a famine in which a quarter of the country's human population died. A haze of dust and gas spread over northern Europe, then more widely across the northern hemisphere, causing many deaths directly from respiratory illnesses, but also a drop in temperatures, crop failures, and famines. The disruption to agriculture in France, and the resulting poverty and famine, is credited with triggering the French revolution in 1789.

THE LAKI FISSURE
This line of volcanic cones in southern Iceland marks the fissure from which the Lakagigar, or Laki, lava flow erupted in 1783–84. Over an eight-month period, an estimated 3 cu miles (14 cu km) of lava poured out over the surrounding terrain.

“THE PECULIAR HAZE OR SMOKY FOG THAT PREVAILED IN THIS ISLAND, AND EVEN BEYOND ITS LIMITS, WAS A MOST EXTRAORDINARY APPEARANCE.”

GILBERT WHITE, NATURALIST, DESCRIBING THE EFFECT OF THE LAKI ERUPTIONS ON BRITAIN

HAWAIIAN-STYLE ERUPTIONS

Hawaiian-type volcanic eruptions are relatively mild events characterized by the emission of fountains and streams of runny lava. Eruptions of this type occur in many parts of the world, not just Hawaii. Larger more explosive eruptions can occur from these volcanoes—in recent history the three active volcanoes on Hawaii's Big Island have mainly erupted this way.

CHARACTERISTICS

Hawaiian-style eruptions occur when there is basaltic (runny) magma that contains little dissolved water and gas. These conditions are common when the magma comes from a hotspot or mantle plume under Earth's crust rather than from a convergent plate boundary. Hawaiian-style eruptions build up broad shield volcanoes, such as Mauna Loa and Kilauea in Hawaii. Before an Hawaiian-style eruption, the volcano may swell or inflate a little. Next there may be an escape of gas. Actual eruptions typically start with a fissure opening on a flank of the volcano and lava spurting out in small fountains; this may progress to the appearance of larger fountains from secondary cones. The hot lava moves downslope in streams. Lava tubes—tunnel-like channels—may form on the flanks of the volcano and facilitate the flow of molten

HAWAIIAN ISLAND CHAIN
This satellite image shows six of Hawaii's volcanic islands. The most prominent volcano on the Big Island,

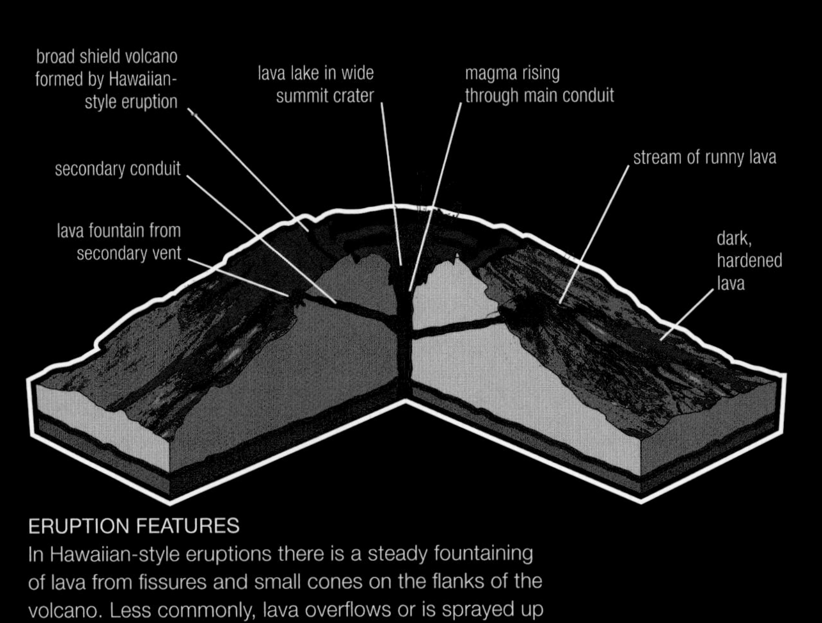

ERUPTION FEATURES
In Hawaiian-style eruptions there is a steady fountaining of lava from fissures and small cones on the flanks of the volcano. Less commonly, lava overflows or is sprayed up

MAUNA LOA ERUPTION
Hot lava bubbles up and flows away from a large vent on the Hawaiian shield volcano Mauna Loa in 1984. During that eruption, the volcano produced about 0.5 million cubic miles (200 million cubic meters) of lava in just three weeks.

DEVELOPMENT OF AN ERUPTION

1 LAVA FOUNTAIN
Voluminous fountains and jets of lava, as here from the Pu'u 'O'o vent on Kilauea, are a hallmark of Hawaiian-style eruptions. The fountains and jets usually develop soon after the eruption starts. They can occur in short spurts or last for hours on end.

2 WIDE LAVA STREAM
Some lava flows away from the eruptive vent in wide streams. These normally consist of the hottest, thinnest type of basaltic lava, called pahoehoe. It can travel for long distances even over the shallow slopes typical of shield volcanoes.

3 LAVA TUBE
Thick flows of lava sometimes develop into channels enclosed by solidified lava. A space can form in the roof of one of these lava tubes—a skylight—allowing observers to look down at the lava flow. Lava tubes can be up to 50ft (15m) below the surface

POTENTIAL HAZARDS

The runniness of the lava and lack of gas in the type of magma associated with Hawaiian-type eruptions means that few explosions, large volcanic bombs, or pyroclastic flows are produced. They are normally relatively safe to witness. In Hawaii, only a handful of fatalities have occurred directly as a result of eruptions on Mauna Loa and Kilauea in recent years, although many houses have been destroyed.

ENTERING THE SEA
Lava on Hawaiian and other island-based volcanoes sometimes cascades right down to the sea, producing large plumes of steam when it enters the water.

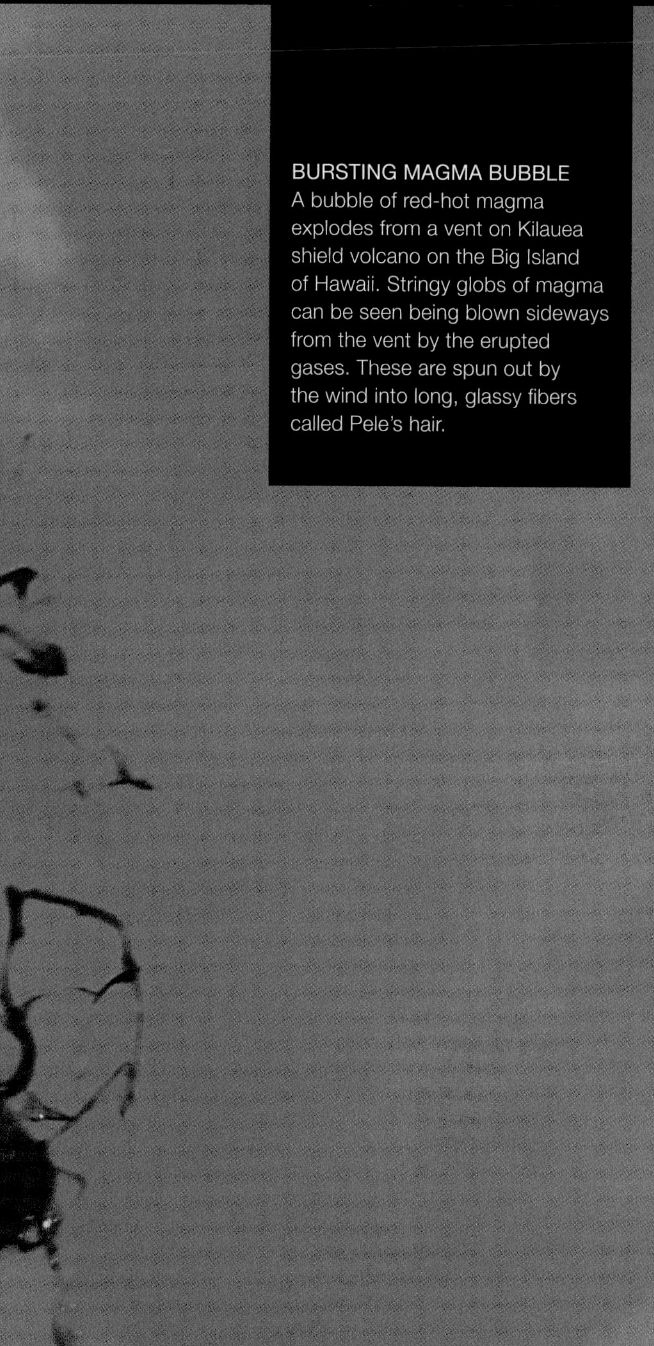

BURSTING MAGMA BUBBLE
A bubble of red-hot magma explodes from a vent on Kilauea shield volcano on the Big Island of Hawaii. Stringy globs of magma can be seen being blown sideways from the vent by the erupted gases. These are spun out by the wind into long, glassy fibers called Pele's hair.

SPRAY OF BOMBS

This long-exposure photograph of a nighttime eruption from Stromboli shows the paths of the small volcanic bombs ejected from the crater. Each falls to the ground in a parabolic arc. Also visible is some erupted steam, which glows red by reflecting light from the incandescent lava.

STROMBOLIAN ERUPTIONS

Low-intensity, episodic eruptions that emit lava as a shower of little "bombs" into the air are called Strombolian eruptions. These are named after the small stratovolcano Stromboli, off the north coast of Sicily.

CHARACTERISTICS

Strombolian eruptions are commonly produced by cinder cone volcanoes (see pp.116–17) and from certain stratovolcanoes (see pp.120–21), notably Stromboli itself, which has been called the "Lighthouse of the Mediterranean." Eruptions occur as a series of short, explosive outbursts that throw showers of lava fragments into the air. Each outburst can be accompanied by noisy bangs but no really large explosions. Some scientists think that Strombolian eruptions are caused by bubbles of gas rising up through the viscous magma in the volcano's conduit and bursting explosively at the top. They may be caused by cyclical gas pressure variations in the volcano's vent. Strombolian eruptive activity can be long-lasting because the eruptive system resets itself.

STROMBOLIAN ERUPTION
These eruptions are characterized by short outbursts in which showers of viscous magma are thrown up as incandescent cinders and lava bombs. Strombolian eruptions never develop a sustained ash column.

significant release of volcanic gases
small, short-lived ash cloud
showers of cinders and lava bombs emitted at regular intervals
main conduit filled with magma
occasional short lava flows

POTENTIAL HAZARDS

Although Strombolian eruptions are much noisier than Hawaiian-style eruptions (see pp.144–45), they are not much more dangerous. Nevertheless, onlookers need to stand well back from the area where volcanic bombs are falling. Although these bombs are usually not large, some of them fall from a height of several hundred feet, so have acquired a considerable speed and potential to injure by the time they reach the ground. Unlike in Hawaiian-style eruptions, there are rarely any sustained lava flows over the ground in Strombolian eruptions, lessening the potential risk. Stromboli itself occasionally erupts in a more violent and dangerous Vulcanian-style eruption (see pp.150–51) and has killed people doing so. An eruption in 1930, for example, produced lava bombs that destroyed several houses and a pyroclastic flow that killed four people.

TWISTED BOMB
Most lava bombs emitted in Strombolian eruptions are no more than 8in (20cm) in diameter.

MOUNT YASUR, VANUATU
This volcano on Tanna Island, Vanuatu, in the southwestern Pacific, has been producing Strombolian eruptions for centuries. Its glowing summit attracted Captain Cook to Tanna in 1774.

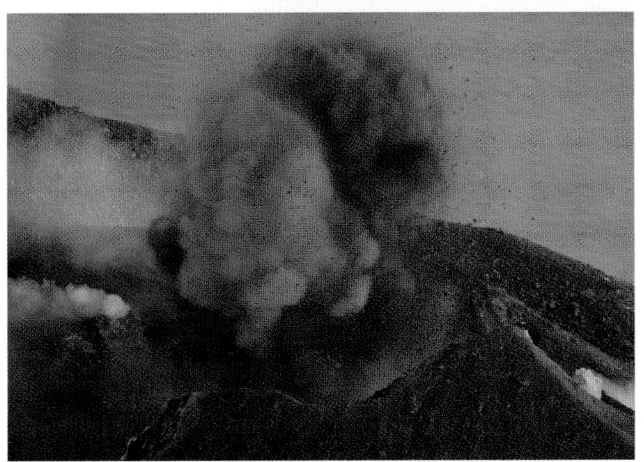

STROMBOLI BY DAY
Some 2,950ft (900m) high, Stromboli has been erupting every 5 to 20 minutes for thousands of years. This makes it a significant tourist attraction in the central Mediterranean region.

VULCANIAN ERUPTIONS

Vulcanian eruptions are moderately violent volcanic events that always start with cannon-like explosions. They are named after the small volcanic island of Vulcano in the Mediterranean, which had a Vulcanian-type eruption in 1890 and also contributed the word "volcano" to most European languages.

CHARACTERISTICS

Vulcanian eruptions usually rate at either level 2 or level 3—described as "explosive" to "severe"—on the Volcanic Explosivity Index (see p.89). Only stratovolcanoes (see pp.120–21) that produce medium-to high-viscosity lava erupt in this way. The explosions that invariably herald a Vulcanian eruption occur as mounting pressure causes a lava plug in the volcano's vent to be blown away. Following the initial volley of explosions and with the production of an ash column, further large bangs occur, at intervals ranging from a few minutes to as much as a day. Also produced are many large volcanic bombs—these are commonly found on the ground after an eruption as "breadcrust" bombs, so-called because their cracked surfaces are reminiscent of the crust of some types of bread loaf.

In most cases, Vulcanian eruptions quieten down within a few hours to days, sometimes ending with a flow of viscous lava. In other cases, activity may continue for years, with occasional explosive eruptions alternating with longer periods of quiet steam emission.

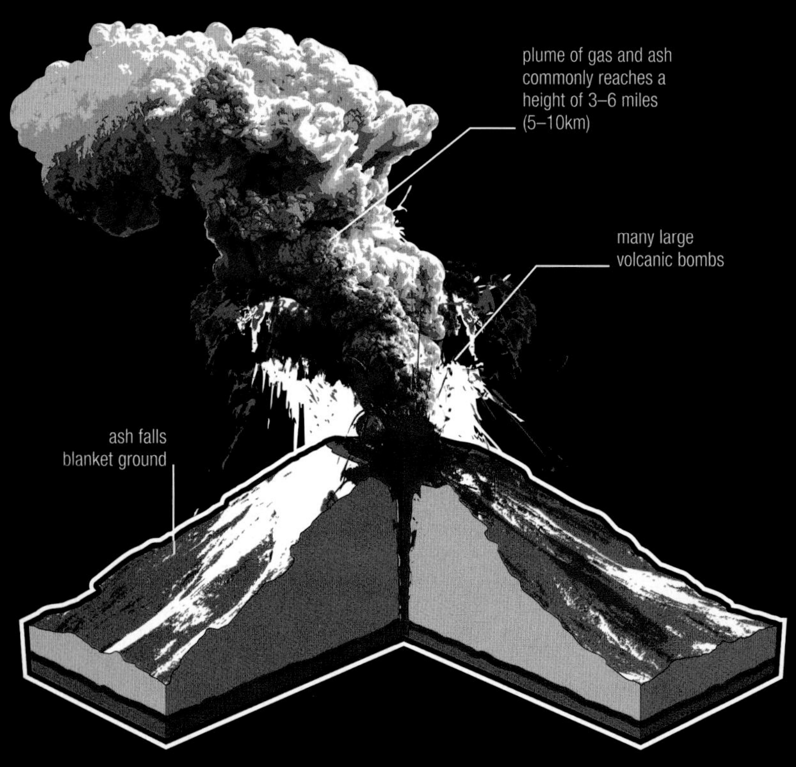

ERUPTION FEATURES
With each explosion, a dense cloud of ash-laden gas is blasted from the volcano's crater and rises high above the peak, the ash forming a fairly large eruptive column before drifting to the ground over a wide area. In addition, there are many high-velocity ejections of large volcanic bombs.

POTENTIAL HAZARDS

Vulcanian eruptions are extremely dangerous for anyone within several hundred feet of the eruption vent because of the volcanic bombs they produce, which can be as much as 7 to 10ft (2 to 3m) in diameter. Despite their name, volcanic bombs rarely, if ever, explode. Nevertheless they can cause severe damage on impact, and because they are often still red-hot or even incandescent when they fall, they can set buildings and vegetation on fire. Volcanoes that produce Vulcanian eruptions also commonly harbor growing lava domes (see pp.136–37), which if they disintegrate, can result in dangerous pyroclastic flows down the sides of the volcano. For these reasons, anyone studying or witnessing a Vulcanian-style eruption is advised to remain a few miles away from its volcanic vent. For example, anyone visiting Anak Krakatau in Indonesia during one of its eruptive phases is kept at a distance of 2 miles (3km) from the active cone and prohibited from landing on Anak Krakatau island itself.

IRAZÚ
This large ash cloud was emitted by Costa Rica's highest volcano, Irazú, during a Vulcanian eruption in 1963. A great quantity of ash later showered down on the Costa Rican capital of San José, 15 miles (24km) away.

BREADCRUST BOMB
The cracked surface of a bomb of this type is caused by the expansion of gas within its still-liquid interior after it has hit the ground.

ANAK KRAKATAU
This volcano in Indonesia has been producing spectacular Vulcanian-style eruptions since 2007. It grew from part of the remains of the volcanic island of Krakatau, which exploded dramatically in 1883.

KILLER COLOMBIAN VOLCANO

The volcano Galeras in Colombia has produced many Vulcanian eruptions over the past 40 years. A sudden eruption in 1993 produced volcanic bombs and poisonous gases that killed nine people, including six scientists. Aside from that incident, Galeras is considered highly dangerous because it lies just 5 miles (8km) from the large city of Pasto with a population of 450,000.

Radar image of Galeras
In this satellite radar image, Galeras is the green area at center, while Pasto is the orange area at bottom right.

PELÉAN ERUPTIONS

In some volcanic eruptions, the key event is an avalanche of hot gas, rock, and ash caused by the collapse of a dome, which feeds a pyroclastic flow that sweeps downhill. These eruptions are named "Peléan" after a notorious example that occurred at Mont Pelée, Martinique, in 1902.

CHARACTERISTICS

Peléan eruptions are among the most dangerous and destructive of all volcanic eruptions. They usually rate at 3 to 4 on the Volcanic Explosivity Index (see p.89). Most of the damage is caused by pyroclastic flows of hot gas, rock, and ash down the sides of the volcano—which may generate even more dangerous pyroclastic surges. Surges contain higher levels of gas than flows and can travel at higher speeds. Only stratovolcanoes that produce high-viscosity lava erupt in this way. Typically, a Peléan eruption is preceded by the development of a lava dome at the volcano's summit. If this collapses, magma may be blasted out sideways, producing a hot, ground-hugging pyroclastic flow, which incinerates all in its path. A new lava dome usually then develops at the summit creating the likelihood of another collapse and pyroclastic flow.

DEVELOPMENT OF A PELÉAN ERUPTION
An eruption of this type is typically triggered when rising pressure within the volcano, or an event such as an earthquake, causes partial disintegration of a lava dome at its summit.

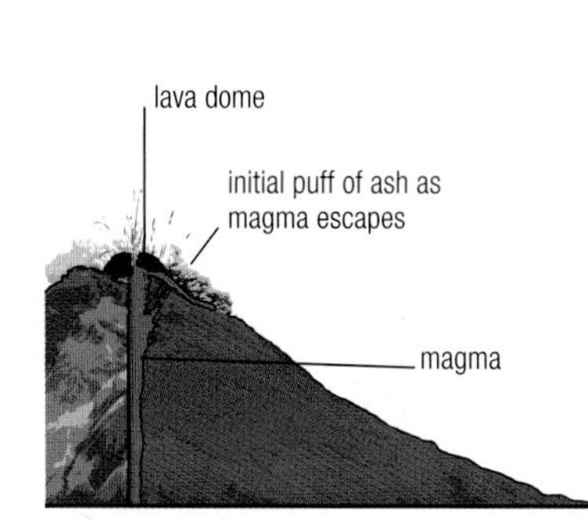

1 PARTIAL COLLAPSE Pressure begins to force gas and magma out from beneath a collapsing lava dome. The magma erupts as ash.

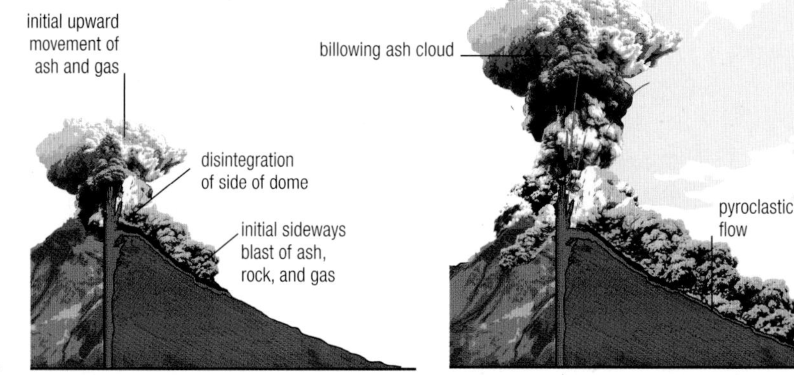

2 ERUPTION Suddenly, as part of the dome disintegrates, large amounts of magma, gas, and rock fragments are blasted out.

3 ASH CLOUD SURGE The ash cloud can billow to a height of 3 to 6 miles (5 to 10km), while a pyroclastic flow rushes to the foot of the volcano.

PYROCLASTIC FLOW ON UNZEN
A small pyroclastic flow is seen moving down Mount Unzen in Japan in May 1993. The volcano underwent some violent Peléan-like eruptions between 1991–1994.

RECENT PELEAN ERUPTIONS

1. Pelée, Martinique 1902
2. Hibok-Hibok, Philippines 1948–51
3. Lamington, New Guinea 1951
4. Bezymianny, Kamchatka 1956, 1985
5. Mayon, Philippines 1968, 1984
6. Soufrière Hills, Montserrat 1995–99
7. Unzen, Japan 1991–95

Mount Mayon, Philippines

Soufrière Hills, Montserrat

THE MONT PELÉE DISASTER

The most lethal volcanic eruption of the 20th century, and one of the five deadliest of all time, occurred from the stratovolcano Mont Pelée, on the Caribbean island of Martinique, on May 8, 1902. It involved a pyroclastic flow that moved at a speed of more than 370mph (600kph). This, in turn, generated a pyroclastic surge that quickly overwhelmed the port city of St. Pierre, about 4 miles (6.5km) from Mont Pelée's summit, killing almost its entire population of about 28,000. There were only three survivors in the direct path of the flow, one of whom was in an enclosed dungeon-like jail cell. A French volcanologist, Alfred Lacroix, who visited Martinique shortly afterward, described these events as *nuée ardentes* ("glowing clouds"). In 1929, another eruption occurred from Mont Pelée, and further eruptions are expected in future, although the mountain is now closely monitored by volcanologists.

MAP OF MARTINIQUE
On this map, made soon after the disaster, the various deposits of lava and pyroclastic material (ash and rock fragments) from Pelée are shown in red. St. Pierre is shown in a small bay at the base of the volcano's southern flank.

PYROCLASTIC SURGE
This photograph, taken in December 1902, shows a pyroclastic surge from Pelée similar to the one that tragically engulfed St. Pierre in May 1902. It was one of several flows and surges that occurred in the months after the original disaster.

> "I SAW ST. PIERRE DESTROYED. THE CITY WAS BLOTTED OUT BY ONE GREAT FLASH OF FIRE."
>
> AN ASSISTANT PURSER ON THE SS *RORAIMA*, WHICH REMAINED TEMPORARILY AFLOAT, ALTHOUGH AFLAME, IN THE HARBOR OF ST. PIERRE

RUINS OF ST. PIERRE
The city burned for several days after the disaster and nearly every building was destroyed. Today the population of St. Pierre is just over 4,500, and although the city was never restored to its former glory, many structures are built upon the foundations of pre-eruption buildings.

PLINIAN ERUPTIONS

The most explosive and violent of all volcanic events are Plinian and ultraplinian eruptions. These eruptions blast a steady, powerful stream of gas and fragmented magma into the air, producing a gas and ash cloud that typically takes the shape of a huge mushroom or cauliflower.

OCCURRENCE

Also called Vesuvian eruptions for their similarity to the deadly eruption of Mount Vesuvius in 79CE, Plinian eruptions are named after the Roman author Pliny the Elder, who died during the event, and his nephew, Pliny the Younger, who observed and later described it in a letter. These eruptions usually stand between 4 and 6 on the Volcanic Explosivity Index, or VEI (see p.89). Those above 6 are called ultraplinian. They only occur in stratovolcanoes or caldera systems that produce very viscous lava, usually rhyolitic (see pp.96–97). Only a few stratovolcanoes have consistently produced such eruptions. These include Vesuvius, which has erupted in a similar fashion about a dozen times since 79CE, and Mount St. Helens, with four or five Plinian eruptions over the past 600 years. Usually, the erupting volcano has shown no activity for hundreds or thousands of years. Chaitén, in Chile, which erupted in 2008, had not seen any volcanic activity in the previous 7,000 years. This indicates that to produce such eruptions, huge amounts of magma and pressure have to build up over time within or under the volcano. In some Plinian eruptions, the amount of magma erupted is so large that the top of the volcano collapses, resulting in a caldera (see pp.126–27).

PLINIAN ERUPTIONS IN HISTORY

VOLCANO	DATE	VEI
Vesuvius, Italy	79CE	5
Tambora, Indonesia	1815	7
Novarupta, US	1912	6
Hekla, Iceland	1947	4
Mount St. Helens, US	1980	5
Pinatubo, Philippines	1991	6
Chaitén, Chile	2008	4
Sarychev, Russia	2009	4
Eyjafjallajökull, Iceland	2010	4

SARYCHEV ERUPTION
In June 2009, astronauts aboard the International Space Station were treated to this view of Sarychev's eruption on Russia's Kuril Islands. The ash plume at one point reached a height of 7 miles (12km).

DEVELOPMENT OF A PLINIAN ERUPTION

Plinian eruptions are marked by columns of gas and ash extending high into the stratosphere. Their key characteristics include powerful blasts of fragmented magma, driven by the thrust of expanding gases, and the ejection of large amounts of pumice (solidified, frothy lava). Short eruptions can end in less than a day, but longer ones may go on for weeks.

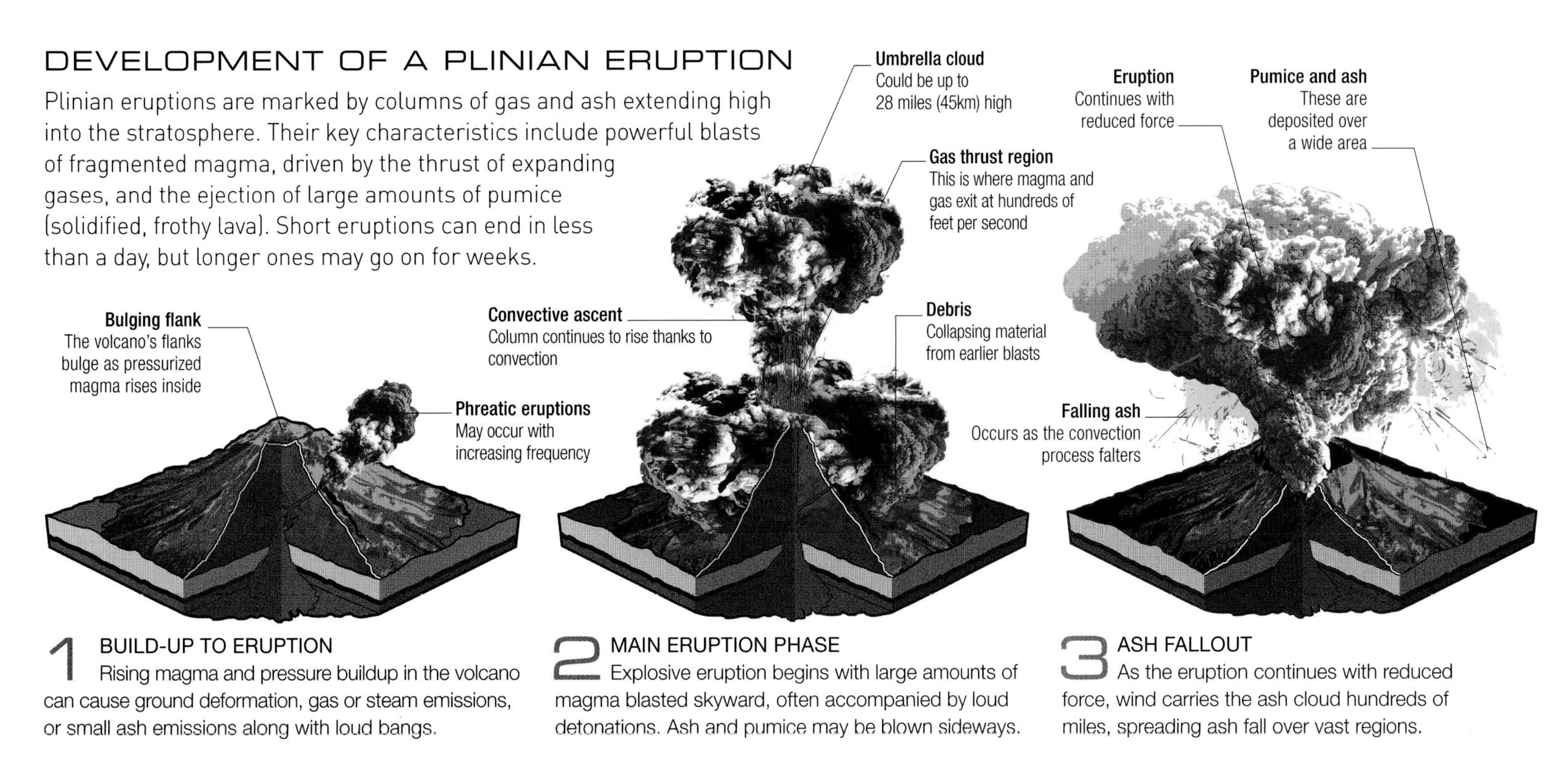

1 BUILD-UP TO ERUPTION
Rising magma and pressure buildup in the volcano can cause ground deformation, gas or steam emissions, or small ash emissions along with loud bangs.

2 MAIN ERUPTION PHASE
Explosive eruption begins with large amounts of magma blasted skyward, often accompanied by loud detonations. Ash and pumice may be blown sideways.

3 ASH FALLOUT
As the eruption continues with reduced force, wind carries the ash cloud hundreds of miles, spreading ash fall over vast regions.

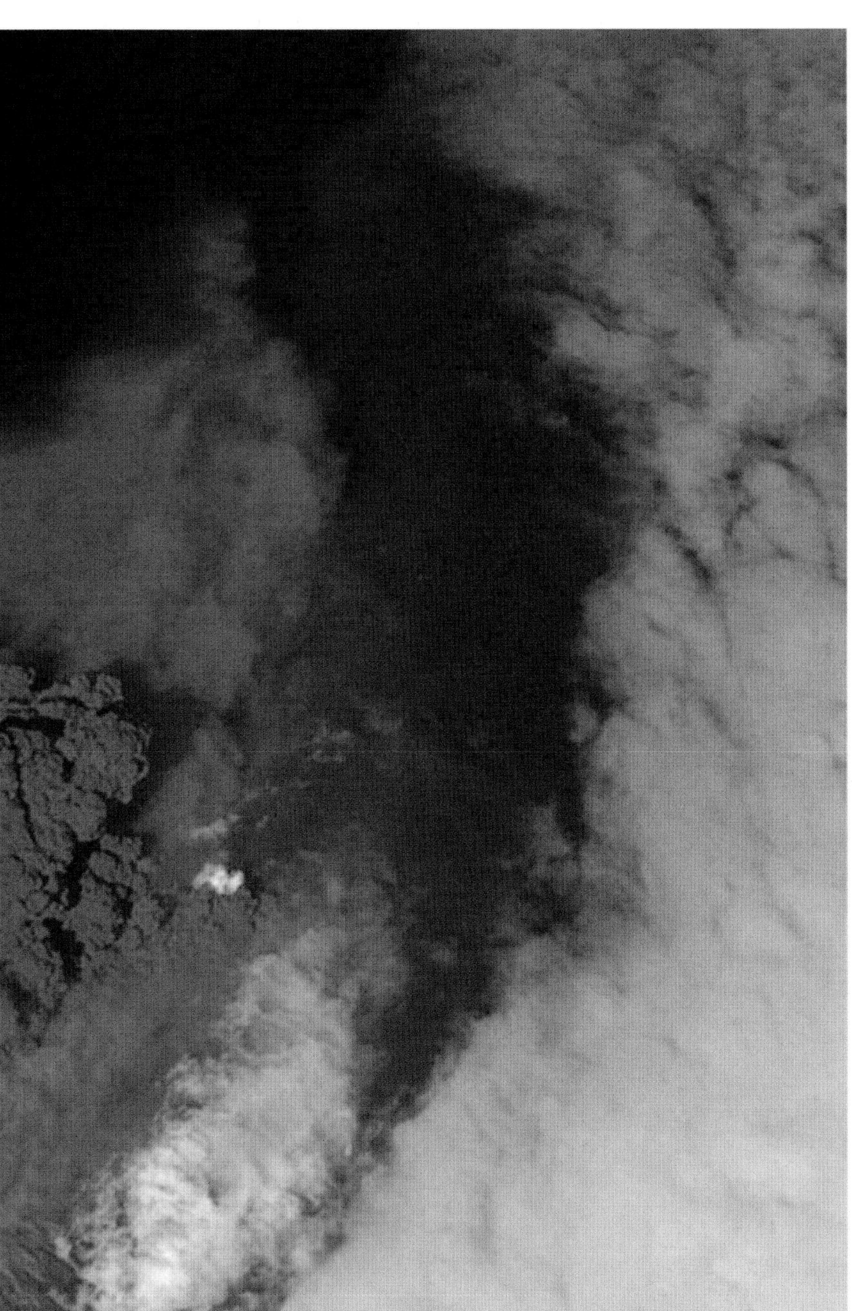

THE TAMBORA ERUPTION

The eruption of Tambora, in Indonesia in 1815 ranks as the largest volcanic eruption in the last 1,800 years. It is also the most lethal of all time in terms of human deaths caused locally. The eruption was so large that it is classified as ultraplinian. During its main phase, three fiery columns were observed rising up into the air and merging. An estimated 38.5 cu miles (160 cu km) of ash and rock were shot into the atmosphere, followed by devastating pyroclastic flows. Close to 12,000 people were killed directly by the eruption; a further 60,000 to 80,000 are estimated to have died subsequently from starvation due to loss of crops and livestock. The ash spewed into Earth's atmosphere lowered temperatures worldwide. In the northern hemisphere, livestock died and crops failed, causing the worst famine of the 19th century.

CALDERA OF TAMBORA
Located on the island of Sumbawa, Tambora has returned to the quiescent state it maintained for 5,000 years before the 1815 eruption. Its caldera is 4 miles (6km) across and 0.7 mile (1.1km) deep. Active fumaroles, or steam vents, still exist on its floor.

VESUVIUS

One of the world's most dangerous volcanoes, Vesuvius is part of a volcanic region of Italy that results from the African Plate pushing under the Eurasian Plate. Although Vesuvius itself is a stratovolcano, it sits in the eroded caldera (collapsed remains) of an older, much larger volcano called Monte Somma. Sitting close to the Bay of Naples, Vesuvius has had various phases of activity and quiescence over the past few thousand years. Many violent eruptions occurred between 1631 and 1944, but there has been hardly any activity since. Today the volcano has to be closely monitored, since a large eruption, with less than a few days warning, could easily kill hundreds of thousands of people.

79CE ERUPTION

Vesuvius's largest eruption of the past 3,500 years, and its most famous, took place in 79CE. The eruption produced a huge ash cloud and ash falls, which together with a pyroclastic surge (see pp.102–103), killed around 2,100 people in the Roman towns of Pompeii and Herculaneum. Excavations of these towns over the past 200 years have revealed much about Roman life at the time.

VESUVIUS FROM SPACE
In this false color infrared satellite image, the cone and crater of Vesuvius appear turquoise at right of center. The surrounding red region is the rest of the Vesuvius/Mount Somma complex. The light blue areas are built-up metropolis, while the Bay of Naples appears black.

SKULL FROM POMPEII
This skull of a victim of Vesuvius's 79CE eruption was discovered during excavations at Pompeii.

> “THE CLOUD WAS RISING FROM... VESUVIUS. I CAN BEST DESCRIBE ITS SHAPE BY LIKENING IT TO A PINE TREE...”
>
> PLINY THE YOUNGER, DESCRIBING THE 79CE ERUPTION THAT KILLED HIS UNCLE, PLINY THE ELDER

PRESERVED EGGS
This wooden bowl, with preserved eggs and eggshells, was unearthed during the Pompeii excavations. Other preserved foods found there include nuts and figs.

CAST OF MAN
This cast is of a man who died lying down at Pompeii while shielding his face—most probably to protect it from the extreme heat of the pyroclastic surge that buried him.

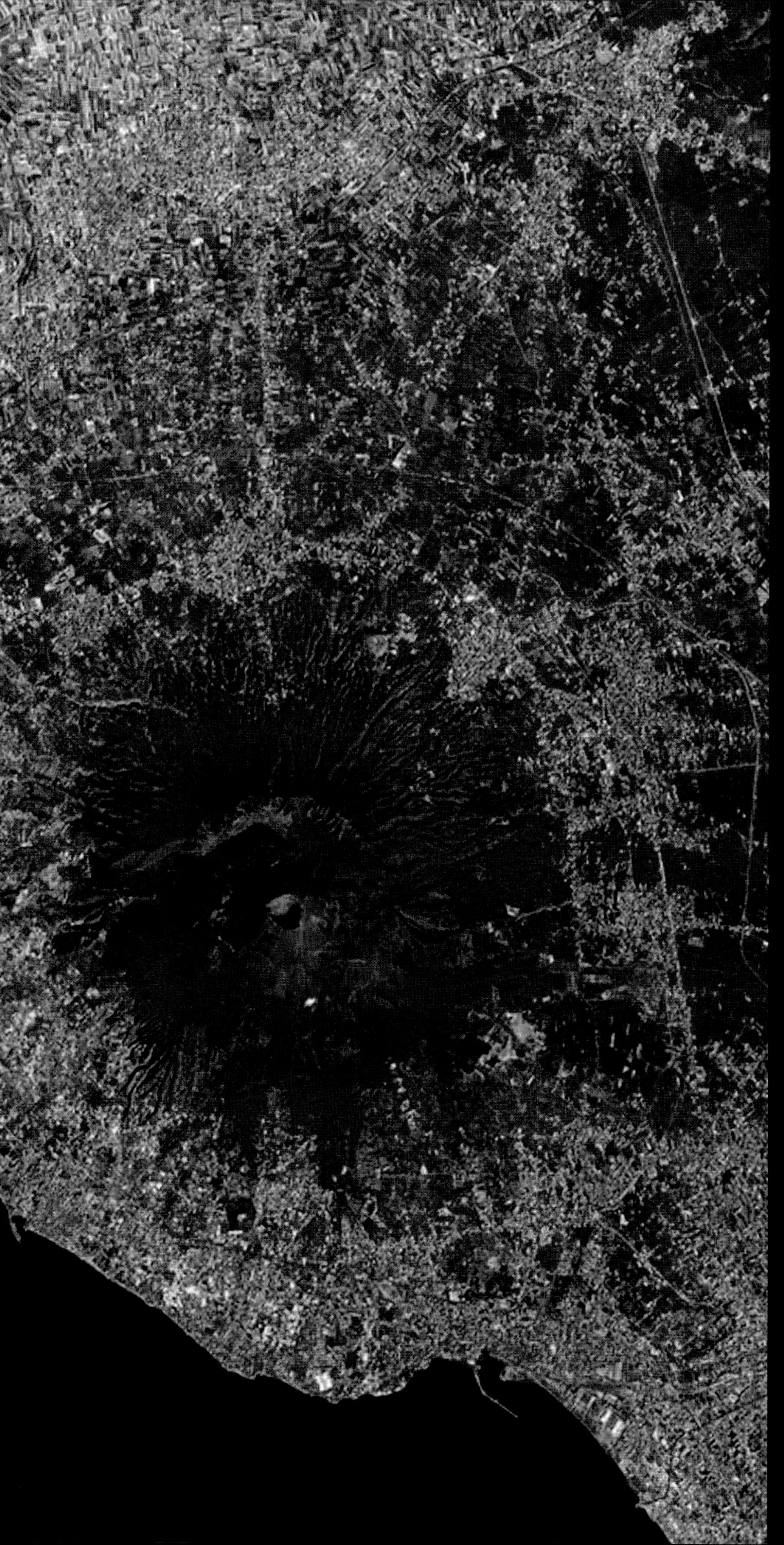

ERUPTION TIMELINE

1631 MASSIVE LAVA FLOWS
A large and sudden eruption occurred, marked by huge lava streams and pyroclastic flows that killed up to 4,000 people. People later mined the lava for building blocks.

1906 ASH CLOUD
A violent eruption, accompanied by earthquakes, produced an 8 mile- (13km-) high ash column, together with more lava than ever recorded before. At least 200 people were killed.

1944 AIRCRAFT DESTROYED
This eruption sent up a 3-mile- (5-km-) high ash plume and produced lava flows that invaded several villages. Heavy ashfalls destroyed US aircraft stationed at a nearby airport.

In June 1912, the most powerful volcanic eruption of the 20th century occurred in a sparsely populated region of Alaska. Over a period of 60 hours, some 3 cu miles (13 cu km) of magma was blasted into the air from a previously unknown, probably new, volcanic vent, later named Novarupta ("new eruption"). A thick layer of ash was deposited over hundreds of square miles of the Alaska Peninsula, and for three days complete darkness prevailed throughout the region. Remarkably, no one is known to have been killed, and few witnessed the eruption, because of its remote location, although the blast was heard up to 746 miles (1,200km) away. The Novarupta vent lies close to a stratovolcano called Mount Katmai. Most of the erupted magma is thought to have drained from a magma reservoir beneath Katmai, which caused the upper half of the stratovolcano to disintegrate. The Novarupta vent itself became filled with a large plug dome.

Location	Alaska Peninsula, Alaska
Eruption type	Plinian
VEI	6

1,060 BILLION

CU FT (30 BILLION CU M) OF ASH, DUST, AND CINDERS DEPOSITED IN THE REGION

VALLEY OF TEN THOUSAND SMOKES

A major feature of the Novarupta eruption was a massive pyroclastic flow. This surged into a valley to the northwest of the eruption site, covering an area of about 40 sq miles (100 sq km) in ash up to 650ft (200m) thick. Thousands of hissing steam plumes rose from the mass of hot ash as it cooled. When the first scientific expedition to the area arrived four years later, the steam plumes were still very much in evidence, prompting one of the investigating scientists, Robert Griggs, to name it the "Valley of Ten Thousand Smokes."

POST-ERUPTION EVENTS

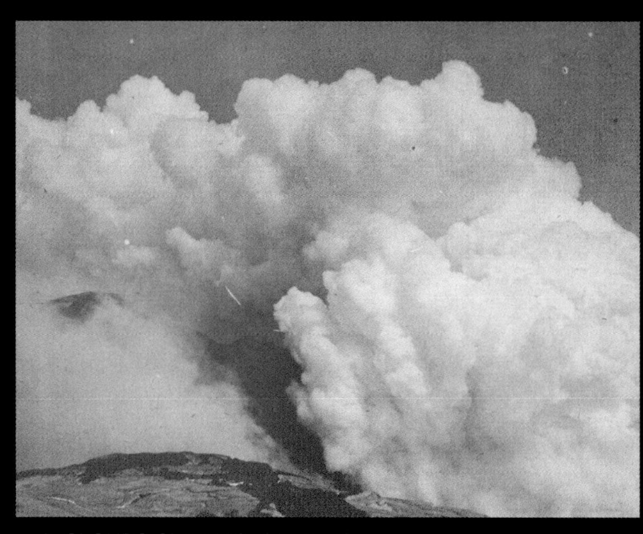

1 CLOUDS OVER MOUNT KATMAI
A photograph taken several months after the eruption showed clouds of steam and ash still rising from the partially collapsed Katmai volcano.

2 SCIENTISTS AT WORK
Surveys of the Valley of Ten Thousand Smokes and the surrounding area took place from 1916–1921.

KATMAI CALDERA ICED OVER
This lake-filled caldera, about 2.8 miles (4.5km) wide, is what remains of the Katmai stratovolcano, which catastrophically subsided during the Novarupta eruption.

3 STEAMING LAVA DOME
A hot, but inactive lava dome, emitting steam plumes, was found in the area between the Katmai stratovolcano and the ash-filled valley. This was later identified as a plug in the eruption vent.

4 ASH DEPOSIT EXPOSED
Over the decades, streams and rivers cut down through the thick deposits of pink volcanic ash. The valley shown here is 130ft (40m) deep.

5 LAVA DOME TODAY
The lava dome that plugs the Novarupta vent site is a black craggy mass some 295ft (90m) high and 1,180ft (360m) wide.

“HAVING REACHED THE SUMMIT OF KATMAI PASS, THE VALLEY OF TEN THOUSAND SMOKES SPREADS OUT... MY FIRST THOUGHT WAS: WE HAVE REACHED THE MODERN INFERNO.”

JAMES HINE, A ZOOLOGIST ON THE FIRST EXPEDITION TO REACH THE ERUPTION SITE

MOUNT ST. HELENS

In May 1980, the most famous volcanic event of the 20th century occurred in Washington State, when the stratovolcano Mount St. Helens lost most of its northern flank. An eruption had been anticipated, because an enormous bulge, caused by injection of magma, had been growing for several weeks near the volcano's summit, accompanied by frequent steam explosions. But what actually happened surprised everyone. At 8:32 am on May 18, an earthquake caused a gigantic chunk of Mount St. Helens' summit and northern flank to suddenly disintegrate and slip away in the largest landslide in history. The landslide exposed the magma body, which exploded in a devastating sideways explosion, called a directed blast.

Location	Cascade Range, Washington State, US
Volcano type	Stratovolcano
Eruption type	Plinian

1,600

MULTIPLE BY WHICH THE ERUPTION BLAST EXCEEDED THAT FROM THE ATOMIC BOMB DROPPED AT HIROSHIMA IN 1945

BEFORE AND AFTER
As a result of the 1980 eruption, the once-graceful cone and summit of Mount St. Helens (top) was scarred with an enormous ugly crater that had opened up on its north side (below).

A MOUNTAINSIDE EXPLODES
At the height of the eruption, a huge plume of ash rose some 14 miles (24km) into the air, driven by expansion of gas in the volcano's vent. Most of the ash eventually settled over a vast region of the northwestern United States.

MONITORING MOUNT ST. HELENS

Since its May 1980 eruption, Mount St. Helens has had further periods of eruptive activity, though it has been quiescent since 2008. The volcano is monitored by scientists from the US Geological Survey and other organizations. A particular watch is kept on earthquake activity close to the volcano, which could indicate new movements of magma, or any signs of surface deformation. Much of the monitoring is performed by automated sensors installed on and around the volcano.

CONSTANT WATCH
One way of remotely monitoring changes in Mount St. Helens' shape is by measuring distances from a fixed point to devices installed on the volcano's flanks. Here a geologist is setting up equipment to perform such measurements.

> "I HAD JUST STARTED TO DRIVE ONTO THE OVERPASS AND THERE IT WAS...IT WAS LIKE WATCHING THE END OF THE WORLD COME SLOWLY."
>
> **LEE HARRIS**, DRIVING NEAR AUBURN, WASHINGTON, 75 MILES (120KM) FROM MOUNT ST. HELENS, WHEN THE ERUPTION BEGAN

THE EFFECTS

The blast from the initial explosion immediately flattened all trees within a fan-shaped area extending for about 20 miles (30km) north from the volcano's summit. Meanwhile, material from the landslide and exploding magma created a gigantic pyroclastic flow, which, moving at 600mph (1,000kph), pulverized and incinerated everything over an area of 230 sq miles (600 sq km). As this impacted a lake in its path, the water flashed to steam, creating a second larger explosion that was heard thousands of miles away, in northern California. Within minutes, millions of tons of melted glacier ice from the summit of the volcano mixed with ash and disintegrated rock to produce several devastating lahars (fluid mudflows). These surged down local rivers, destroying everything in their path, including bridges, trees, and buildings. By the time everything had settled, 57 people had perished—most by asphyxiation or incineration by the pyroclastic flow—and more than a billion dollars in damage had been done.

MUDFLOW DEPOSITS
The volcanic mudflows, or lahars, that surged down local rivers, sweeping away everything in their path, eventually slowed to a trickle, leaving thick deposits of mud and ash.

MOUNT ST. HELENS
Smashed, burnt, and blasted trees covered the terrain surrounding Mount St. Helens after its eruption on May 18, 1980. The initial sideways blast of hot gas, ash, and rock from the volcano knocked down some 230 sq miles (600 sq km) of forest, killing every tree in this area within 60 seconds.

PHREATIC ERUPTIONS

When heated volcanic rocks encounter groundwater or other surface water, the result may be a phreatic eruption—an explosion of steam and rock fragments. Another distinct type of eruption, termed phreatomagmatic, occurs when magma (molten rock) contacts water.

CAUSES AND FEATURES

Phreatic eruptions, also known as steam blast or ultravulcanian eruptions, can occur in any situation in which groundwater or other surface water comes into contact with volcanically heated rock or freshly deposited volcanic ash. These eruptions vary considerably in size and strength. Some herald larger eruptions—for example, several phreatic eruptions preceded the famous Plinian-style eruption at Mount St. Helens in May 1980. Usually, the eruptive column is predominantly white due to the high steam content. Phreatic eruptions do not produce any fountains or streams of lava (see pp.96–99), but flying volcanic bombs (see pp.100–01) are fairly common.

Phreatomagmatic eruptions occur when quantities of groundwater, seawater, or other surface water come in contact with magma. A common scenario leading to a phreatomagmatic eruption occurs when magma rises into a layer of rock that has become saturated with groundwater. The 1883 eruption of Krakatau in Indonesia, one of the largest in recorded history, is often suspected to have been a phreatomagmatic eruption. A wall of the volcano ruptured, allowing seawater to flood into its magma chamber.

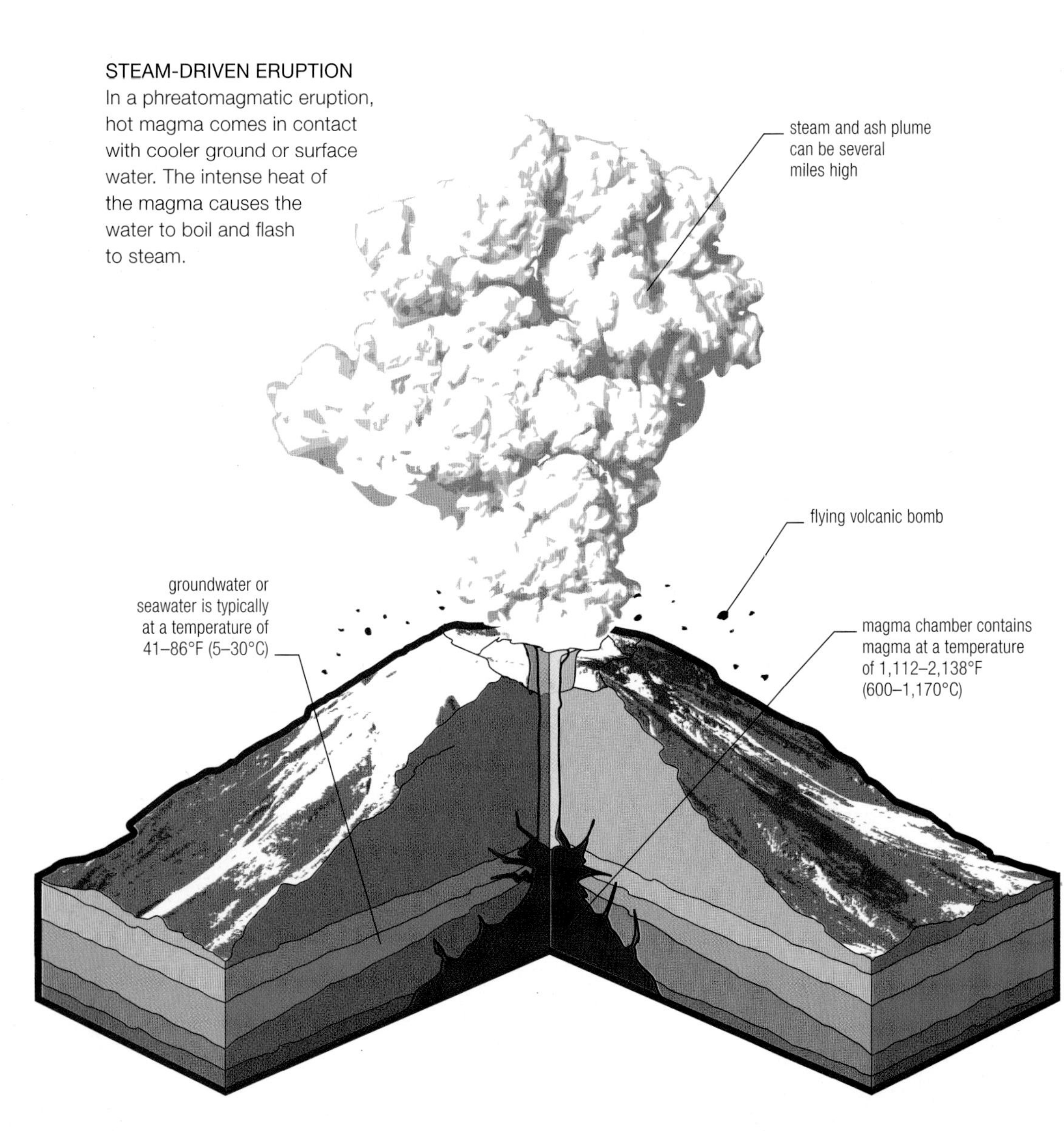

STEAM-DRIVEN ERUPTION
In a phreatomagmatic eruption, hot magma comes in contact with cooler ground or surface water. The intense heat of the magma causes the water to boil and flash to steam.

TOXIC WATER
Phreatic eruptions commonly occur from this crater lake—the world's largest acidic volcanic lake—within the Ijen volcano complex in Indonesia.

GUAGUA PICHINCHA
Located close to Ecuador's capital, Quito, this volcano has produced many large phreatic eruptions in recent decades. Its October 1999 eruption (seen here) caused a sizeable ashfall, which blanketed Quito.

HAZARDS

A number of dangers arise from phreatic and phreatomagmatic eruptions. A major hazard is the emission of large quantities of carbon dioxide, which can asphyxiate at high concentrations, and hydrogen sulphide, which is poisonous. In February 1979, for example, a phreatic eruption from the Dieng volcano in the central highlands of Java, Indonesia, killed 149 people due to carbon dioxide asphyxiation. Crater lakes on volcanoes that experience repeated phreatic eruptions are often highly acidic due to the presence of sulphuric acid produced by reactions between water and sulphurous volcanic gases. During eruptions from such lakes, highly acidic rain may fall from the sky. Another hazard with these eruptions is volcanic bombardment. In 1924, an eruption from the shield volcano Kilauea in Hawaii hurled massive boulders up to half a mile from the crater.

VOLCANOES PRONE TO PHREATIC ERUPTIONS

POÁS VOLCANO
A highly active stratovolcano in Costa Rica, the Poás volcano produces frequent phreatic eruptions, most recently in 2009. The floor of its crater contains one of the world's most acidic lakes.

MOUNT TARUMAE
Located within the Shikotsu caldera in Hokkaido, Japan, this active stratovolcano has produced a number of phreatic explosions in the last few centuries.

SUBGLACIAL VOLCANOES

A few of the world's volcanoes lie buried under enormous glaciers, in the form of ice caps or ice-sheets. Most are found either in Iceland, lying under one of its large ice caps, or in Antarctica. Eruptions of these volcanoes can have spectacular, but occasionally catastrophic, results.

ERUPTIONS UNDER ICE

Although some huge volcanoes lie under the West Antarctic ice sheet (see pp.172–73), none have erupted in thousands of years, so recent subglacial eruptions have been confined to Iceland. The heat from hot gases and magma coming up through the volcano causes the overlying ice to melt. The meltwater quickly cools the erupting magma, resulting in pillow lava (see pp.96–97). Once a hole has been melted in the ice all the way up to the glacier's surface, the eruption becomes visible in the form of a huge plume of steam and ash—produced by the explosive interaction of hot magma and water—shooting up into the atmosphere.

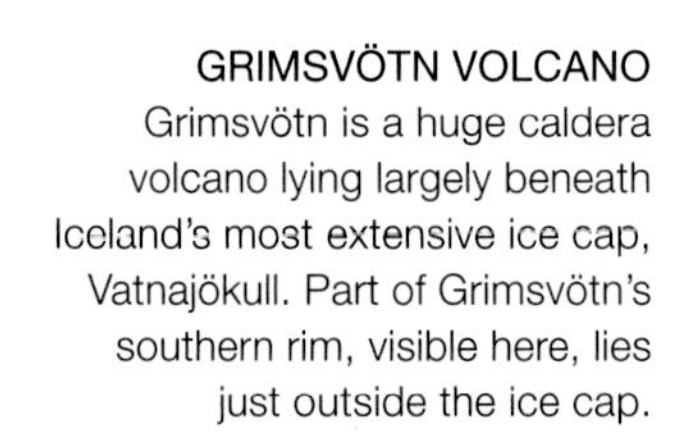

GRIMSVÖTN VOLCANO
Grimsvötn is a huge caldera volcano lying largely beneath Iceland's most extensive ice cap, Vatnajökull. Part of Grimsvötn's southern rim, visible here, lies just outside the ice cap.

VATNAJÖKULL ICE CAP
This satellite photograph shows an ash plume rising from the western half of Vatnajökull as a result of a 2004 eruption of Grimsvötn.

PLUME FROM GRIMSVÖTN
The plume of steam and ash from the 2004 eruption of Grimsvötn rose to a height of 6 miles (10km). The eruption is thought to have started from a fissure in the subglacial caldera. Within a few days, it had melted a hole through 660ft (200m) of ice.

ERUPTION AND FLOOD

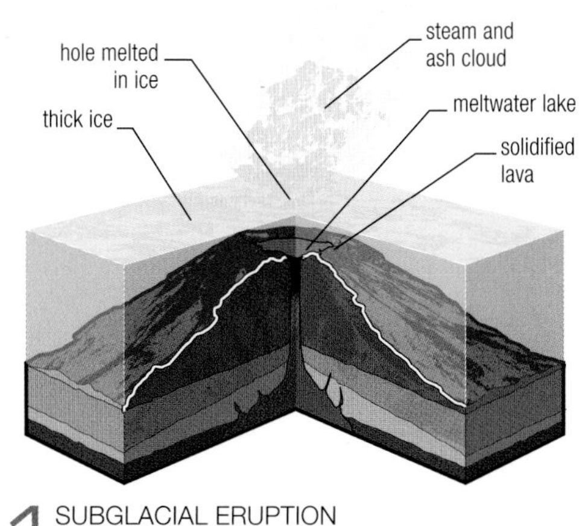

1 SUBGLACIAL ERUPTION Heat from hot gases and magma coming up through the volcano melts a hole in the overlying ice. The released water cools the rising magma to form pillow lava.

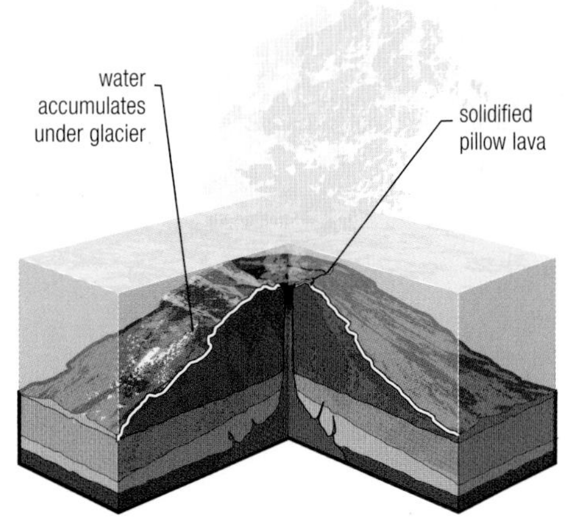

2 WATER ACCUMULATION The released water can become trapped as a lake under the glacier. As the eruption continues and the lake enlarges, it may even start to slightly lift the glacier, even though this may weigh billions of tons.

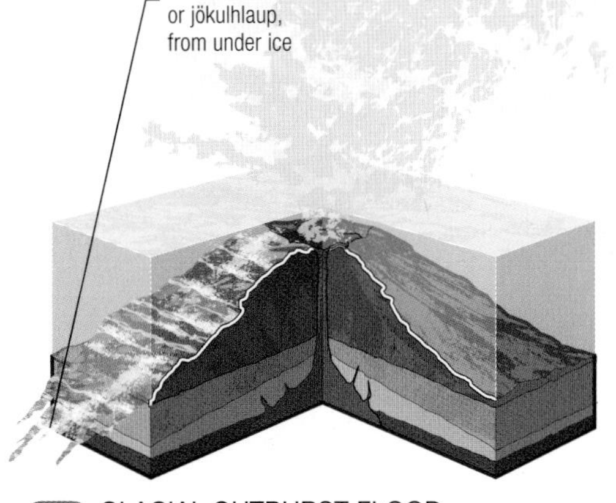

3 GLACIAL OUTBURST FLOOD Eventually the lake becomes so large, and the buildup of pressure so great, that the water suddenly bursts out, either by flowing out under the glacier (as shown here) or exploding out through its side.

JÖKULHLAUPS

In most subglacial eruptions, the water produced from melted ice becomes trapped as a lake between the volcano and the overlying glacier. Eventually, this may be released in a violent and dangerous flood. Events of this type are so common in Iceland that Icelanders have a specific term for them—"jökulhlaup," or "glacial outburst flood." One of the most dramatic jökulhlaups in Iceland of all time occurred in 1996 as a result of an eruption of Grimsvötn. Over three or four weeks, more than 0.7 cu miles (3 cu km) of meltwater accumulated beneath the Vatnajökull ice cap. The subglacial lake suddenly burst out, some of the water escaping beneath the ice cap and some blasting out through a fissure in its side. The resulting flood was temporarily the second biggest flow of water in the world (after the Amazon River). It caused $14 million of damage and left numerous 33-ft-(10-m-) high icebergs—chunks of the Vatnajökull ice cap—scattered across Iceland's coastal plain.

TUYA

HERÐUBREIÐ TUYA, ICELAND

A subglacial volcano develops steep sides and a flat top due to the way lava builds up at its summit, constrained by ice. After the volcano becomes extinct, and the ice cover retreats, and a landform with this shape, called a tuya, is left.

RUSH OF MELTWATER

During the eruption of the Eyjafjallajökull volcano in Iceland in 2010, meltwater rushed out from under the edges of the ice cap overlying the volcano to produce lakes and some jökulhlaups.

VOLCANIC LIGHTNING

Eyjafjallajökull's ash plume was frequently illuminated by dramatic electrical discharges, thought to be the result of untold numbers of collisions between ash and ice particles generating static electricity.

EYJAFJALLAJÖKULL

For several weeks in April and May 2010, a previously little-known Icelandic volcano became a focus of attention in Europe as it belched huge quantities of ash into the atmosphere, bringing air traffic across a large part of the continent to a standstill. The eruption of the volcano—which lies partly under the small Icelandic ice cap of Eyjafjallajökull (Icelandic for "island mountain glacier")—had started off in a low-key way with the appearance of fissures near a mountain pass next to the glacier, together with some lava fountaining. But a few weeks later, on April 14, the eruption moved up to the summit crater underneath the ice itself and entered a new, explosive phase with the ejection of fine, glass-rich ash to a height of 5 miles (8km) in the atmosphere. This was a Plinian-style eruption, though in Iceland only considered moderately big compared with some of the monstrous eruptions the country has seen in the past. What was worrying Icelanders, and others, was that it might soon be followed by activity at Eyjafjallajökull's larger and more dangerous neighbor, a huge caldera volcano called Katla.

APRIL–MAY 2010	
Location	Southern Iceland
Volcano type	Stratovolcano
Eruption type	Plinian
VEI	4
People displaced	500 families

95,000
SCHEDULED PASSENGER FLIGHTS CANCELED

AIR TRAFFIC DISRUPTIONS

The location of the eruption meant that the ash was carried into the heavily used airspace over northwestern Europe, and from April 15, 2010, aviation authorities closed much of the airspace in this region. This situation continued for the next eight days, stranding 10 million passengers and causing damage to some European economies. After April 23, the affected airspace was reopened, but closures continued to occur intermittently in different parts of Europe as the ash cloud was monitored using technology such as lidar (see right). In Iceland, the eruption of Eyjafjallajökull had stopped by the end of June 2010, and as of May 2011, there had been no follow-up eruption of Katla.

STUDYING THE ASH
An engineer demonstrates a lidar (light detection and ranging) device—laser-based technology that can be used to gather data on atmospheric particles. Lidar proved particularly useful in studying the ash plume.

AVIATION HAZARD
Even when it has been dispersed to the point of being invisible, volcanic ash poses grave dangers to aircraft. Ash particles can melt inside jet engines and force engine shutdown. They also abrade windshields and clog vital sensors.

> "SUDDENLY...YOU REALIZE THAT ALL ALONG YOU HAVE BEEN INHABITING THE EARTH."
>
> **FRENCH ANTHROPOLOGIST BRUNO LATOUR,**
> *A PLEA FOR EARTHLY SCIENCES*, KEYNOTE SPEECH AT THE BRITISH SOCIOLOGICAL ASSOCIATION, 2007

THE PLUME SPREADS

1 APRIL 15, 2010
The plume had spread to cover much of Norway and northern England and was impinging on other Scandinavian countries. Most airspace in the British Isles, Norway, and parts of Sweden had by now been closed.

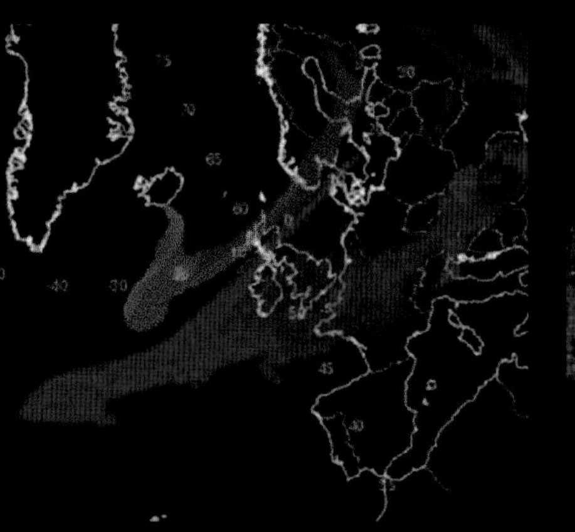

2 APRIL 18, 2010
There was now an extensive cloud of two- to three-day-old ash over much of the British Isles, France, and central Europe, with a new pulse heading in. Airspace was now closed, or partially closed, in 20 countries.

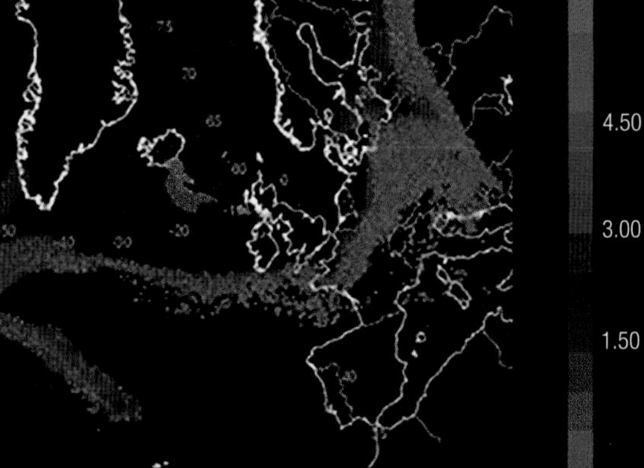

3 APRIL 20, 2010
Following a period of relative quiescence at the volcano, the area of new ash was much reduced, though a cloud of older ash covered eastern Europe. A few days later most of the formerly closed airspace was reopened.

MEAN AGE OF ASH PARTICLES IN DAYS

ERUPTION OF EYJAFJALLAJÖKULL

An angry, billowing plume of hot ash and steam rises up through a fissure in the Eyjafjallajökull ice cap, from the volcanic vent lying underneath, and is blown horizontally toward the sea. All around the fissure, a grubby deposit of ash lies on the ice. The erupted ash was of a fine, abrasive, glass-rich type, making it particularly dangerous to aviation.

ANTARCTIC VOLCANOES

It might seem odd to think of volcanoes in a place as chilly as Antarctica, but for red-hot magma rising up toward Earth's surface, the cold of Antarctica is no barrier. This continent has about 30 volcanoes, only a few of which have recently been active.

MOUNT EREBUS

Antarctica's most active volcano, Mount Erebus, is also the most southerly on Earth to have erupted in recorded history. The 12,447-ft- (3,794-m-) high stratovolcano is located on Ross Island, off the coast of East Antarctica, along with three other, apparently inactive, volcanoes. All four are thought to sit over a hotspot under the Antarctic Plate. Erebus is one of a few volcanoes to have a lava lake in its summit crater. It was erupting when first sighted by Captain James Ross in 1841, and is still erupting today. It produces frequent minor explosions in its lava lake and occasional larger Strombolian explosions (see pp.148–49).

TOWER OF ICE
On the flanks of Erebus are fumaroles that have formed strange ice towers. As steam emerges from a vent, some of it condenses to form water. As this flows toward the ground, it quickly freezes, adding more ice to the tower.

STEAM PLUME
A small steam plume emanates from the summit of Erebus, which is almost entirely covered by ice. On its more active days, Strombolian explosions may hurl small lava bombs onto its upper slopes.

WEST ANTARCTIC VOLCANOES

In an area called Marie Byrd Land in West Antarctica lies a group of large volcanoes almost completely buried in ice. This group is thought to be the result of a 2,000-mile- (3,200-km-) long continental rift opening up under the West Antarctic ice sheet. Some West Antarctic volcanoes, such as Mount Hampton, are probably extinct, but others, such as Mount Takahe—which is thought to have erupted about 7,000 years ago—are potentially active. In 2008, scientists found evidence of a relatively recent eruption, about 2,000 years ago, beneath the Hudson Mountains in West Antarctica. Using ice-sounding radar, they discovered a layer of ash from the eruption lying beneath the ice sheet.

MOUNT HAMPTON
The only visible part of this massive shield volcano is its summit area. This is largely occupied by a 4-mile- (6-km-) wide crater, which is completely filled with ice.

VOLCANOES IN THE FROZEN CONTINENT

Antarctica's volcanoes fall into three main groups—one around the tip of the Antarctic Peninsula, a second in West Antarctica, and a third near the coast of East Antarctica. The massive stratovolcano Mount Melbourne is the only recently active volcano on the Antarctic mainland. Mount Erebus is on an island. The Seal Nunataks, a group of nunataks (mountain tops poking above the ice) near the Antarctic Peninsula, are thought to be separate volcanoes or remnants of a large shield volcano.

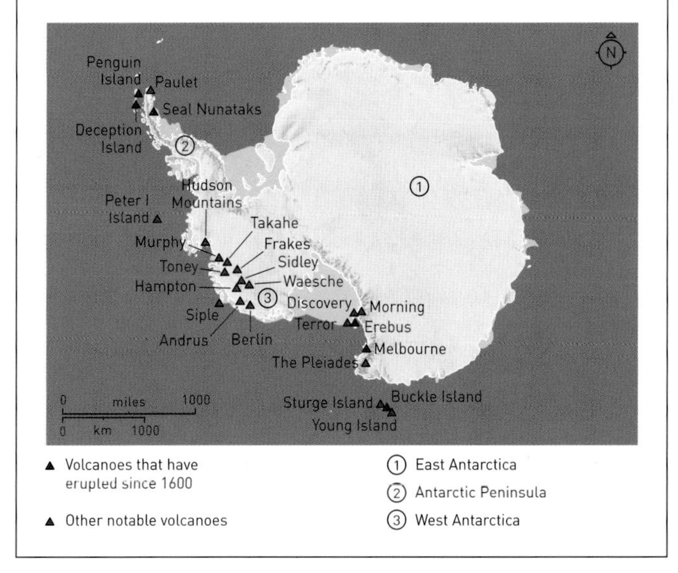

DECEPTION ISLAND
Consisting of a horseshoe-shaped flooded caldera (see pp.126–27), Deception Island last erupted between 1969 and 1970, destroying two scientific stations. It is one of the South Shetland Islands, an arc of volcanic islands that lie along a convergent plate boundary near the Antarctic Peninsula.

AFRICAN RIFT VOLCANOES

Continental rifts are regions of Earth's surface that are pulling apart. The rifting is associated with plumes of hot mantle rising up from the interior. As a result, volcanoes often form in such regions. Those found in the East African Rift Zone are some of the most spectacular volcanoes on the planet.

EASTERN RIFT VOLCANOES

The eastern part of the rift system, called the Eastern Rift Valley, contains a handful of volcanoes, including Ol Doinyo Lengai. It is the only volcano in the world to erupt natrocarbonatite—the most fluid and coolest of all lavas, with a temperature of about 930°F (500°C). Although black or brown, it turns white on contact with moisture. Near Lengai lies Kilimanjaro, which consists of a complex of three stratovolcanoes. Although no eruptions have been recorded, it contains a magma chamber 1,310ft (400m) below its crater, and future eruptions are not ruled out. Other volcanoes include The Barrier in Kenya composed of four overlapping shield volcanoes, and a stratovolcano, Meru.

KEY

Fault lines

Major volcanoes that have erupted since 1800

Other notable volcanoes

Afar Depression

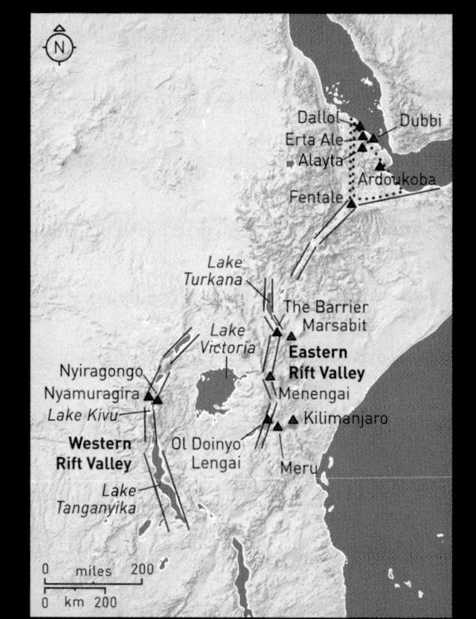

EAST AFRICAN RIFT SYSTEM
This region has three main volcanically active areas—the Eastern Rift Valley, the Western Rift Valley, and the Afar Depression in the north.

OL DOINYO LENGAI, TANZANIA
Known as "Mountain of God" in the language of the native Masai people, Ol Doinyo Lengai produces a unique form of lava that glows a vivid orange at night. Here, lava can be seen oozing from one of the steep-sided cones that have formed in the volcano's summit crater.

SUMMIT CRATER
Ol Doinyo Lengai's crater is filled with solidified lava, which appears white by day. The volcano's eruptive activity is usually centered on small lava flows from the cones on the crater's floor. Occasionally, this pattern changes to more violent eruptions with ash plumes and larger lava flows.

THE VIRUNGA VOLCANOES

In the Western Rift Valley, within the Virunga Mountains west of Lake Victoria, lie two dangerous volcanoes, responsible for about 40 percent of Africa's volcanic eruptions and most volcano-related deaths. Mount Nyamuragira, Africa's most active volcano, is a shield volcano that has erupted more than 40 times since 1885, most recently in 2010. It is noted for producing large amounts of sulphur dioxide gas and toxic ash. Mount Nyiragongo, nearby, has a lava lake at its summit that has been described as a "boiling cauldron of hell." From time to time, the lava either erupts from the summit or bursts out of fissures on its flanks, flowing downhill at great speeds and killing anyone in its path. One of its most disastrous eruptions occurred in 2002 (see pp.176–77).

MOUNT NYIRAGONGO
This stratovolcano in the Democratic Republic of the Congo is 11,384ft (3,470m) high, with a 1-mile- (2-km-) wide summit crater, visible in this satellite photograph. It has erupted at least 34 times since 1892, and is dangerous due to its steep flanks and fluid lava.

VOLCANOES OF THE AFAR TRIANGLE

The Afar Triangle, or Afar Depression, is a barren region of northeastern Africa. It is situated well below sea level and is one of the hottest, driest, and most inhospitable places on Earth. It contains some notable volcanoes, the best known being Erta 'Ale (the Fuming Mountain or Devil's Mountain), which is a huge active shield volcano. At its summit lies a large, elliptical crater, measuring 2,297 by 5,249ft (700 by 1,600m), which contains two smaller steep-sided "pit" craters, one holding a spectacular lava lake. This lake occasionally produces dangerous eruptions, but

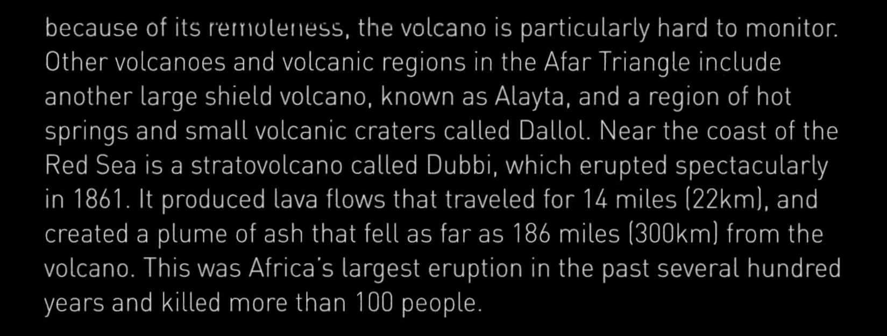

because of its remoteness, the volcano is particularly hard to monitor. Other volcanoes and volcanic regions in the Afar Triangle include another large shield volcano, known as Alayta, and a region of hot springs and small volcanic craters called Dallol. Near the coast of the Red Sea is a stratovolcano called Dubbi, which erupted spectacularly in 1861. It produced lava flows that traveled for 14 miles (22km), and created a plume of ash that fell as far as 186 miles (300km) from the volcano. This was Africa's largest eruption in the past several hundred years and killed more than 100 people.

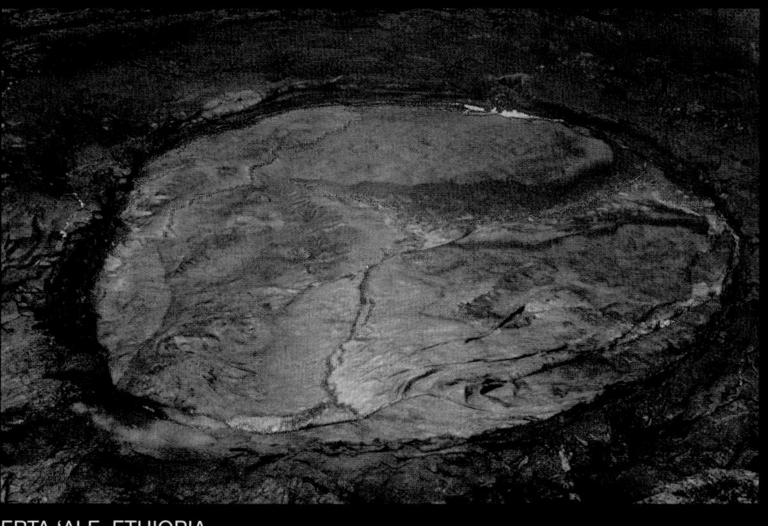

ERTA 'ALE, ETHIOPIA
The most famous feature of Erta 'Ale is a fiery, circular lava lake with a 500ft (150m) diameter. When fully molten, it produces lava fountains and gives off large amounts of heat. At other times, a solid crust forms over much of the lake's surface.

ALAYTA VOLCANO
This large shield volcano in the western part of the Afar Depression covers more than 1,040 sq miles (2,700 sq km). In this satellite image, much of its surface looks black due to erupted lava. An eruption in 1907 led to considerable damage to land, property, and human life. Alayta last erupted in 1915.

NYIRAGONGO DISASTER

The 2002 eruption of Mount Nyiragongo was the most destructive flow of lava in modern history. Nyiragongo's lava is highly fluid, and when it escapes from a large lava lake at the summit, it races down the volcano's steep sides. In 2002, fissures opened up on the volcano's flanks and streams of lava, 655 to 3,280ft (200 to1,000m) wide and up to 6½ft (2m) deep, oozed into Goma City, creating fires and explosions. Forty-five people died, some due to asphyxiation by volcanic gases, others following the explosion of a gas station. About 12,000 people were left homeless.

Location	Democratic Republic of the Congo
Volcano type	Stratovolcano
Eruption type	Hawaiian
Explosivity index	1
350,000	THE NUMBER OF PEOPLE EVACUATED

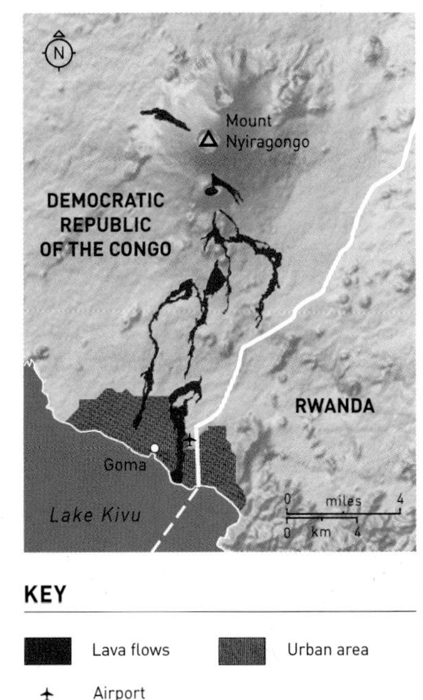

ERUPTION CHRONOLOGY
Streams of lava burst from fissures on Mount Nyiragongo. One cut across the runway at Goma Airport, igniting aviation fuel, while others poured into the city and reached Lake Kivu.

THE 2002 DISASTER

1 LAVA LAKE
Nyiragongo is the only steep-sided volcano that holds a lake of red-hot, extremely fluid lava at its summit. On this occasion, the lava suddenly drained from the lake through fissures that appeared on the volcano's flanks.

2 GOMA AIRPORT
Streams of hot fluid lava entered Goma Airport, covering the northern end of its runway, and then swept into Goma, destroying 45,000 buildings, including 90 percent of the town's business district.

3 LAKE KIVU
Lava entered Lake Kivu, raising concerns that it might cause gas-saturated waters to rise from the bottom of the lake and release lethal amounts of carbon dioxide. Thankfully, the lava did not penetrate far enough.

"THIS MORNING, AN APOCALYPTIC SCENE. IT'S AS THOUGH A GIANT BULLDOZER HAS SWEPT THROUGH GOMA. THE FUMES...ALMOST OVERPOWERING."

ANDREW HARDING, BBC CORRESPONDENT, SURVEYING THE SCENE OF THE DISASTER

LAVA CASCADES
In this 2010 photograph, streams of lava cascade down Nyiragongo's flanks, in a similar, though slightly more subdued, manner to the 2002 eruption.

VOLCANIC REMNANTS

Volcanic activity leaves its mark on our planet in many ways. In addition to volcanoes and their eruptive products are the remains of ancient magma bodies that solidified inside volcanoes or deep underground and were later exposed at the surface through erosion.

SHIP ROCK
Made of hard rock types called lamprophyre (also known as minette) and volcanic breccia, Ship Rock is a volcanic plug with radiating dikes. It stands more than 1,700ft (518m) above the surrounding high-desert plain.

VOLCANIC PLUGS

These landscape features are the result of magma solidifying within the vent and other internal spaces in a volcano, to form a hard plug of igneous (magma-derived) rock. Later, when the rest of the volcano has eroded away, this plug is uncovered on the surface. A classic example is Ship Rock, in New Mexico, the remains of a volcano that existed about 27 million years ago. When the magma that formed Ship Rock solidified, the plug was probably about 2,800ft (850m) below the surface. Subsequently, weathering and erosion removed the lava and ash layers of the volcano, together with underlying soft shales, to expose Ship Rock as it is today.

lens-shaped laccolith

IGNEOUS INTRUSIONS
Magma that pushes up into any cracks and other spaces it can find in existing rock layers, sometimes forcing them aside, is called an igneous intrusion. Intrusions eventually cool to form solid bodies, which erosion may later expose at the surface as a variety of landscape features.

DIKES AND SILLS

Two relatively small types of intrusion that are often exposed at the surface are dikes and sills. Both usually form from runny basaltic magmas, which can squeeze through thin gaps and cracks in country rock (rock that predates the intrusions). These magmas solidify to typically form a rock called dolerite. Dikes are thin, vertical intrusions that cut through horizontally bedded rock layers. They can take various forms, including those that radiate from a central point, ring dikes (encircle volcanic centers and are formed when magma is squeezed up through a ring fracture), or a series of parallel structures, called a dike swarm. Sills are horizontal structures that lie along bedding planes (boundaries between layers of sedimentary rock). An example is Whin Sill in England, which is exposed in a few places as outcrops about 100ft (30m) thick.

WHIN SILL
Hadrian's Wall was built on part of this exposed sill in northeastern England. The sill was intruded some 295 million years ago.

BATHOLITHS AND LACCOLITHS

Larger igneous intrusions include batholiths, laccoliths, and some minor types such as lopoliths and stocks. All of these are usually made from granitic magmas—these have a different composition from, and are more viscous than, basaltic magmas. As a result, they form huge masses underground instead of penetrating cracks and crevices. Batholiths are the biggest of the large igneous intrusions and are typically made of granite or related rocks. Once exposed at the surface, batholiths have an area of at least 40 sq miles (100 sq km), and many are much larger than this. Similar structures with areas less than 40 sq miles (100 sq km) are called stocks. An example of a batholith is the Sierra Nevada batholith in California, which is some 370 miles (600km) long and consists of more than 100 individual bodies, called plutons. These were formed from the cooling of separate blobs of magma some time between 225 and 80 million years ago. Of the other types of large intrusion, laccoliths are large, lens-shaped intrusions that bulge upward and are often composed of a rock called gabbro. Lopoliths are similar but sag downward. The Pine Valley Mountains laccolith in Utah is one of the largest in the world.

EL CAPITAN
A part of the Sierra Nevada batholith, El Capitan is a 3,000-ft- (910-m-) high rock formation, made of solid granite in Yosemite National Park, California. It formed underground about 100 million years ago.

volcanic plug with radiating dikes

ring dike erodes to form circular outcrop patterns

batholith exposed at surface

dike in parallel swarm

sill that has intruded at sufficiently high pressure may produce dikes that rise up vertically from it

massive batholith

stock is a smaller version of a batholith or a small upward bulge in one

country rock

sill forms between bedding planes

dike forms vertically through rock layers

ARRIGETCH PEAKS
These rugged spires in Alaska started as a granite laccolith within limestone country rock. After exposure, the laccolith was eroded by glaciers to produce the peaks visible today.

MONITORING VOLCANOES

Many of the world's most dangerous volcanoes, especially those near cities, are monitored scientifically. Although volcanologists cannot predict precisely when or how a volcano will erupt, they can warn of an increased risk of an eruption and give an idea of its likely character.

METHODS AND TOOLS

Volcanologists use several different methods to monitor volcanic activity. For example, they analyze the gases emitted by a volcano, use seismometers to detect tremors in the ground underneath it, and employ instruments called tiltmeters to measure any bulging on its flanks. Increased emission of certain gases, a rise in the frequency and intensity of earthquakes, or ground bulging, can all indicate that magma is welling up inside the volcano and are fairly reliable signs of an imminent eruption. Satellite surveillance is a more expensive alternative to measuring ground deformation. Other methods include monitoring for changes in magnetic field strength and gravity near a volcano, which may be used to trace magma movement. Once scientists have thoroughly studied a volcano, they produce hazard maps. These maps may show, for example, the most likely routes of future lava flows or lahars (volcanic mudflows). When the risk of eruption rises to a significant level, people can be advised to move away temporarily from the main danger zones or even to evacuate the entire area.

MONITORING STATIONS
Technicians set up equipment to measure ground deformation near the Soufrière Hills Volcano on the Caribbean island of Montserrat. This is one of the world's most heavily monitored volcanoes.

GAS SAMPLING
Volcanologists often measure and analyze the gases emitted by a volcano. A number of these gases are highly noxious so gas masks are worn.

REMOTE SENSING
Satellites are used to monitor erupting volcanoes, such as this explosion of Augustine volcano in Alaska. They track the extent and movement of ash plumes, which are a hazard to aviation.

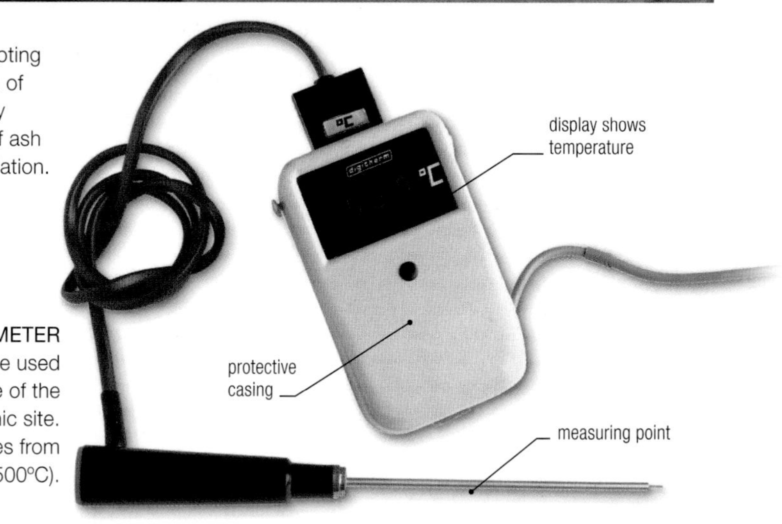

SPECIAL THERMOMETER
Thermocouple thermometers are used to measure the temperature of the ground, lava, or gases at a volcanic site. They can withstand temperatures from -358°F to 2,732°F (-200°C to 1,500°C).

FIELD WORK
Wearing a protective heat suit, this volcanologist is taking samples from a lava flow on Mount Etna. Changes in the temperature and composition of lava provide clues about the future course of an eruption.

VOLCANOES FOR SPECIAL OBSERVATION

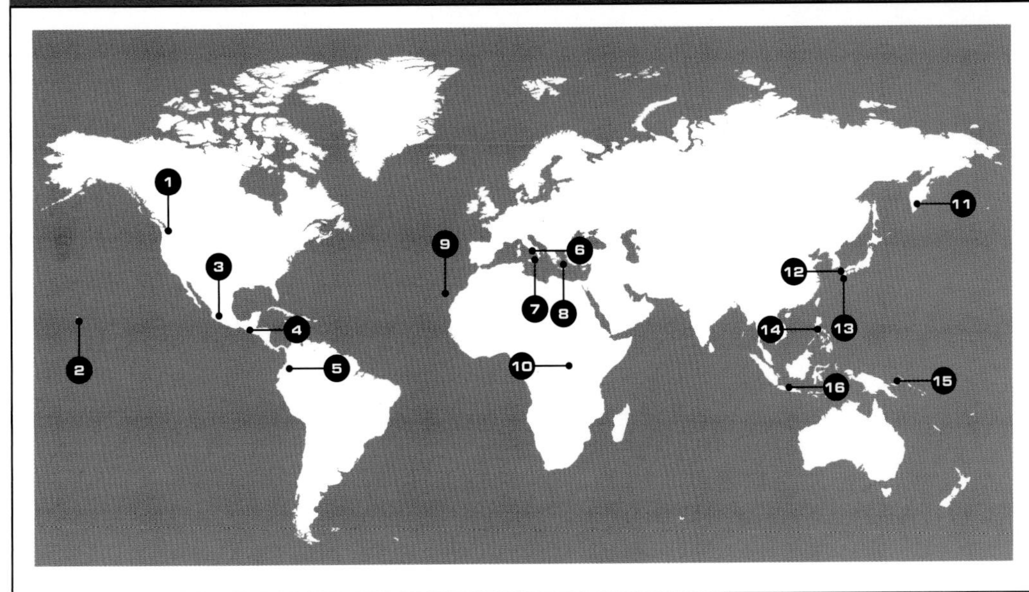

The International Association of Volcanology has identified a group of 16 volcanoes known as "decade" volcanoes for particular study and monitoring because of their history of destructive eruptions and proximity to heavily populated areas. When selecting these volcanoes, additional criteria included the existence of multiple dangers, such as the risk of pyroclastic flows, lahars, and various types of collapse. Not all have received special monitoring so far, mainly due to funding difficulties.

1. Rainier
2. Mauna Loa
3. Colima
4. Santa María
5. Galeras
6. Vesuvius
7. Etna
8. Santorini
9. Teide
10. Nyiragongo
11. Avachinsky-Koryaksky
12. Unzen
13. Sakurajima
14. Taal
15. Ulawun
16. Merapi

LIVING WITH VOLCANOES

About 8 percent of the world's population lives near a volcano, which is surprising given their reputation for danger. Despite the risks, millions of people put up with them, mainly for economic reasons.

HAZARDS

Analysis of the causes of deaths from volcanic eruptions over hundreds of years shows that the greatest dangers come from pyroclastic flows (see pp.102–03) and, for a few volcanoes, mudflows (see pp.106–07). Some volcanoes can cause gas poisoning and large eruptions near coasts can set off tsunamis. Heavy ashfalls can also be dangerous, but lava flows are generally more of an economic nuisance than a lethal hazard.

PICKING UP THE PIECES
A man recovers part of a zinc roof three days after an eruption of the Nyiragongo Volcano (see pp.176–77) caused widespread destruction.

MINING AND TOURISM

Against the risks they pose, volcanoes can bring some advantages. In the volcanic regions, the amount of heat flow coming out of Earth (geothermal heat) is high and this can be exploited as a carbon-free energy source (see pp.34–35). A few volcanoes are a source of other natural resources, such as sulphur and diamonds. They can also bring economic benefits to a region through tourism. The beauty and excitement of erupting volcanoes attracts hundreds of thousands of sightseers every year.

MINING SULPHUR
The Ijen volcano in Java, Indonesia, supports an industry in which volcanic gases are artificially converted to sulphur, which is then carried out of the volcano's crater. This provides employment, though the work is extremely arduous and involves health hazards.

AGRICULTURAL BENEFITS

An erupting volcano usually produces a great amount of ash, lava, or both. In the short term these can be harmful to the environment, but in the long term they decompose to form a fertile soil, containing many useful minerals. In the region around an active volcano, there is nearly always a sizeable group of people farming the soil. Even after a large and possibly lethal eruption, they return to reclaim their livelihood. This explains the high density of populations in volcano-dominated islands like Java and the settlements around dangerous volcanoes, such as Merapi, also in Indonesia.

CULTIVATED VOLCANO
Terraced wheat fields adorn the slopes of a small, probably extinct, volcanic cone in Rwanda. The fact that every bit of the volcano's surface is being used provides a good indication of the soil fertility here.

MAYON VOLCANO
The symmetrical cone of Mount Mayon in the Philippines is surrounded by a fertile agricultural region. People live near this dangerous volcano for the livelihood it provides, and also for the beautiful setting.

WITNESSING AN ERUPTION
Tourists admire and photograph the erupting Icelandic volcano, Eyjafjallajökull, in April 2010. More than 100,000 people, many from outside Iceland, went to see the eruption. The tourist influx boosted Iceland's economy.

ASH AND LAVA RECOLONIZATION

Despite the devastation they can cause, lava flows and ashfalls are usually quite rapidly recolonized by plants. Recolonization usually starts within 10 years for lava flows and within 3 to 4 years for ashfalls, emphasizing the additional fertility they can bring to soils.

FERNS IN LAVA
This crack in a pile of pahoehoe lava, in which some ferns have recently started growing, was spotted on the Puna coast of Hawaii's Big Island.

FERTILE CINDER CONE
Terraced rice paddies and fields filled with vegetables geometrically cover the slopes of a volcano in central Java, Indonesia. Thanks to the fertile soil formed from breakdown of the volcano's eruption products, the fields provide three harvests each year.

VOLCANIC HOT SPRINGS

Hot springs form when large amounts of groundwater is heated by magma below the volcano. The presence of minerals and microbes can make these pools look colorful, but the water can often be at an extreme pH due to the volcanic gases that are in the system.

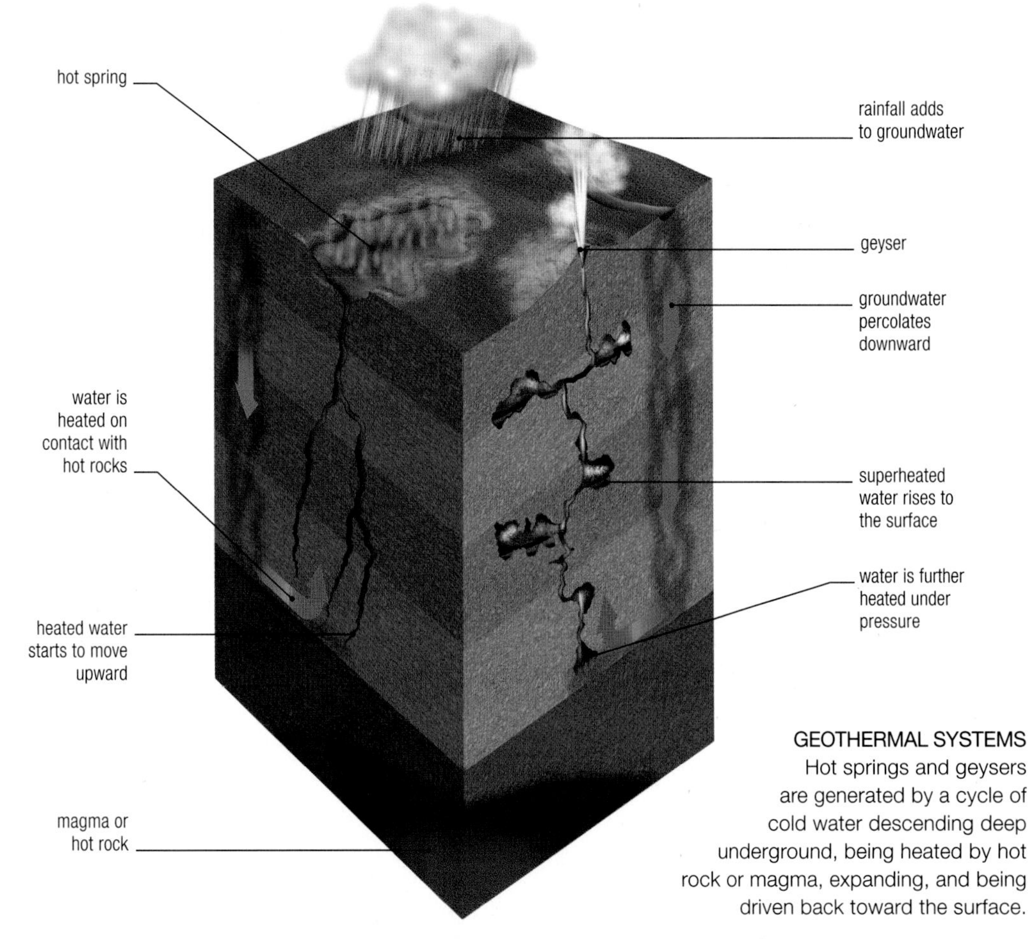

GEOTHERMAL SYSTEMS
Hot springs and geysers are generated by a cycle of cold water descending deep underground, being heated by hot rock or magma, expanding, and being driven back toward the surface.

SPRING GENERATION

Volcanic hot springs occur when surface water trickles down through the crust until it comes into contact with the rocks surrounding a magma chamber, or the magma itself. As the water warms, typically close to boiling point, it becomes less dense, and rises up through fissures and chambers in the rock until it reaches the surface, where it forms a pool of hot water. Often, the water temperature in such a pool or spring is very high—many springs have temperatures in the range from 158°F (70°C) to as high as 206°F (97°C).

THERAPEUTIC POTENTIAL

Some volcanic hot springs contain warm rather than hot water, making them safe to bathe in. The bathing pool in the Blue Lagoon spa in Iceland uses water processed by the nearby geothermal plant, with water temperature averaging 98 to 102°F (37 to 39°C). Bathing in the mineral-enriched waters of a hot spring is reputed to have therapeutic value, for example, in alleviating some skin ailments, as well as psychological benefits.

FLOW RATE AND MINERAL CONTENT

Volcanic hot springs usually produce an uninterrupted flow of hot water every second at an average temperature of 206°F (97°C). They vary from insignificant seeps to strong plumes, such as the hot spring at Deildartunguhver, Iceland, which flows at a rate of 50 gallons (180 liters) every second. During its underground journey, the hot water dissolves minerals out of the rock. As well as coloring some springs, minerals often precipitate out when the water cools at the surface, producing hard deposits that can develop into spectacular formations, such as the travertines at Mammoth Hot Springs in Wyoming.

BOILING LAKE, DOMINICA
This scaldingly hot body of water, on a Caribbean island, is one of the world's largest hot springs. Its surface is permanently enveloped in a cloud of steam.

CHAMPAGNE POOL, NEW ZEALAND
Formed 900 years ago in the Wai-o-tapu geothermal area, the Champagne Pool contains salts of arsenic and antimony, which give it a colorful appearance.

DALLOL HOT SPRINGS, ETHIOPIA
Located in the hot, highly volcanic Danakil Depression, these springs are notable for the various strangely shaped mineral deposits found around them.

RIOT OF COLORS
Hot springs may be brightly colored due to the presence of microbes. Differently colored microbes with different temperature preferences can lead to a spectrum of colors in one pool, as seen here at Grand Prismatic Spring in Yellowstone Park, Montana.

FUMAROLES

Found in many of the world's volcanic regions, fumaroles are openings in the planet's crust that emit steam and a variety of volcanic gases, such as carbon dioxide, sulphur dioxide, and hydrogen sulphide. They often make loud hissing noises as the steam and gases escape. Many fumaroles are extremely foul smelling.

CHARACTERISTICS

Fumaroles are similar to hot springs (see pp.186–87). However, unlike in hot springs, the water in fumaroles gets heated to such a high temperature that it boils into steam before reaching the surface. The main source of the steam emitted by fumaroles is groundwater heated by magma lying relatively close to the surface. Other gases, such as carbon dioxide, sulphur dioxide, hydrogen sulphide, and smaller amounts of water, are released from the magma. Fumaroles are often present on active volcanoes during periods of relative quiet between eruptions. They may last for decades or centuries if they are above a persistent heat source, or disappear within weeks if they occur atop a fresh volcanic deposit that cools quickly.

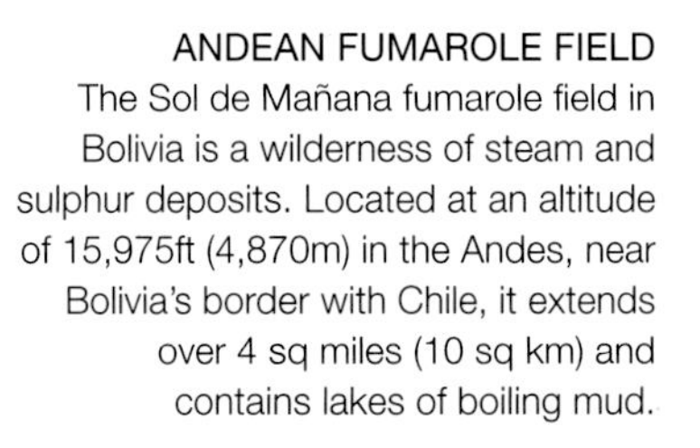

ANDEAN FUMAROLE FIELD
The Sol de Mañana fumarole field in Bolivia is a wilderness of steam and sulphur deposits. Located at an altitude of 15,975ft (4,870m) in the Andes, near Bolivia's border with Chile, it extends over 4 sq miles (10 sq km) and contains lakes of boiling mud.

SULPHUROUS FUMAROLES

A fumarole that emits a lot of hydrogen sulphide, which smells of rotten eggs, or pungent-smelling sulphur dioxide, is called a solfatara. Hydrogen sulphide often reacts with other gases to make sulphur, which is deposited on the ground as yellow crystals. The steam and sulphurous gases also combine to make weak sulphuric acid. In some locations, this dissolves nearby rocks into a mass of hot mud (see pp.190–91) and may also be a health hazard for downwind communities.

MOUNT ASAHI-DAKE, JAPAN
Numerous snow-melting fumaroles punctuate the sides of this stratovolcano on the island of Hokkaido, Japan. They create a remarkable sight in winter when the slopes are used as a skiing area.

DALLOL FUMAROLES AND ACID POND
The Dallol region of Ethiopia, one of the lowest and hottest spots on the Earth, consists of many hot springs, fumaroles—seen here at the top of conical mounds—and sulphur deposits. The liquid on the ground is weak sulphuric acid.

FUMAROLE GASES

The illustration shows the breakdown of gases given off by fumaroles in one large study (in Taiwan) and is considered fairly typical. While water and carbon dioxide predominate, another major component—the poisonous gas hydrogen sulphide—is what gives most fumaroles their unpleasant smell.

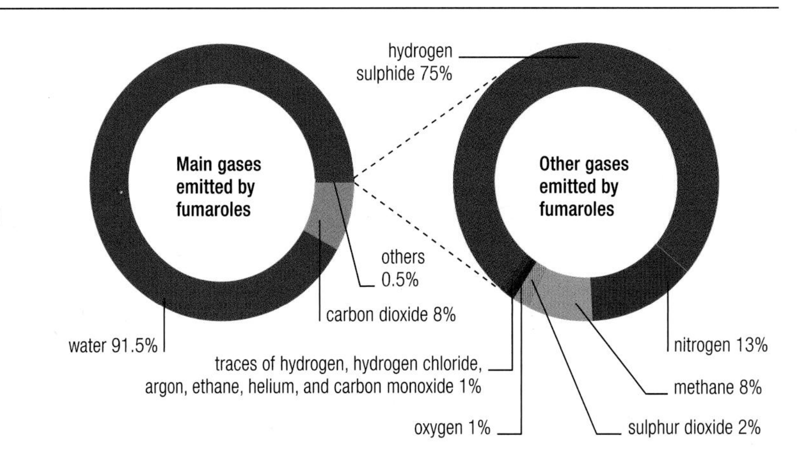

SULPHUR CRYSTALS

GRAN CRATERE, VULCANO
Extensive sulphur deposits cover the rim of the Gran Cratere ("big crater") on the small volcanic island of Vulcano, north of Sicily, Italy. Numerous active fumaroles exist within the crater, and visitors are advised to avoid breathing in the foul-smelling, poisonous gases they give off.

BUBBLING MUD POTS
This pool of hot mud, at Whakarewarewa in New Zealand, formed when sulphurous gases in a fumarole combined with steam to produce chemicals that dissolved the nearby rock. The swirly patterns are caused by successive gas bubbles welling up through the mud and bursting.

GEYSERS

Geysers are extraordinary natural fountains that intermittently shoot boiling-hot water and steam into the air in violent eruptions. They form under rare conditions and exist in only a few locations worldwide.

ACTIVITY AND TYPES

Like hot springs (see pp.186–87), geysers result from increased heat flow from Earth's interior meeting fluids in permeable or fractured rocks. However, there is a crucial difference between the two. In a hot spring, water flows freely from below the ground, so no pressure builds up. A geyser, in contrast, has a narrow opening near the top of its plumbing system that restricts movement of water. As a result, pressure can intermittently build up, aided by the presence of geyserite, a pressure-sealing mineral that lines all of a geyser's underground chambers and channels. It is a form of silica that precipitates at hot springs and geysers.

As pressure builds inside a geyser's subterranean chambers, the water is unable to change into steam, even though the temperature of the water may rise to 480°F (250°C). Eventually, the high pressure blasts water out from the constricted opening. As this happens, pressure falls further down in the plumbing system, allowing some of the hot water to flash into steam, which then rapidly expands. This sustains the eruption, which continues until pressure in the geyser has dropped close to zero. The whole cycle then starts again. There are two main types of geysers: cone and fountain. In a cone geyser, geyserite forms a cone-shaped nozzle at the surface, which directs the flow of water. By contrast, the surface opening of a fountain-type geyser is a crater filled with water, so its eruptions are more diffuse. Geysers are highly variable in their behavior—some erupt frequently and regularly, others do not. While most eruptions last for only a few minutes, some may persist for hours.

> "THAT WHICH THE FOUNTAIN SENDS FORTH RETURNS AGAIN TO THE FOUNTAIN."
>
> **HENRY WADSWORTH LONGFELLOW**, AMERICAN POET, REFERRING TO CYCLICAL PHENOMENA

CASTLE GEYSER
This cone geyser in Yellowstone National Park erupts every 10 to 12 hours, sending a fountain to a height of 89ft (27m) for 20 minutes. Carbon dating suggests that it is 5,000 to 15,000 years old.

540
The approximate number of active geysers in Yellowstone, which represents about half the global total. Most other geysers are found either in New Zealand's North Island, in northern Chile, Iceland, or in a small area in Kamchatka in eastern Russia.

WORLD'S TALLEST GEYSERS

NAME	LOCATION	HEIGHT (MAX)	ERUPTION CYCLE
Steamboat	Yellowstone, US	295ft (90m)	Irregular
Giant	Yellowstone, US	245ft (75m)	Irregular
Splendid	Yellowstone, US	245ft (75m)	Irregular
Geysir	Iceland	230ft (70m)	Irregular
Great Fountain	Yellowstone, US	220ft (67m)	9–15 hours
Beehive	Yellowstone, US	200ft (60m)	8–24 hours
Grand	Yellowstone, US	200ft (60m)	7–15 hours
Giantess	Yellowstone, US	200ft (60m)	Irregular

STROKKUR

One of Iceland's most famous spectacles, Strokkur is a fountain geyser that erupts every 4 to 8 minutes, sending steam and water to a maximum height of 98ft (30m). Eruptions were first reported in 1789, after an earthquake.

FLY GEYSER

Located in the Black Rock Desert of Nevada, this arresting feature consists of three colorful cones, each continuously spouting hot water. It is not entirely natural, being the accidental result of an oil well drill in 1916.

MUD VOLCANOES

The less-famous cousins of ordinary magmatic (lava- and ash-erupting) volcanoes, mud volcanoes are channels through which large amounts of pressurized gas, mud, and salty water are expelled from deep underground onto Earth's surface. Once dry, the mud builds up into cones that can be up to several hundred feet high.

GEOTHERMAL VOLCANO
Although a separate phenomenon from the mud pools (see pp.190–91) that are often seen in geothermal areas, mud volcanoes sometimes occur in such areas, as here in Rotorua, New Zealand.

CAUSES AND DISTRIBUTION

Mud volcanoes are caused by pressure building up underground, in regions where gas and water are trapped. As the gas and water are driven up toward the surface through lines of weakness in the crust, they soften some of the rock layers they encounter, turning these into mud. Pressure exerted by gas deposits deep in the crust can be a major precipitating factor, so many mud volcanoes occur near gas fields. They may also develop near tectonic plate boundaries and fault lines. More than 1,000 mud volcanoes have been identified on land and in shallow water—in eastern Azerbaijan, Romania, Venezuela, and elsewhere. They also exist underwater—about 10,000 are thought to be scattered across continental shelf areas and the deep sea floor.

BARREN LANDSCAPE
Where mud volcanoes occur, they create a strange lunar-like landscape. Vegetation is normally sparse or absent, because most plants cannot tolerate the amount of salt deposited in the soil along with the mud.

SHALLOW MUD VOLCANO
This unusual, small mud volcano with gently sloping, mineral-stained sides, formed in Glen Canyon, Utah. It erupts watery mud from its vent and has developed a shallow summit crater.

MUD CONES
In Azerbaijan, southwest Asia, mud volcanoes commonly form lines of small cones. These emit cold mud, water, and gas almost continually from a deep underground reservoir.

SMALLER SCALE
This mud volcano in Berca, Romania, has the same conical shape as a typical magmatic volcano, but is much smaller in scale.

CAULDRON OF GASES
At Yellowstone National Park is a steam-belching, mud-filled hole in the ground that is referred to as a mud volcano. However, it is really a fumarole (see pp.188–89)—an opening through which steam and other gases escape.

ERUPTION CHARACTERISTICS

The mud spewed out by a mud volcano varies in its temperature and viscosity (runniness), depending on the volcano. It is usually cold, since it comes from the crust and not the mantle, but some volcanoes release warm mud. Where the underground pressure causing a volcano is especially high, it may break rock formations, throwing out chunks of rock with the mud. Mud volcanoes also emit large amounts of gases, mainly hydrocarbons such as methane. A large methane plume can ignite at the vent of a mud volcano. In Azerbaijan, mud volcanoes have erupted explosively, hurling flames from burning methane hundreds of feet into the air, and depositing tons of mud on surrounding areas.

THE LUSI DISASTER

In May 2006, one of the largest mud volcano eruptions ever recorded was triggered in the Sidoarjo region of eastern Java, Indonesia. The eruption, called Lumpur Sidoarjo (*lumpur* is Indonesian for "mud") or Lusi is continuing and is expected to do so until about the year 2040. It has already flooded an area of about 4 sq miles (10 sq km) with mud and forced more than 40,000 people out of their homes. In addition, 13 people were killed in November 2006 when subsidence near the eruption site ruptured a gas pipeline and caused an explosion. Faulty procedures during the drilling of a nearby gas well have been blamed for triggering the disaster. Experts believe that the well accidentally punched into an aquifer deep below the surface, allowing hot water under high pressure to escape, rise, and mix with a layer of volcanic ash, causing the blowout. But the company that drilled the well claims an earthquake was the cause. Legal proceedings connected to the Lusi disaster continue.

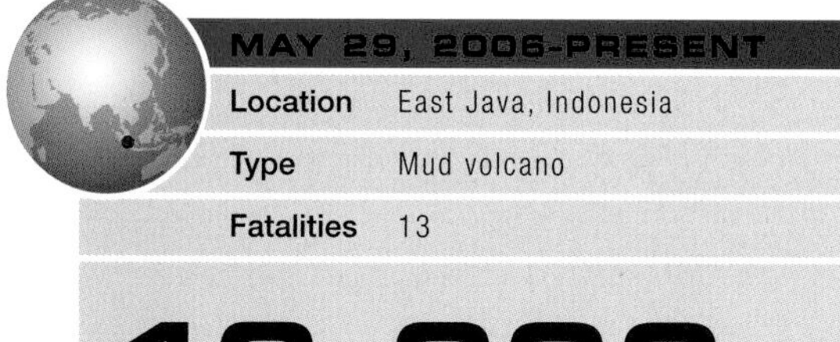

MAY 29, 2006–PRESENT

Location	East Java, Indonesia
Type	Mud volcano
Fatalities	13

40,000

THE NUMBER OF PEOPLE DISPLACED

68

The number of Olympic-sized swimming pools that the mud flow could have filled each day, during peak flow rates. It reached about 250,000 cu ft (7,000 cu m) per hour at times during the first 18 months after the eruption started.

CATASTROPHE UNFOLDS

1 BLOW OUT
A day after a new gas borehole was drilled nearby, plumes of steam and hot sulphurous mud suddenly appeared from a ground fissure near the village of Porong.

LARGEST MUD VOLCANO
A plume of steam rises from one of the mud volcano's vents, which is essentially a simple circular hole in the ground. Efforts to plug the flow have only resulted in more vents appearing that continue to maintain the mud flow.

DISASTER RESPONSE

So far, all efforts to stem the mud flow have failed, though its flow rate has diminished naturally. Huge dikes were built to surround and contain the affected area, but there have been occasional breakouts from these. Also, there are major concerns that the whole area inundated will subside, perhaps as much as 500ft (150m), due to the ever-increasing weight of the mud.

Pumping the mud
Some efforts are being made to pump the mud out of the flooded area into a nearby river. But as the river carries the mud down to the coast and deposits it as sediment, there are concerns about the wider impact on coastal ecosystems.

Plugging holes
To slow the mud flow, engineers tried dropping hundreds of huge concrete balls—each weighing 88lb (40kg) and linked together by chains—into the largest vent. However, it had a barely noticeable effect.

2 SLUDGE COVER
Within two years, a dozen nearby villages, containing many thousands of homes, had been submerged in mud to a depth of several feet.

> "...IT WILL TAKE 26 YEARS FOR THE ERUPTION TO DROP TO A MANAGEABLE LEVEL AND FOR LUSI TO TURN INTO A SLOW-BUBBLING VOLCANO."
>
> **RICHARD DAVIES**, DIRECTOR OF DURHAM ENERGY INSTITUTE, UK, 2011

3 LAND DEVOURED
This false color near-infrared satellite image from November 2008 clearly shows the roughly rectangular area created by the containment dikes built to hold the gray mud back. The surrounding vegetation appears red.

4

EARTHQUAKES

<< **Earthquake damage**
The 6.9-magnitude earthquake that struck Kobe, Japan, in 1995 caused several portions of an elevated highway to collapse.

WHAT IS AN EARTHQUAKE?

The ground may feel solid but an earthquake shows that this is not the case—the Earth can shake so violently that buildings collapse, cracks open up in the surface, and mountainsides tumble down. Yet these terrifying events, occurring without warning, are part of the natural workings of our planet.

WHY EARTHQUAKES HAPPEN

An earthquake (also known as a quake, tremor, or temblor) is the elastic vibration of rocks caused by the sudden release of energy in the Earth's interior. At the surface this becomes violent shaking, which may last from a few seconds to several minutes. Almost all earthquakes are triggered by the sudden breaking or fracturing of rocks, and the most earthquake-prone parts of Earth are the tectonically active regions near the boundaries of the tectonic plates. Profound geological shifts can occur during an earthquake, changes that remain imprinted in the landscape long after the shaking has stopped. The location of the earthquake in the interior is called the hypocenter and the projection of the hypocenter at the surface is called the epicenter. When a large earthquake's epicenter is located offshore, the seabed may move sufficiently to cause a tsunami. Earthquakes can also trigger landslides, and occasionally volcanic activity. But the consequences to man-made structures near the epicenter can be devastating, leading to huge loss of life.

EARTHQUAKE FAULT LINE

Earthquakes occur below fault lines that cut through the landscape, most commonly near tectonic plate boundaries. As the plates are moving in different directions, stress builds up in the rock. Movement begins at the epicenter, deep within the crust. Waves of energy radiate outward, shaking the ground above. Blocks of land are displaced either horizontally or vertically.

Fault line
A crack in Earth's crust where two plates are moving past one another

Area of displacement
Land displaced (horizontally or vertically) by the earthquake

Epicenter
Lies immediately above the hypocenter on the surface

Hypocenter
This is the focus, the point where the movement takes place and most energy is released

STRUCTURAL DAMAGE

The ground accelerations caused by shaking during even a moderate earthquake can easily exceed gravity and are enough to topple free-standing structures. Buildings and other structures that lack internal bracing, or with poorly supported roofs or walls, will collapse.

TSUNAMI

If the seabed is uplifted or dropped suddenly as a consequence of ground movements, then large volumes of water will be displaced, creating a tsunami, a series of waves, that travels at more than 400mph (700kph) across the ocean. Waves can be more than 33ft (10m) high.

MAGNITUDE AND INTENSITY

The intensity of shaking during an earthquake is indicated on the modified Mercalli scale, which measures the effect of shaking on people or buildings. Intensity is measured at a point at the surface and depends on the depth of the earthquake and on the nature of the ground—hard rock shakes less than soft ground. The moment magnitude (M) scale, like its predecessor the Richter scale, is a measure of the amount of energy released during an earthquake. It is an estimate of the energy at the source, which may be deep underground. It is a logarithmic scale so, for example, a M-6 earthquake releases roughly 30 times more energy than a M-5 earthquake. However, the moment magnitude scale cannot be directly related to the destructive effects of the quake, because this depends on the location. An M-8 earthquake at a depth of several hundred miles is unlikely to do much damage at the surface, whereas the same earthquake at a depth of only a few miles could destroy a city. Most earthquakes each year have a small moment magnitude, with fewer larger ones. In general, for earthquakes at depths of less than 30 miles (50km), large moment magnitude events will be far more damaging than smaller ones, but the destructive power also depends on the length of the fault that breaks.

MAGNITUDE Measured on the moment magnitude scale	**INTENSITY** Typical maximum on the modified Mercalli scale	
The amount of energy released at the source of the earthquake.	**The intensity or strength of the shaking measured by the effect on people and buildings.**	
1.0 – 3.0	I	(not felt)
3.0 – 3.9	II – III	(felt on upper floors of buildings)
4.0 – 4.9	IV – V	(rocking motion felt by most)
5.0 – 5.9	VI – VII	(some damage to buildings)
6.0 – 6.9	VII – IX	(moderate to considerable damage)
7.0 and above	VIII +	(slight damage to total devastation)

Tectonic plates
Huge blocks of crust slide past or over each other. These movements lead to earthquakes

Seismic waves
Waves of energy radiate out from the epicenter, on the surface as well as underground

SHAKING
One of the most obvious effects of the waves of seismic energy released during an earthquake is the shaking experienced at the surface. One of the safest places to be during an earthquake is braced in a door frame or under a table.

LANDSLIDES
The most earthquake-prone parts of Earth also tend to be mountainous. The shaking during an earthquake can destabilize hill slopes, causing rock falls and landslides (see pp.236–37). Landslides can be one of the greatest causes of death in an earthquake.

FIRES
Fires are not a direct consequence of earthquakes, however, when an earthquake severely shakes the ground beneath a city, electricity cables, gas pipes, and oil installations can be damaged or broken. Electrical faults or overheating can lead to explosions and fires.

EARTHQUAKE ZONES

10 LARGEST EARTHQUAKES SINCE 1950

1 VALDIVIA
Country Chile
Date 1960
Magnitude M–9.5

2 PRINCE WILLIAM SOUND
Country Alaska, US
Date 1964
Magnitude M–9.2

3 SUMATRA
Country Indonesia
Date 2004
Magnitude M–9.1

4 KAMCHATCKA
Country Russia
Date 1952
Magnitude M–9.0

5 TOHOKU
Country Japan
Date 2011
Magnitude M–9.0

6 ECUADOR
Country Ecuador
Date 1906
Magnitude M–8.8

7 MAULE
Country Chile
Date 2010
Magnitude M–8.8

8 RAT ISLANDS
Country Alaska, US
Date 1965
Magnitude M–8.7

9 ASSAM
Country India
Date 1950
Magnitude M–8.6

10 SUMATRA
Country Indonesia
Date 2005
Magnitude M–8.6

10 DEADLIEST EARTHQUAKES SINCE 1900

11 TANGSHAN
Country China
Date 1976
Deaths 255,000

12 GANSU
Country China
Date 1920
Deaths 235,502

13 INDIAN OCEAN
Country Off the coast of Sumatra
Date 2004
Deaths 230,210+

14 HAITI
Country Haiti
Date 2010
Deaths 222,570

15 GREAT KANTO
Country Japan
Date 1923
Deaths 142,800

16 ASHGABAT
Country Turkmenistan
Date 1948
Deaths 110,000

17 SICHUAN
Country China
Date 2008
Deaths 87,587

18 AZAD, KASHMIR
Country Pakistan
Date 2005
Deaths 86,000

19 MESSINA
Country Italy
Date 1908
Deaths 72,000

20 CHIMBOTE
Country Peru
Date 1970
Deaths 70,000

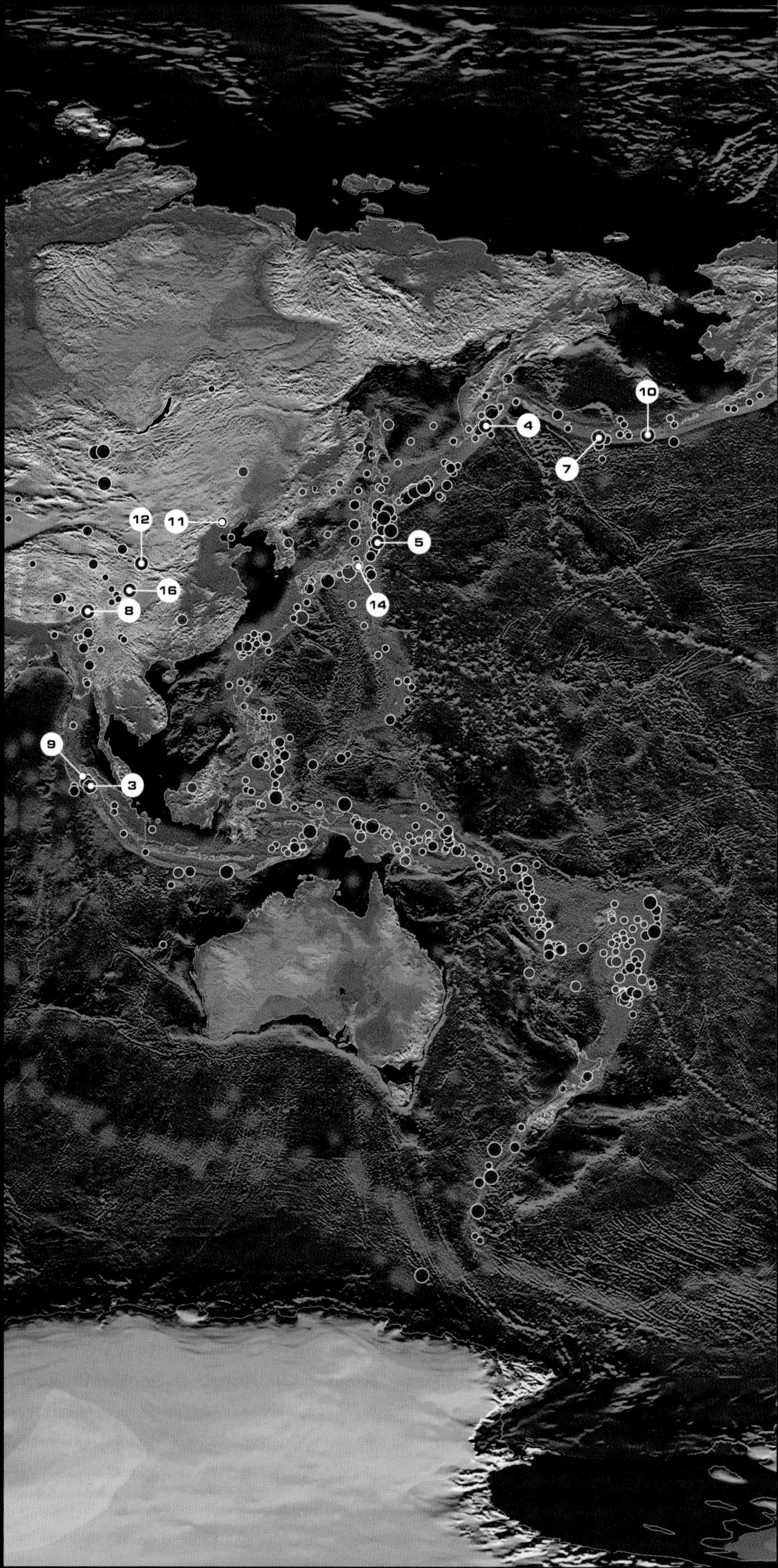

One of the great geological discoveries about our planet is that it is alive with earthquakes. Almost all of these earthquakes occur in relatively narrow zones along the edges of the tectonic plates. The area known as the Pacific "Ring of Fire" is the most tectonically active.

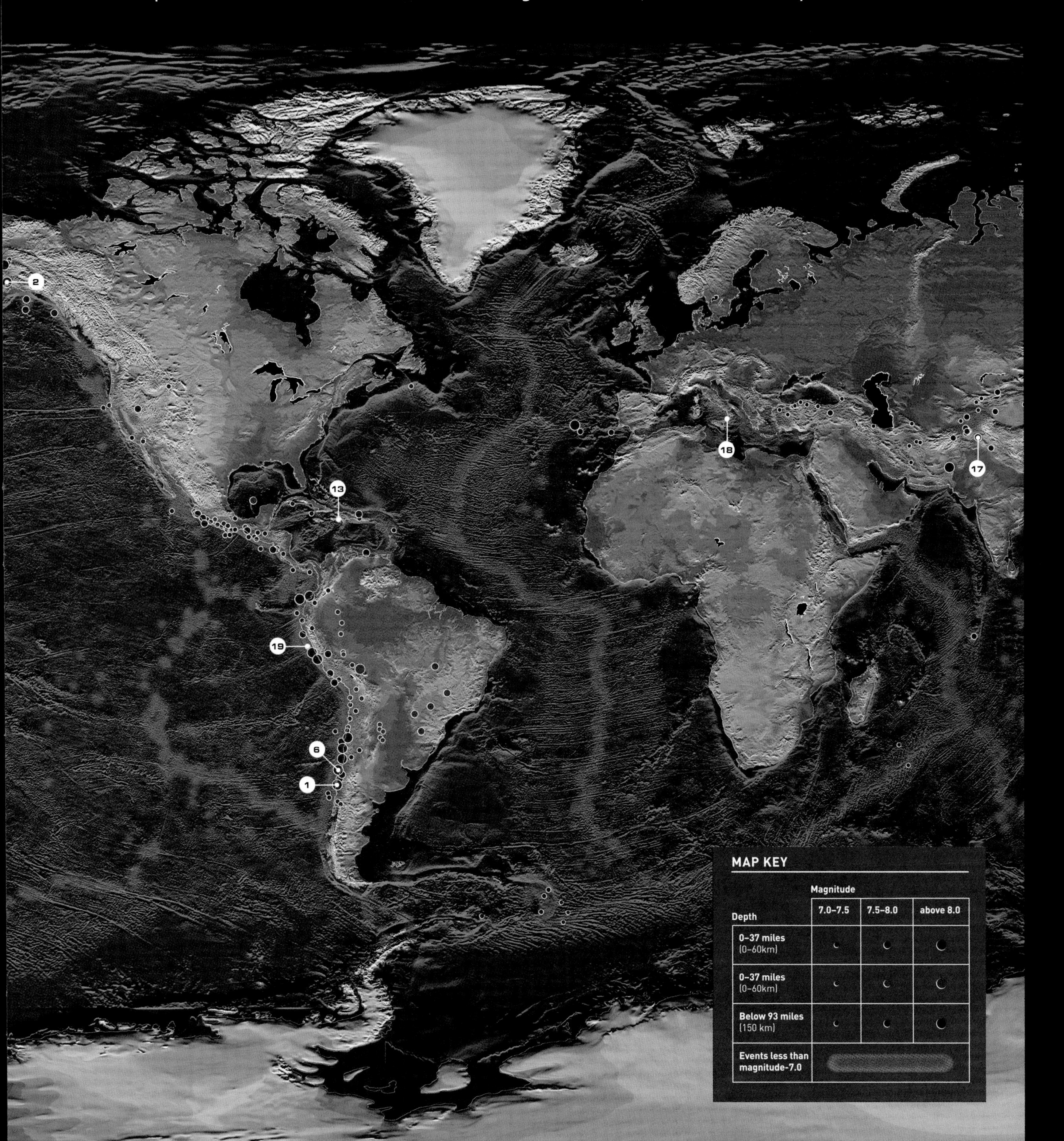

CAUSES OF EARTHQUAKES

Along the boundaries of the tectonic plates, rocks are squeezed and stretched like a spring by the huge forces inside Earth. Nearer the surface, rocks are sufficiently cold and strong that they eventually break along faults. If this occurs suddenly, the rocks snap, generating the violent vibrations of an earthquake.

FAULT LINES

Geologists have long observed that Earth's crust is intensely broken and fractured, cut through by fault lines. Some of these fault lines extend for hundreds or even thousands of miles, traversing great tracts of the continents. Others are only visible under a microscope. The link between movement along faults and earthquakes was realized by geologists in the late 19th century, but it was not until the great 1906 San Francisco earthquake that scientists began to study the connection closely. Man-made structures, including railway lines and roads, were offset across the fault line during the earthquake.

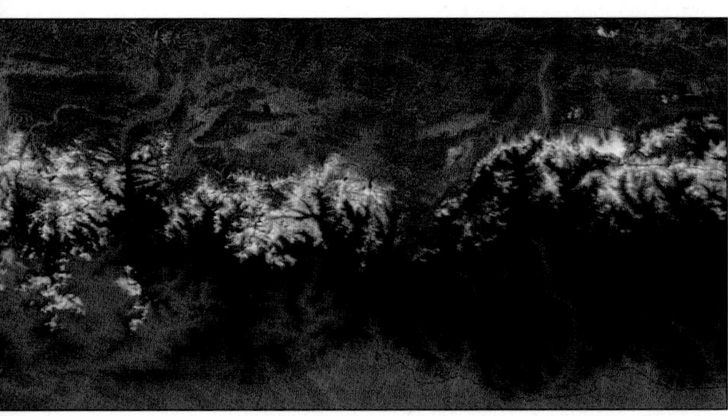

HIMALAYAN MOUNTAINS
The Himalayas lie above a giant, gently inclined fault line, which ruptures every few hundred years during earthquakes. Movement on this fault is pushing up the mountains.

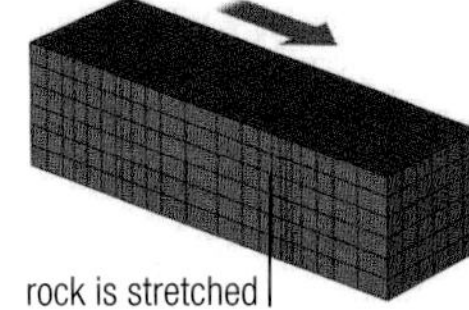

rock is stretched and then compressed

P WAVES
P waves pass through any material including liquids such as molten rock.

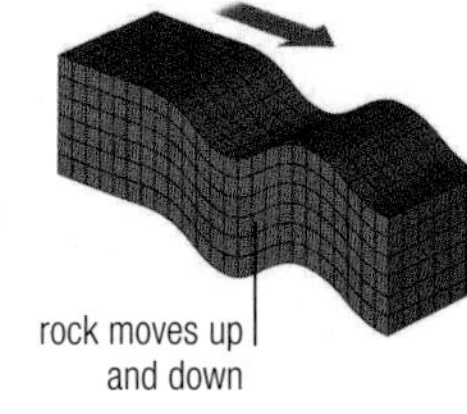

rock moves up and down

S WAVES
S waves can only move through solid rock. Their absence in the outer core is key evidence that it is liquid.

liquid outer core
no S waves detected here
lower mantle
upper mantle
epicenter of earthquake from where seismic waves radiate
P waves pass through solid as well as molten rock
S waves only travel through solid rock
Earth's inner core of solid iron
P waves refracted as they pass through layers
P waves detected up to 140° from origin around the surface

SEISMIC WAVES
Waves are detected by seismometers in different parts of the world during an earthquake, and are used by seismologists to unravel the sequence of events, including the direction and amount of slip on the earthquake-generating fault. The time delay between the P and S waves can be used to locate the earthquake.

P WAVES AND S WAVES

The vibrations that occur during an earthquake have distinct patterns. Some travel right through the Earth's interior, and are called elastic body waves with two distinct types of motion. P waves (primary), or compressional waves, are the first to arrive at a seismometer, traveling through the crust at about 3.5 miles per second (6km per second). They have a back-and-forth motion, like a vibrating spring, and are identical to sound waves, but at too low a frequency to be heard by the human ear. S waves (secondary), or shear waves, have a distinctive snakelike motion and travel relatively slowly—typically at a speed of about 2.5 miles per second (4km per second) in the crust—arriving after the P waves.

FAULT LINES AND LANDSCAPES

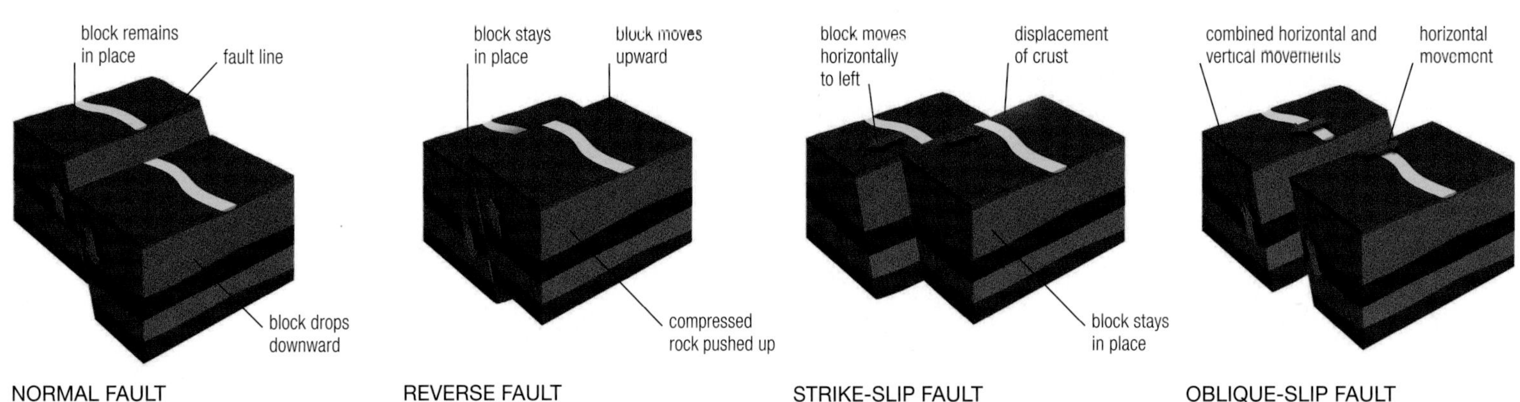

NORMAL FAULT
When rocks are stretched horizontally, they break along normal faults. Here, the overlying rocks slide down the fault and away from the underlying rocks.

REVERSE FAULT
The overlying rocks slide upward giving rise to uplifted regions or mountain ranges. Faults inclined at less than 45° are referred to as thrust faults.

STRIKE-SLIP FAULT
Two sides of the fault slide past each other horizontally. If this is to the right, it is referred to as right-lateral. If it is to the left, it is left-lateral.

OBLIQUE-SLIP FAULT
When a strike-slip fault is combined with either extension or compression, the blocks can slide diagonally producing an oblique-slip fault.

OFFSET RAILWAY LINES
Elastic rebound either side of the fault buckled these railway lines after the 1987 magnitude 6.3 Edgecombe quake in New Zealand.

ELASTIC REBOUND

US geologist Harry Fielding Reid studied the 1906 San Francisco earthquake and suggested that the rocks were behaving rather like a piece of elastic. Before the earthquake there was no movement on the fault itself, but the surrounding rocks were being slowly distorted. Eventually, this distortion exceeded the limit that rocks could undergo and they snapped, breaking along the fault. It was this sudden snapping, or "elastic rebound", that generated the violent shaking during the earthquake. Movement on the fault during the 1906 earthquake was strike-slip (see pp.224–25). However, this pattern of elastic distortion, then sudden breaking when an earthquake occurs, can be observed on all types of faults, resulting in the characteristic ground motions that geologists call the "earthquake cycle."

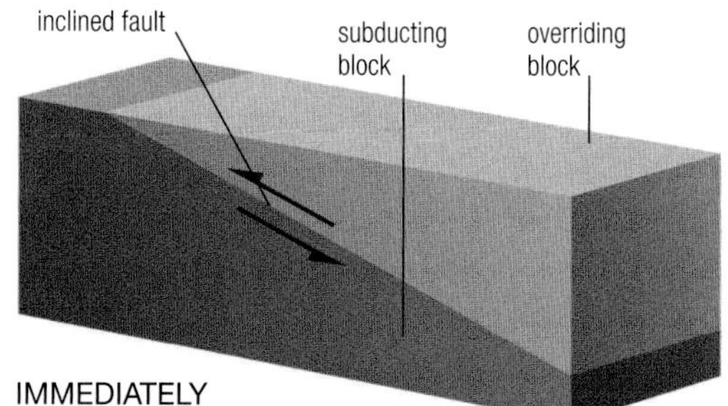

IMMEDIATELY AFTER EARTHQUAKE
After a sudden slip along the fault, the earthquake cycle soon begins again as the rocks and land surface settle into a new arrangement.

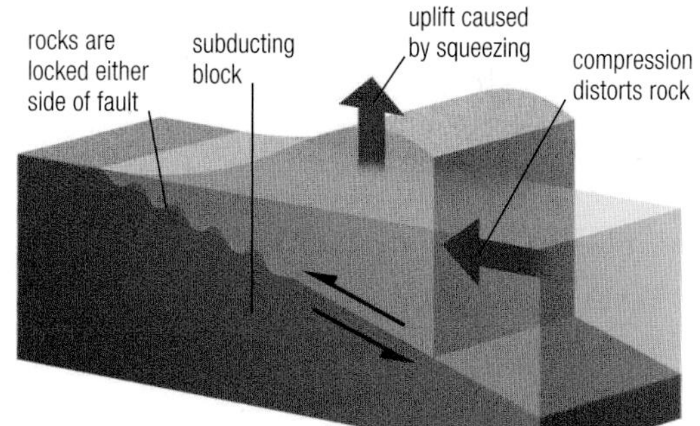

BETWEEN EARTHQUAKES
The motion of the tectonic plates causes stress to build up. There is no movement on the fault, but the rocks are distorted as stress accumulates.

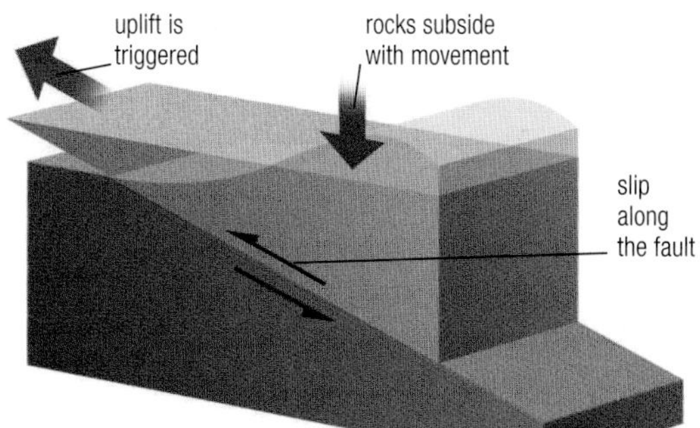

DURING EARTHQUAKE
The accumulated strain that has built along the fault eventually overcomes the strength of the rock and unlocks it. The rock plates then spring back, releasing the displacement and producing an earthquake.

HAITI 2010

On Tuesday, January 12, 2010, at 16:53 local time, a magnitude-7.0 earthquake struck close to Port-au-Prince, the capital of Haiti, in the Caribbean. The earthquake was felt as far afield as parts of the Bahamas, Puerto Rico, and the US Virgin Islands, and even in southern Florida, northern Colombia, and northwestern Venezuela. It occurred in a geologically active region at the boundary between the Caribbean and the North American plates, where the Caribbean Plate is moving about 0.8in (20mm) per year relative to North America, mainly along left-lateral strike-slip faults. However, the main shock did not produce any observable surface displacement on the main fault lines in this region, including the nearby Enriquillo-Plaintain Garden Fault, but rather on previously undetected faults. The earthquake was followed by as many as 52 aftershocks measured at magnitude-4.5 or above between January 12 and January 24.

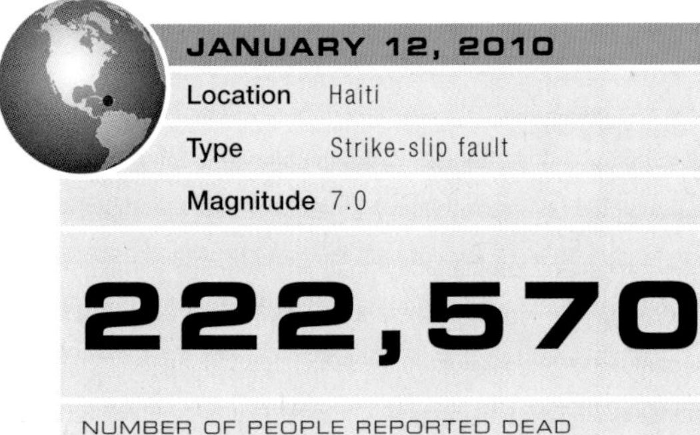

JANUARY 12, 2010	
Location	Haiti
Type	Strike-slip fault
Magnitude	7.0

222,570

NUMBER OF PEOPLE REPORTED DEAD

INTERNATIONAL RESCUE
Search-and-rescue teams from all over the world, including the US, Britain, Japan, and New Zealand, came to the aid of the Haitians.

THE PRESIDENTIAL PALACE BEFORE AND AFTER
A significant proportion of the city of Port-au-Prince was destroyed in the earthquake, including the Presidential Palace, the National Assembly, and Port-au-Prince cathedral. Haiti is a poor country, and inadequate design codes for earthquake-proof buildings led to much of the damage.

AFTER THE QUAKE
A shortage of clean drinking water was a major problem for the population of Port-au-Prince and diseases such as diarrhea spread easily. Aid organizations made water-filtration systems a priority.

IMPACT AND CASUALTIES

According to official estimates, 222,570 people were killed, 300,000 injured, 1.3 million displaced, 97,294 houses destroyed, and 188,383 houses damaged in the Port-au-Prince area of southern Haiti. This total includes at least four people killed by a local tsunami. The earthquake destroyed the main infrastructure in Port-au-Prince—a town of nearly 750,000 people—especially the water and sewage systems. Most of the population were displaced into temporary shelters, with poor sanitation and a lack of drinking water and food, resulting in the spread of diseases such as cholera and dystentery.

> “WE HAD A LOT OF HOUSES DESTROYED, A LOT OF PEOPLE DEAD... A LOT OF PROBLEMS.”
>
> **JEAN**, MAYOR IN SOUTHERN HAITI

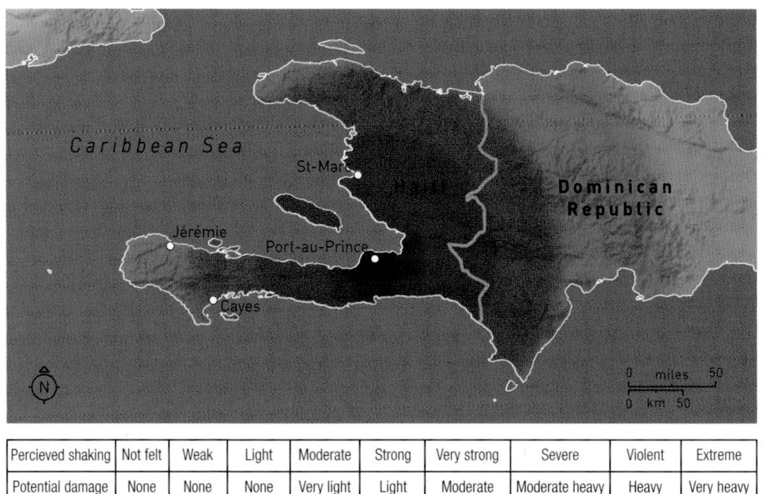

Percieved shaking	Not felt	Weak	Light	Moderate	Strong	Very strong	Severe	Violent	Extreme
Potential damage	None	None	None	Very light	Light	Moderate	Moderate heavy	Heavy	Very heavy
Intensity	1	2 – 3	4	5	6	7	[illegible]	[illegible]	[illegible]

HAITI SHAKE MAP

This color-coded map shows the intensity of the shaking experienced, from red (most intense) to green (weak). The most intense shaking is close to the epicenter, where a fault ruptured at depth. However, the intensity of shaking was also determined by the nature of the local sediments beneath Port-au-Prince.

WATER PURIFICATION

Damage to the water and sewage systems in Port-au-Prince led to wells and water mains becoming contaminated with dirty water and sewage, which eventually became a major source of waterborne diseases. To deal with this problem aid teams distributed water purification kits. Designed for personal use, a simple tube device, called a LifeStraw, consists of a series of filters: some are fabric and some chemical. Mesh filters trap dirt and sediment and an iodine layer destroys most bacteria and viruses, then active carbon granules purify water using a process called adsorption. This also helps remove the unpleasant taste of the iodine.

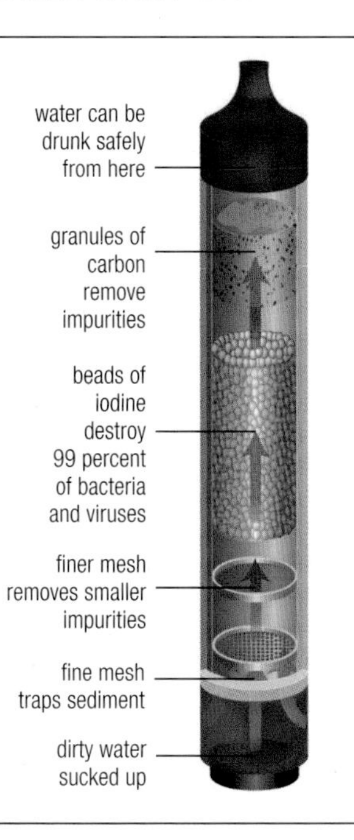

LifeStraw

This specially designed tube of filters can save lives when it is used as a straw to safely drink contaminated water.

MOVEMENTS AND FAULTS

Earthquakes are an expression of a more profound process—the long-term shifting of rocks and the landscape caused by the motion of tectonic plates. Geologists measure and monitor these movements, either directly using accurate positioning systems, or by studying geological changes in the landscape.

LATERAL AND VERTICAL MOVEMENTS

The permanent changes in the landscape that occur during an earthquake extend over a wide region, up to hundreds of miles from the earthquake epicenter. Close to the epicenter, if the earthquake is relatively shallow, there will be an abrupt break in the ground surface, revealing a fault that has broken during the earthquake. In the largest earthquakes (magnitude-8 and above on the moment magnitude scale), displacements on the fault up to nearly 66ft (20m) have been observed, though a displacement up to a few feet is more typical of an earthquake of magnitude-6 and above. These displacements may be both vertical and horizontal shifts of the land surface, which over a span of geological time, shape our Earth.

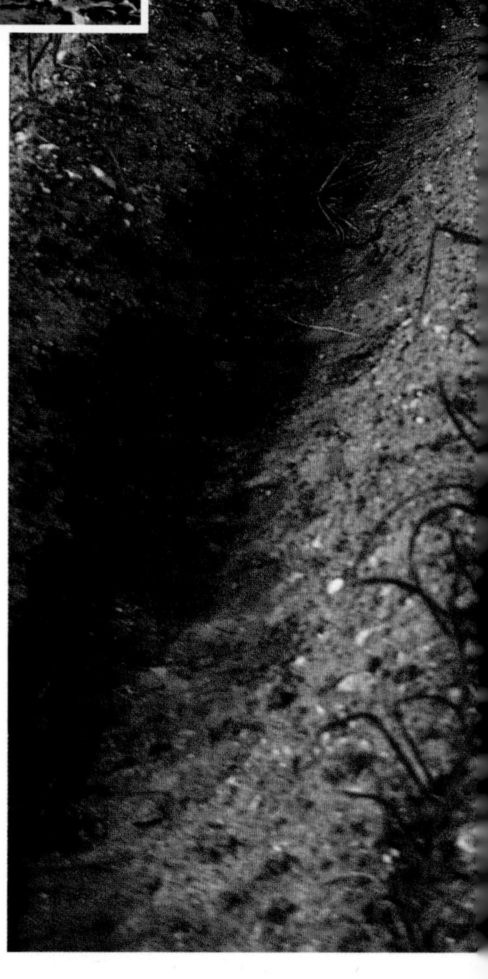

VERTICAL MOVEMENT
Along the Hanning Bay Fault on Montague Island in Prince William Sound, Alaska, vertical movement of about 13ft (4m) created this new scarp in the landscape during the 1964 magnitude-9.2 earthquake. Repeated over many earthquakes, vertical motion like this can create a mountain range.

1–6in
(2.5–15cm) The range of distance that each tectonic plate can move relative to Earth's interior each year.

SHIFTING STADIUM
The exterior walls of Berkeley's Memorial Football Stadium, California, are slightly offset. The Hayward Fault runs across the field, and the two halves of the stadium have shifted since it was constructed.

EARTHQUAKE FAULT CREEP

Not all faults in the landscape are inactive between earthquakes. Though tremors may not be felt at the surface, the fault may still be moving. Several fault lines in California and Turkey have been observed to "creep" or make slight but continuous movements. In some cases, these creeping motions seem to decrease with time, and are likely to represent continued longer term readjustments of the fault after an earthquake. The Hayward Fault in California is still moving, offsetting buildings, drains, and roads that cross its path. In Turkey, part of the North Anatolian Fault, near Ismit Pasha, was observed to creep in the 1970s and 80s, continually distorting a railway line that crossed its path. But at depths of a few tens of miles, where the rocks are sufficiently hot and weak, all faults undergo continuous creep along aseismic, ductile, shear zones. The earthquake-generating part of the fault is confined to the colder and stronger rock at shallower depths, where the movement of the fault occurs in a series of earthquake-triggering jerks. Between earthquakes the motion at this depth is absorbed by distortion of the rocks over a wide area.

“ HUGE AMOUNTS OF STRESS BUILD UP. WHEN THE STRESS IS TOO GREAT THE ROCKS FORCE APART LIKE A... SPRING. ”

PROFESSOR GEOFFREY KING, INSTITUT DE PHYSIQUE DE GLOBE

KASHIWAZAKI, JAPAN
This right-lateral shift across a field in Kashiwazaki, Japan indicates a fault line which appeared following the earthquake in Kobe, in 1995.

FAULT LINES AROUND THE WORLD

When the first satellite images of the continents became available in the early 1970s, scientists were stunned to see long, knifelike scars in the landscape, in some cases running for hundreds or even thousands of miles. A French geologist, Paul Tapponnier, who had been studying these images of Central Asia, was one of the first to realize that what he was looking at were fault lines in Earth's crust, along which there had been large shifts in the landscape. Tapponnier's work revealed one of the largest fault lines on land—the Altyn Tagh Fault in northern Tibet. Many other giant fault lines are now recognized, such as the San Andreas Fault in California, the Denali Fault in Alaska, and the Alpine Fault in New Zealand. All these faults occur where the tectonic plates are moving relative to each other.

For example, coastal California is sliding northward along the San Andreas Fault, with a right lateral sense of motion, taking up the motion of the Pacific Plate relative to the North American Plate. Beneath the world's oceans, even bigger fault lines lie hidden, forming where the sea floor is spreading apart or slipping beneath a neighboring continent. Earthquakes provide the key to understanding the significance of these faults, because the seismic waves generated when a fault ruptures tell us the amount and direction of movement.

COMPLEX FAULT LINES IN TURKEY

The countries of the Eastern Mediterranean are geologically very active, with many earthquakes triggered by movement along the large number of fault lines in the region. The African and Arabian tectonic plates are moving northward, colliding with Eurasia. Turkey is forming a "micro-plate" caught between the bigger plates and pushed out westward along two major strike-slip faults—the North and South Anatolian Faults—toward the subduction megathrust along the Hellenic Arc. There are many smaller fault lines too, especially in the Aegean region of Greece. Nowhere escapes the danger of a deadly quake.

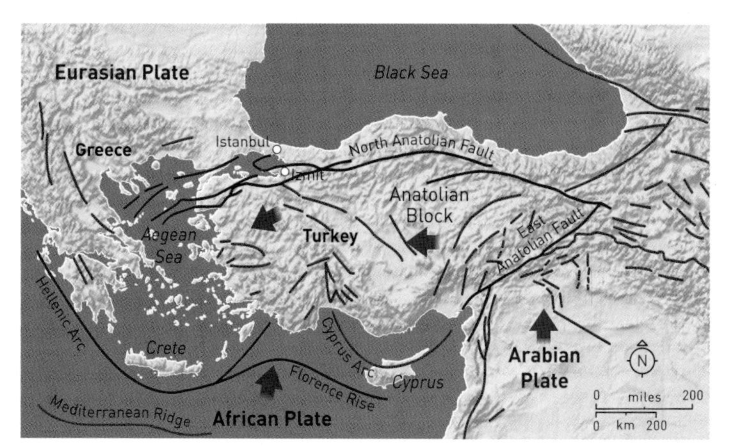

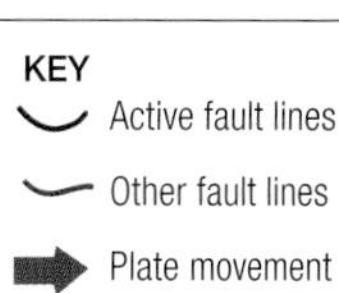

Faults in the Eastern Mediterranean
The red lines show the major active fault lines. However, there are many other faults, shown in green, which may or may not be active. Movement on all these faults is part of the motion of the African Plate, pushing northward into Eurasia at a rate of about 0.4in (10mm) per year.

Kaynaslie, Turkey
Rescue workers sift through the rubble of houses and vehicles destroyed by the wreckage of apartment buildings by a powerful earthquake which measured 7.2 on Richter scale in Northern Turkey, on November 12, 1999.

MEASURING EARTHQUAKES

Anybody close to an earthquake will feel the ground shaking violently for a few moments. However, these felt vibrations are just part of a pattern of seismic wave vibrations associated with the earthquake, which can be detected by sensitive instruments around the globe called seismometers. The details of this wave pattern provide seismologists with a rich source of data that can be used to investigate both the causes and consequences of an earthquake.

MAGNITUDE SCALES

Seismologists have come up with various ways to describe the amount of energy released during an earthquake. The original Richter scale was based on the magnitude of displacement recorded on a particular type of seismometer at a known distance from the epicenter. These days it is often quoted as moment magnitude, which is similar but is calculated from seismic vibrations on any seismometer. The moment magnitude can be interpreted in terms of the amount of slip that has occured on the earthquake-triggering fault multiplied by the size (area) of the fault. Both the Richter and moment magnitude scales are logarithmic. An increase of one unit on the scale corresponds to about a 30-fold increase in energy.

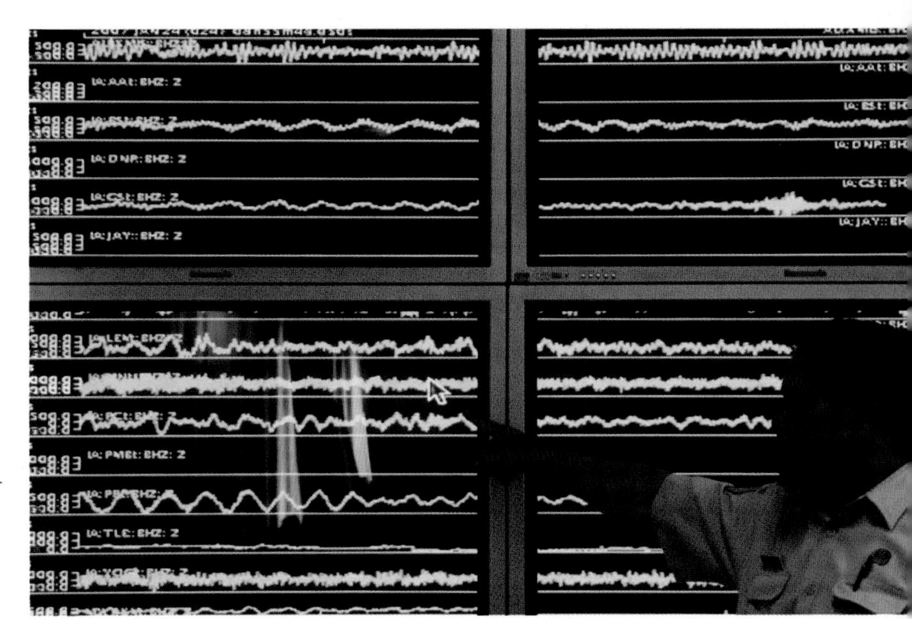

DIGITAL SEISMOGRAPHS
Data from seismometers around the world is fed into computers belonging to earthquake research stations, such as this one in the Philippines.

> "THE HIGHEST MAGNITUDES ASSIGNED SO FAR... ARE ABOUT 9, BUT THAT IS A LIMITATION IN THE EARTH, NOT IN THE SCALE."
>
> **CHARLES RICHTER**, US SEISMOLOGIST, 1980

SEISMOMETERS

A seismometer is a delicate instrument designed to measure seismic vibrations during an earthquake. It works on the simple principle that the recording part is free to move with the Earth while the bulk of the instrument, which is securely anchored to the ground by a heavy weight, remains stationary. A great step forward in the early 19th century was a neat portable model that could be carried by scientists to areas of seismic activity. Until recently, these vibrations were recorded by a vibrating pen, scratching out a trace called a seismograph on a rotating drum. Modern digital instruments record the vibrations electronically, and the data are processed and stored in a small computer. Readings from several seismometers are passed to a central recording station where the data is analyzed. Warnings of serious earthquake activity may then be transmitted to radio or television stations to alert the public. GPS devices are also used to measure the extent of shifts in Earth's surface during earthquakes. Individual, solar-powered devices transmit information from out in the field to research stations.

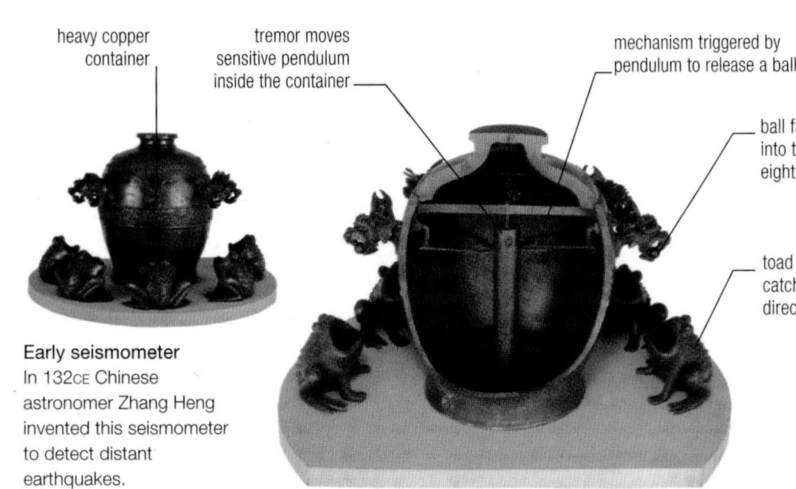

Early seismometer
In 132CE Chinese astronomer Zhang Heng invented this seismometer to detect distant earthquakes.

solid, rotating drum is held steady with weights

paper wrapped around the drum

pen records any movements the ground makes

solid base to hold it still

Recent seismometer
This 1965 model is a Willmore Portable seismometer, capable of picking up and recording much more detailed seismic activity.

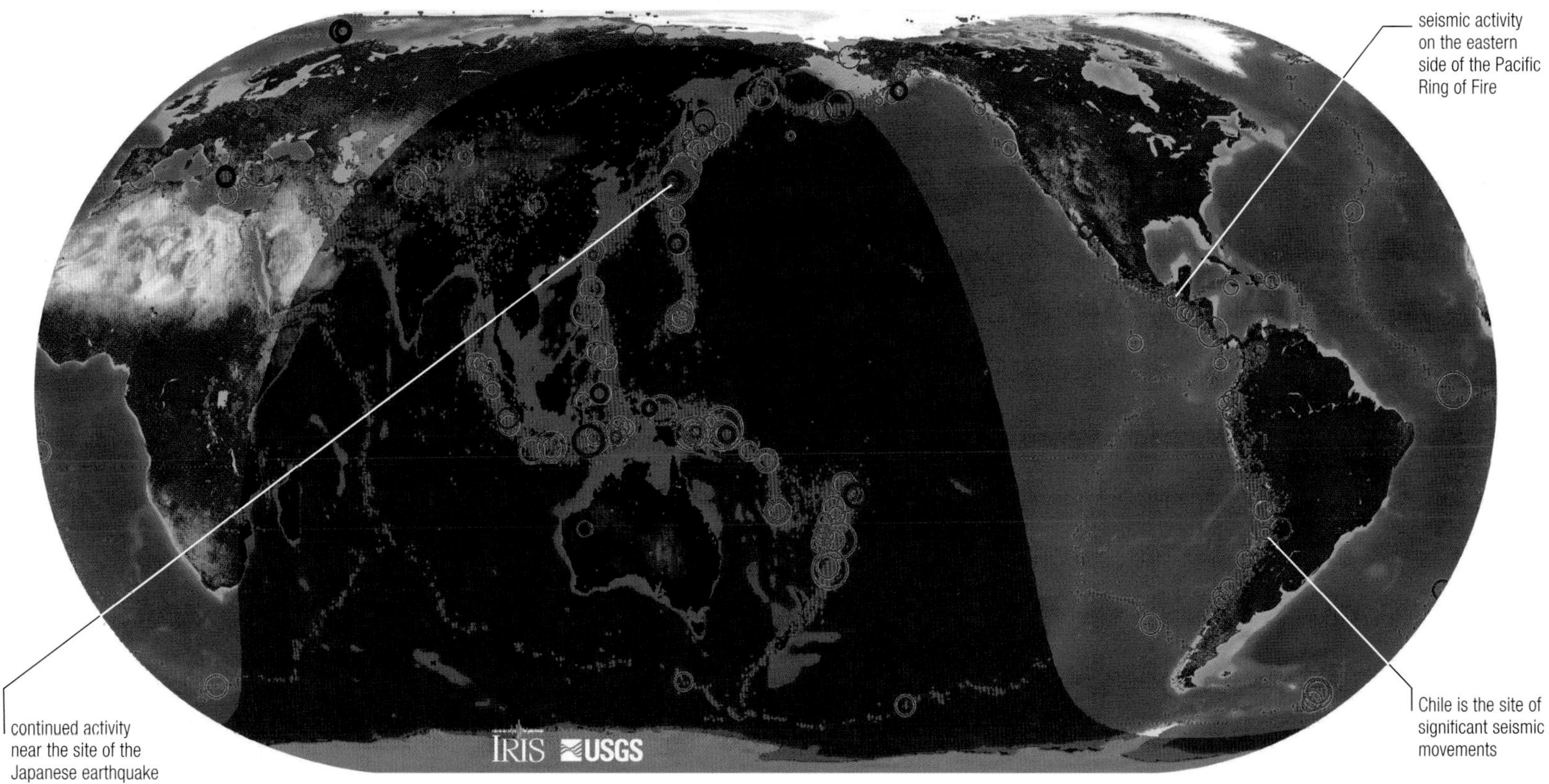

KEY—GLOBAL MAGNITUDE ACTIVITY

8
6
4

- Today
- Yesterday
- Past 2 weeks
- Past 5 years

GLOBAL SEISMIC MONITORING
This snapshot of seismic activity on April 14, 2011 was recorded by the Global Seismographic Network. The location and magnitude of earthquakes is calculated from this information.

TRACKING GLOBAL SEISMIC ACTIVITY

The Global Seismographic Network, with contributions from a number of countries, is a global state-of-the-art, digital seismic network of seismometers, providing free, real-time data. The instruments are capable of measuring and recording all seismic vibrations, ranging from high-frequency, strong ground motions near an earthquake to the slowest vibrations from a distance away. With the USGS (US Geological Survey), this is the principal source of data for earthquake locations.

GPS INSTRUMENTS
The Global Positioning System (GPS) is used to accurately measure displacements of marker points before, during, and after an earthquake.

FOCAL MECHANISM

The focal mechanism is a method to indicate the nature of the fault movement that triggered an earthquake. The angle of slip and type of fault movement—such as normal, strike-slip, thrust fault, or reverse thrust (see pp.206–07)—can be inferred from graphical representations termed "beach balls." The colored areas indicate where the first vibrations of P waves (see pp.206–07) pushed or pulled, and this makes the distinctive beach ball pattern. Seismologists use these icons when they produce maps of fault lines to show the orientation of movements on actual faults.

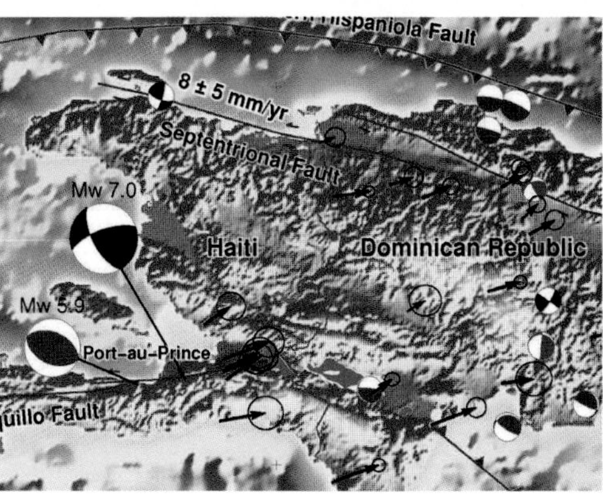

FOCAL MAP
This segment of a map shows fault lines with focal mechanisms and moment magnitude measurements around the site of the 2010 Haiti earthquake.

BEACH BALLS
The "beach ball" diagrams describe the pattern of seismic waves for distinct types of fault movement. The blocks show the type of fault movement.

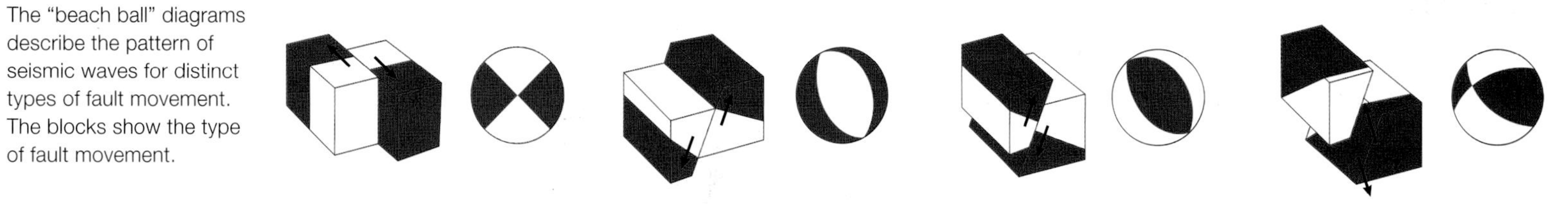

SIRVIENDO
LA CIUDA

MEXICO CITY EARTHQUAKE
The 1985 Mexico City earthquake measured 8.0 on the Richter scale and caused about 9,500 deaths. The earthquake was located off the Mexican Pacific coast, in a subduction zone more than 220 miles (350km) away. Due to its strength and the fact that Mexico City sits on an old lakebed, the city suffered major damage with 400 buildings collapsing.

SUBDUCTION EARTHQUAKES

In a subduction zone, especially where the sea floor slides beneath a continent, movement on large thrust faults slice and pile up rocks, building a mountain range such as those found along the Pacific Ring of Fire. These faults are capable of rupturing suddenly, triggering the largest earthquakes on the planet.

SUBDUCTION ZONES AND MEGATHRUSTS

The Pacific Plate is sinking back into Earth's interior in a series of subduction zones that follow the edges of the surrounding continents. Here, the ocean floor rubs against the overlying plate along a giant, gently inclined fault known as a megathrust. Between earthquakes, where the rocks are relatively cold and strong, the two plates are locked together and the motion of the plates is absorbed by squeezing and distortion of the rocks. Eventually, the forces become too great and the fault suddently ruptures in a great earthquake. The resulting slip raises or drops the seabed or ground surface. The largest earthquakes of this type—such as the one in Japan in 2011 (see pp.266–69)—can reach magnitude-9, and the displaced water generates devastating tsunamis.

RING OF FIRE
This map shows the depth of the sea floor and elevation of the continents along the margins of the Pacific Ocean. A deep ocean trench and a line of active volcanoes follow the Pacific Rim, defining a series of subduction zones in the so called Ring of Fire.

KEY
- Earthquakes since 1980 at magnitude 5 and above
- Plate boundaries

PACIFIC PLATE
The Pacific Plate dives down in the ocean trench offshore and slides beneath the continental margin. The crust is pushed up to create a chain of mountains. At depths of less than 30 miles (50km), the rocks are cool enough to break only during earthquakes. Molten rock rises and erupts in volcanoes.

Volcano
Eruptions form distinctive conical volcanoes, typical of subduction zones

Ocean trench
A deep trench forms where the ocean floor descends into the subduction zone

Lithosphere
The crust plus the very top part of the mantle

Continental plate
The edge of the continent is squeezed and thickened, built up by further volcanism

Magma
A source of magma feeds the volcano

Oceanic plate
The oceanic plate together with the very top part of the mantle sinks below the continental crust

Megathrust
The relatively cold, earthquake-generating area of rocky crust

Asthenosphere
The soft, upper part of the mantle

THE GREAT ALASKAN EARTHQUAKE
On March 27, 1964, a sudden slip on a megathrust under Alaska triggered the largest recorded earthquake in North America, severely damaging the city of Anchorage. A devastating tsunami followed the magnitude-9.2 quake.

THRUST FAULTS

Thrust faults are defined as faults that are inclined at angles less than 45 degrees, where the overlying rocks slide up and over the underlying rocks. Faults like this tend to occur in layered rocks, sometimes inclined at only a few degrees where they follow the rock layers. A megathrust is merely a very large thrust, and can extend for thousands of miles along its length. The biggest megathrusts are found in subduction zones. Large thrusts also form the edges of the world's largest mountain ranges.

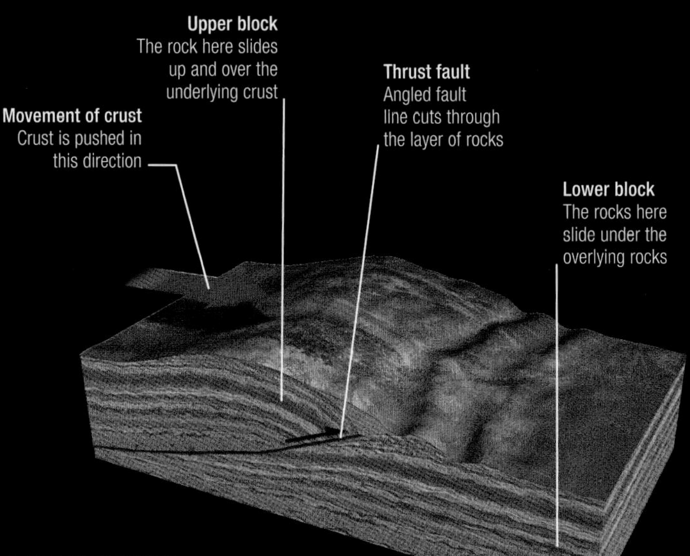

THRUST MOVEMENT
This cutaway view shows a typical thrust fault in a layered sequence of rocks. This fault places deeper older rocks above younger rocks.

WADATI-BENIOFF ZONE

In a subduction zone, where the oceanic plate dives into the Earth's interior, there is intense earthquake activity. However, the earthquakes do not occur only along the megathrust, where the oceanic plate rubs against the overlying plate. There are also earthquakes in the oceanic plate itself, down to depths of nearly 435 miles (700km). This phenomenon was first described in the early 20th century by seismologists Kiyoo Wadati from Japan and Hugo Benioff from the US. They related the activity of many small earthquakes in the Pacific to a subducting oceanic plate, as it was stretched and squeezed deep in the Earth.

Deep earthquakes
This illustration shows the depth of earthquakes that originate in the slablike Wadati-Benioff zone. In the Pacific Rim, this slab plunges to a depth of 435 miles (700km).

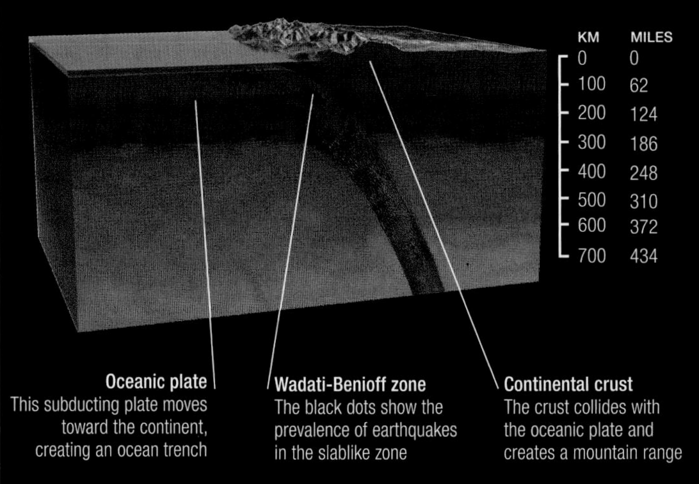

CONCEPCIÓN 2010

In 1835, while voyaging on the *Beagle* the famous English naturalist Charles Darwin witnessed a massive earthquake near Concepción, Chile. Many years later, in 1960, the residents of this same city experienced a magnitude-9.5 earthquake—the largest ever recorded with modern instruments. On Saturday, February 27, 2010 history repeated itself, as a magnitude-8.8 earthquake struck the coastal region of central Chile, again near the town of Concepción. The earthquake was felt across a large area of South America, including Brazil, Bolivia, and Argentina. It triggered a Pacific-wide tsunami, and killed at least 521 people, injured 12,000, and left hundreds of thousands displaced. The tsunami damaged or destroyed many buildings and roads along the coast in the vicinity of Concepción, and even damaged boats and a dock as far away as San Diego, California. More than 6.5ft (2m) of uplift along the coast was observed near Arauco. The earthquake was the result of 16 to 49ft (5 to 15m) of slip on the megathrust along the plate boundary between the South American Plate and the subducting Nazca Plate, rupturing a zone around 310 miles (500km) long.

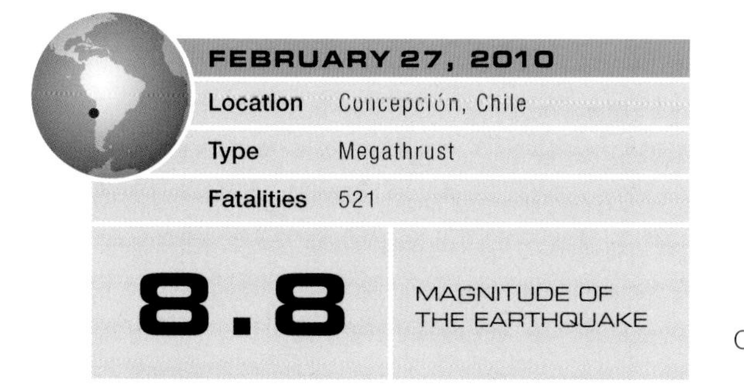

FEBRUARY 27, 2010	
Location	Concepción, Chile
Type	Megathrust
Fatalities	521
8.8	MAGNITUDE OF THE EARTHQUAKE

MASS DAMAGE
The considerable force that was unleashed by the earthquake caused these large apartment buildings in Santiago, Chile, to buckle and lean at precarious angles.

DISASTER STRIKES

1 THE GROUND SHAKES
The violent shaking during the earthquake was enough to overturn these cars traveling along a road in Santiago, Chile. A maximum ground acceleration of 0.65g was recorded at Concepción.

2 THE TSUNAMI STRIKES
Within 30 minutes of the main shock, a tsunami washed onshore, created by the sudden movement of the sea floor up to 60 miles (100km) from the coast. This picture shows the destruction caused by the tsunami in the Chilean coastal town of Talcahuaro.

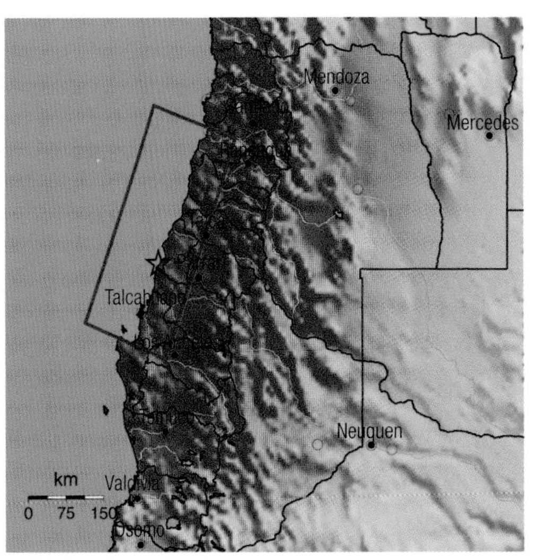

SHAKE MAP
The greatest shaking is clearly defined in a region about 310 miles (500km) long, following the coast of Chile. The region is underlain by the gently inclined megathrust where the ocean floor (Nazca Plate) is sliding beneath part of South America. During the earthquake, a sudden slip on the megathrust released a vast amount of seismic energy that violently shook the region.

Percieved shaking	Not felt	Weak	Light	Moderate	Strong	Very strong	Severe	Violent	Extreme
Potential damage	None	None	None	Very light	Light	Moderate	Moderate heavy	Heavy	Very heavy
Intensity	1	2–3	4	5	6	7	8		

EARTHQUAKE RESCUE TECHNOLOGY

In high-risk areas, buildings are constructed to control vibration and remain stable in the event of an earthquake. However, the destructive power of an earthquake can leave buildings in ruins, so rescue technology is crucial in the search for people trapped under rubble. Highly trained sniffer dogs take minutes to locate people buried deep in the debris. Rescuers then operate heat imaging and listening devices to confirm the prescence of, and communicate with, any survivors. More recently, robots have been developed that are equipped with wheels to negotiate obstacles, cameras to relay images back to the controllers, and infrared sensors to detect people in the rubble.

SURVIVOR-DETECTING ROBOT
This device is a flexible robot developed in Japan to aid the search for survivors among the debris of earthquake and tsunami ravaged regions. It rotates with a snakelike motion, enabling it to slide over the ground and glide through water. Sensors detect the presence of survivors.

3 LOOKING FOR SURVIVORS
Rescue workers climb into a collapsed building in Santiago, looking for survivors trapped inside. An estimated 500,000 homes in Chile were severely damaged by the earthquake and ensuing tsunami.

"THREE MINUTES IS AN ETERNITY. WE KEPT WORRYING THAT IT WAS GETTING STRONGER, LIKE A TERRIFYING HOLLYWOOD MOVIE."

DOLORES CUEVAS, SANTIAGO HOUSEWIFE

SICHUAN, CHINA 2008

The magnitude-7.9 earthquake that struck eastern Sichuan, 55 miles (90km) from the city of Chengdu in May 2008 was one of the most destructive earthquakes to strike China in more than 30 years, with strong aftershocks continuing for several months. At least 69,000 people were killed, hundreds of thousands were injured or missing, and many tens of millions displaced or made homeless by the disaster. More than five million buildings collapsed, with landslides and rockfalls blocking roads and railway lines. The deadly earthquake was felt throughout much of China and as far afield as Bangladesh, Taiwan, Thailand, and Vietnam, where buildings swayed with the tremor. The earthquake occurred when a reverse fault slipped between 7 and 30ft (2 and 9m), extending for about 124 miles (200km) along the southeastern edge of the Longmen Shan Mountains. This region lies at the eastern edge of the vast collision zone between India and Central Asia. Here, the convergence of the two plates has also resulted in the uplift of the Himalayas and Tibetan Plateau. The Longmen Shan region has previously experienced destructive earthquakes. In August 1933, a magnitude-7.5 earthquake killed more than 9,300 people.

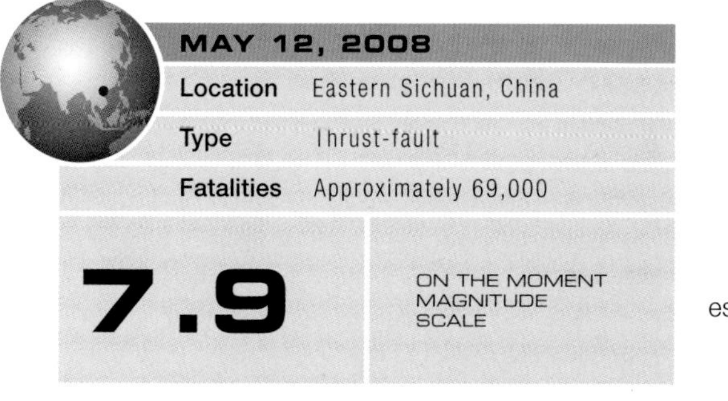

MAY 12, 2008

Location Eastern Sichuan, China

Type Thrust-fault

Fatalities Approximately 69,000

7.9 ON THE MOMENT MAGNITUDE SCALE

DESTRUCTIVE FORCE The earthquake annihilated cities and villages in Sichuan Province. In the city of Beichuan alone, it is estimated that 80 percent of the buildings were reduced to rubble.

EASTERN SICHUAN SHAKE MAP

This color-coded map shows the intensity of shaking, from red (most intense) to green (weak). The most intense shaking occurred in a long zone, extending from the earthquake epicenter at the far southwestern end and following the line of where the Longmen Shan Fault ruptured at depth, which was more than several hundred miles in length.

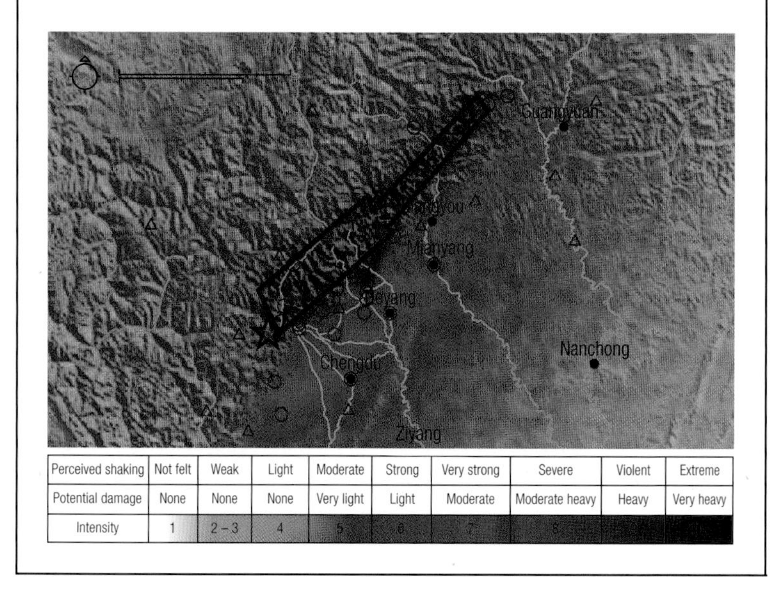

Perceived shaking	Not felt	Weak	Light	Moderate	Strong	Very strong	Severe	Violent	Extreme
Potential damage	None	None	None	Very light	Light	Moderate	Moderate heavy	Heavy	Very heavy
Intensity	1	2–3	4	5	6	7	8	[illegible]	[illegible]

DISASTER AND RESCUE

1 RUSHING WATER
"Quake lakes" formed when landslides triggered by the earthquake blocked rivers. Here, rushing water from a breached quake lake flows through devastated Beichan.

2 RESCUE EFFORTS
Search and rescue teams helped locate and dig out survivors trapped in the debris of collapsed buildings in Beichan.

3 TEMPORARY SHELTERS
Prefabricated houses for quake victims were hastily erected in hard-hit Wenchuan, one of the cities closest to the epicenter.

> "THIS IS THE MOST DESTRUCTIVE EARTHQUAKE SINCE THE PEOPLE'S REPUBLIC OF CHINA WAS FOUNDED [IN 1949] AND HAS AFFECTED THE WIDEST AREAS."

WEN JIABAO, CHINESE PREMIER

BIRTH OF A FAULT LINE

This aerial view of fields shows a new fault line created during the 2011 earthquake near Christchurch, New Zealand. An area of ground has been torn for a distance of nearly 18.5 miles (30km) across farmland. Where the fault line crosses a water-filled ditch, the right-lateral displacement is clear, offsetting the ditch horizontally by several feet.

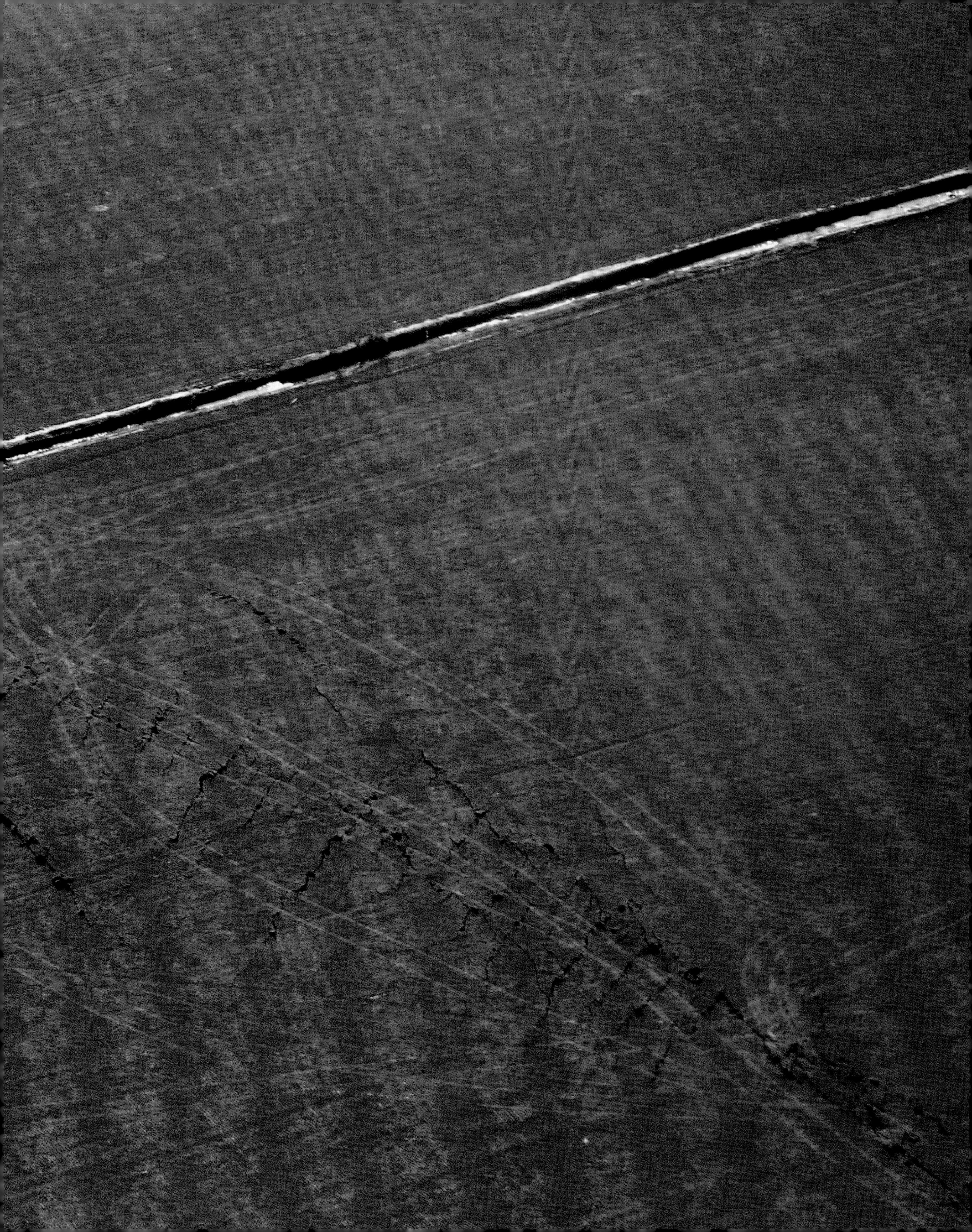

STRIKE-SLIP EARTHQUAKES

When two plates slide past each other, strike-slip fault lines form in the crust, which can suddenly rupture during earthquakes. From space, these fault lines can clearly be seen across the continents as giant scars on the landscape.

SLIPPING AND SLIDING

A strike-slip fault forms a deep break in the crust, extending to depths of tens of miles. As one side moves relative to the other, either to the left (left-lateral displacement) or the right (right-lateral displacement), the rocks become intensely fractured, sometimes ground down into a fine, claylike powder. The actual movement during an earthquake usually occurs as a series of jerks because the blocks stick and then slip past each other. For example, the San Andreas strike-slip fault in California broke during the 1906 earthquake, moving about 17ft (5m) with a right-lateral displacement. Measured over geological periods of time, this fault slips at a rate of about 80ft (25m) every thousand years while it absorbs motion between the Pacific and North American plates. An earthquake on this part of the fault occurs on average every 200 years.

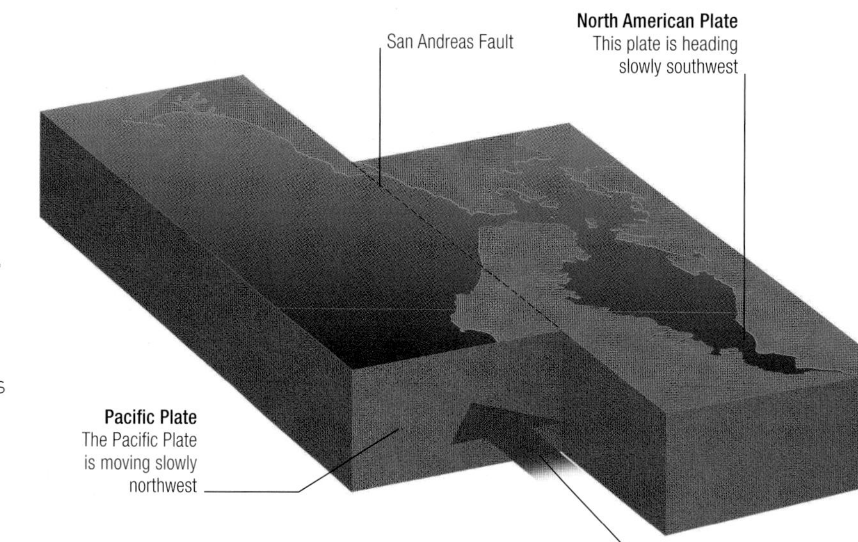

FAULT MOVEMENT
The San Andreas Fault splits San Francisco, California. The two sides of the area form blocks of crust that are sliding past each other. The fault itself is a nearly vertical break in the crust.

OFFSETTING LANDSCAPE

Over millions of years, the displacements during individual strike-slip earthquakes build up. On major strike-slip faults, these displacements can reach hundreds of miles, slicing up and offsetting the bedrock by staggering amounts. Geologists used to believe that such large offsets on strike-slip faults were impossible, because nobody could understand what happened at the ends of the fault lines. Now, with the theory of plate tectonics, it is clear that the faults ultimately connect with other plate boundaries, where the ocean floor is sinking, or is being created at mid-ocean ridges. Thus, over the past 20 million years, the roughly 310-mile- (500-km-) strike-slip offset of the rocks along the San Andreas Fault, or a similar amount for New Zealand's Alpine Fault, is part of the destruction or creation of oceans (see pp.28-31).

1906 EARTHQUAKE, SAN FRANCISCO, CALIFORNIA
These multistory wooden houses on Howard Street in San Francisco tilted over, but remained more or less intact during the powerful 1906 earthquake that shook the city.

KOBE EARTHQUAKE

In January 1995, a strike-slip earthquake of magnitude 6.8 struck the city of Kobe, Japan, killing thousands and causing major structural damage. Japan lies on the fault line between the Philippine and Pacific plates.

IZMIT 1999

On August 17, 1999 at 3:00am, a magnitude-7.4 earthquake struck northwestern Turkey, about 7 miles (11km) from the town of Izmit. This devastating earthquake was felt more than 310 miles (500km) away on the south coast of Crimea, in the Ukraine. It was caused by a sudden movement on the major fault line in the region—the North Anatolian Fault—with up to 16ft (5m) of right lateral strike-slip displacement on a segment of the fault extending for about 75 miles (120km) within 30 miles (50km) of Turkey's largest city, Istanbul, where 13 million people live. The main shaking lasted for 37 seconds, with maximum ground accelerations of up to 0.4g. With at least 17,118 people killed, and the estimated cost of damage put at $6.5 billion, the earthquake was both a major humanitarian tragedy and an economic disaster. The Izmit earthquake is just the latest in a sequence of earthquakes over the past 60 years that have been moving closer to Istanbul. Each earthquake was caused by the sudden breaking of segments of the North Anatolian Fault. The 1999 disaster filled in an earthquake gap on this fault that had been identified by geophysicists as likely to rupture.

A CITY IN RUINS

1 DESTRUCTION AND MASS DAMAGE
Many cities in the region, including Izmit, Adapazari, and Istanbul, suffered major damage from shaking, with the destruction of apartment buildings, mosques, and historic monuments.

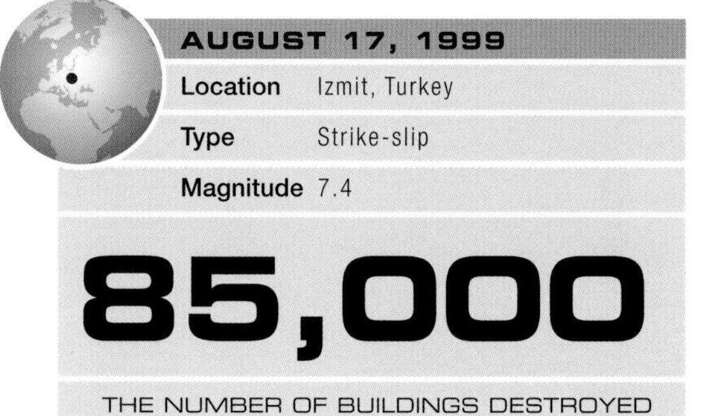

AUGUST 17, 1999

Location	Izmit, Turkey
Type	Strike-slip
Magnitude	7.4

85,000

THE NUMBER OF BUILDINGS DESTROYED

A HUMAN TRAGEDY
Relatives search for belongings and survivors among the debris of concrete apartment blocks, which crumbled all around them during the earthquake.

> "WHEN IT HIT, I FELT HELPLESS—LIKE BEING THROWN EVERY WHICH WAY IN A FRYING PAN."
>
> **ERCÜMENT DOĞUKANOĞLU**, NAVAL CAPTAIN

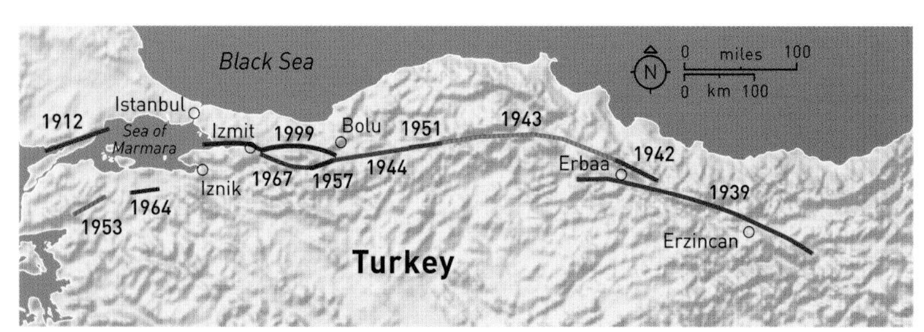

THE NORTH ANATOLIAN FAULT
The North Anatolian Fault is a major strike-slip fault line that runs nearly 930 miles (1,500km) through northern and western Turkey. It marks the southern edge of the Eurasian Plate, slipping today at about 1in (25mm) per year, with a right-lateral displacement.

2 OIL REFINERY FIRES
A tower at the Tupras oil refinery collapsed, starting a major fire that took several days to bring under control. More than 772,000 tons of oil were stored at the refinery.

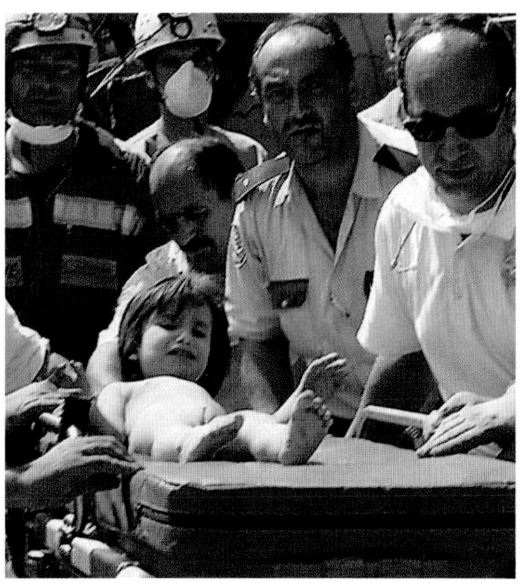

3 INJURIES AND SURVIVAL
Nearly 50,000 people were injured during the earthquake, trapped in their houses and apartments when the earthquake struck during the early hours of the morning.

4 HOMELESSNESS AND REBUILDING
Few buildings in Izmit were constructed to withstand earthquakes. Whole districts collapsed and about 500,000 people lost their homes during the 37-second tremor.

CHRISTCHURCH 2011

On February 22, 2011, at about midday, an earthquake measuring 6.3 on the Richter scale struck the major city of Christchurch on South Island, New Zealand, with its epicenter close to the port of Lyttleton. Ground accelerations of 2g rocked the central business district, destroying more than a third of the buildings. Extensive liquefaction (see p.232) in the eastern suburbs of Christchurch badly damaged residential properties. There were landslides in the Port Hills, to the south of Christchurch, and dislodged boulders rolled through the suburbs, leaving a trail of destruction. More than 100 people were killed, while 1,500 were injured.

This earthquake was part of the protracted sequence of aftershocks following an earthquake on September 4, 2010 about 12 miles (20km) to the west of Christchurch in Darfield. There was no loss of life after this magnitude-7.1 earthquake, although it damaged historic buildings and caused ground ruptures on the fault line.

BUILDINGS REDUCED TO RUBBLE
The Darfield earthquake in September 2010 was far less fatal than its February 2011 aftershock, but it destroyed several buildings in Christchurch.

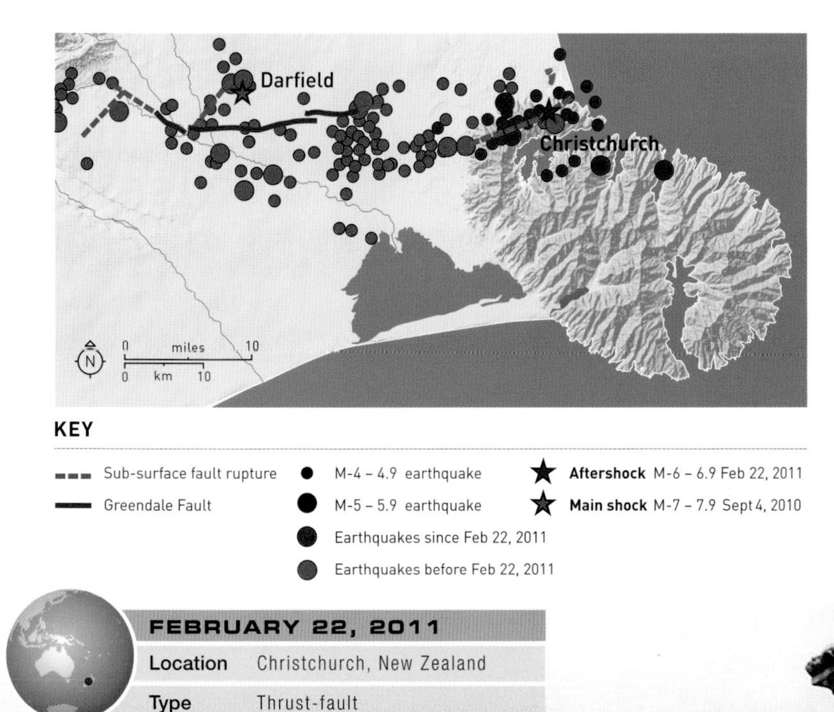

THE GREENDALE FAULT
The Christchurch and Darfield events, and the numerous aftershocks, all cluster along the newly recognized Greendale fault line, and various fault ruptures, running roughly east–west.

FEBRUARY 22, 2011	
Location	Christchurch, New Zealand
Type	Thrust-fault
Fatalities	Approximately 166

100,000

APPROXIMATE NUMBER OF BUILDINGS DESTROYED

A DEADLY AFTERSHOCK

Although the Christchurch earthquake was a relatively moderate earthquake having nearly 30 times less energy than the magnitude-7.1 Darfield earthquake in September 2011, it was far more destructive.

Seismologists believe a number of factors led to this. Firstly, fault movement during the February 22 earthquake was much closer to the city of Christchurch, and was less than a mile below the surface of its southern suburbs. In addition, the way the fault ruptured focused much of the energy of the earthquake in the central business district of Christchurch, which sits on relatively poorly consolidated sediments prone to ground shaking and liquefaction. Finally, the earthquake happened on a weekday during office hours, when the center of Christchurch was full of people. However, many of the deaths occurred in two downtown buildings that failed to withstand the shaking and collapsed almost completely, crushing the people inside.

> “THE PHONE LINES AND WATER ARE NOT WORKING. THE ROADS ARE GRIDLOCKED AS THE CITY IS BEING EVACUATED.”
>
> **CHRISTOPHER STENT**, EYEWITNESS

SEARCHING FOR SURVIVORS
The February 22 earthquake struck during office hours and trapped scores of people underneath mounds of rubble. Rescue workers had an enormous task searching for survivors amid the debris.

15,000
The approximate number of earthquakes that occur in New Zealand every year. Most of these are minor and only 10 or fewer of these would have a magnitude in excess of 5.0. The country is located on the border of the Australian and Pacific plates, which are colliding, making it prone to earthquakes.

THE CHRISTCHURCH CATHOLIC CATHEDRAL
The February 2011 earthquake caused irrevocable damage to several iconic cathedrals in Christchurch, including the Christchurch Catholic cathedral. Two bell towers in the front of the cathedral collapsed, bringing down most of its front façade.

SEARCHING THE RUBBLE
Rescue workers search for survivors near the Canterbury Television (CTV) building in the heart of Christchurch, New Zealand. The CTV building completely collapsed during the February 2011 earthquake and was the focus for intense efforts to find survivors. Ninety-four bodies were recovered from the ruins.

SEISMIC DESTRUCTION

Earthquakes wreak destruction on a vast scale. Not only are man-made structures damaged but the very ground itself is altered, liquifying soil and creating quake lakes. This does not just occur during the main shock. Numerous aftershocks can be sufficiently violent to cause damage during the vulnerable recovery period. In some cases, foreshocks can be a warning of what is to come.

FORESHOCKS AND AFTERSHOCKS

Earthquakes are not isolated events. This is because rocks do not just break along one fracture, but on numerous fractures or faults—and each break will trigger its own small earthquake. The main shock is triggered by the largest break, but there may be precursor breaks—triggering so-called foreshocks that have the potential to be large events themselves—and many subsequent breaks, triggering a long sequence of aftershocks that can continue for years. It was hoped that the detection of foreshocks could be used to predict an imminent main shock, but in practice foreshocks are only recognized after the main shock has occurred. The 2002 earthquake in Sumatra is now recognized as a foreshock, occurring two years before the massive magnitude 9+ Indian Ocean earthquake in 2004. Back in 2002, this was not known. The September 4, 2010 magnitude 7.3 Darfield earthquake, which was focused just outside Christchurch in New Zealand, had numerous magnitude 5+ aftershocks in the following months, culminating in a magnitude 6.3 earthquake on February 22, 2011, which destroyed much of the central business district of Christchurch and killed 166 people. The whole earthquake sequence is now recognized as the result of a propagating fault line, moving toward the east. Although an earthquake can relieve the accumulation of stress in one part of a fault line, it can mean that there is more likelihood of another quake in a different part of the fault.

SUMATRA AFTERSHOCK
A view of the destruction left by the massive earthquake that hit the island of Nias off the coast of Sumatra in March 2005. This was an aftershock following the huge earthquake in the region in December 2004.

12 COSTLIEST EARTHQUAKES IN LAST 100 YEARS

NO	YEAR	COUNTRY	COST OF DAMAGE/ REBUILD IN US$
1.	2011	Tohoku, Japan	more than 300 billion
2.	1995	Kobe, Japan	131.5 billion
3.	1994	Northridge, California, USA	20–40 billion
4.	2004	Niigata, Japan	28 billion
5.	1988	Armenia	14.2–20.5 billion
6.	1980	Irpinia, Italy	10–20 billion
7.	1999	Taiwan	9.2–14 billion
8.	1999	Izmit, Turkey	6.5–12 billion
9.	1994	Kuril Islands, Russia	11.7 billion
	1994	Hokkaido, Japan	11.7 billion
11.	2010	Christchurch, New Zealand	11 billion
12.	2004	Indian Ocean	7.5 billion

LIQUEFACTION PROCESS

In water-saturated ground, shaking during an earthquake can force the water out of pore spaces, turning the ground into a liquid. The water then erupts at the surface—known as sand boiling—bringing up with it mud and sandy material, and causing local flooding. Heavy buildings or cars may subside into the liquefied ground.

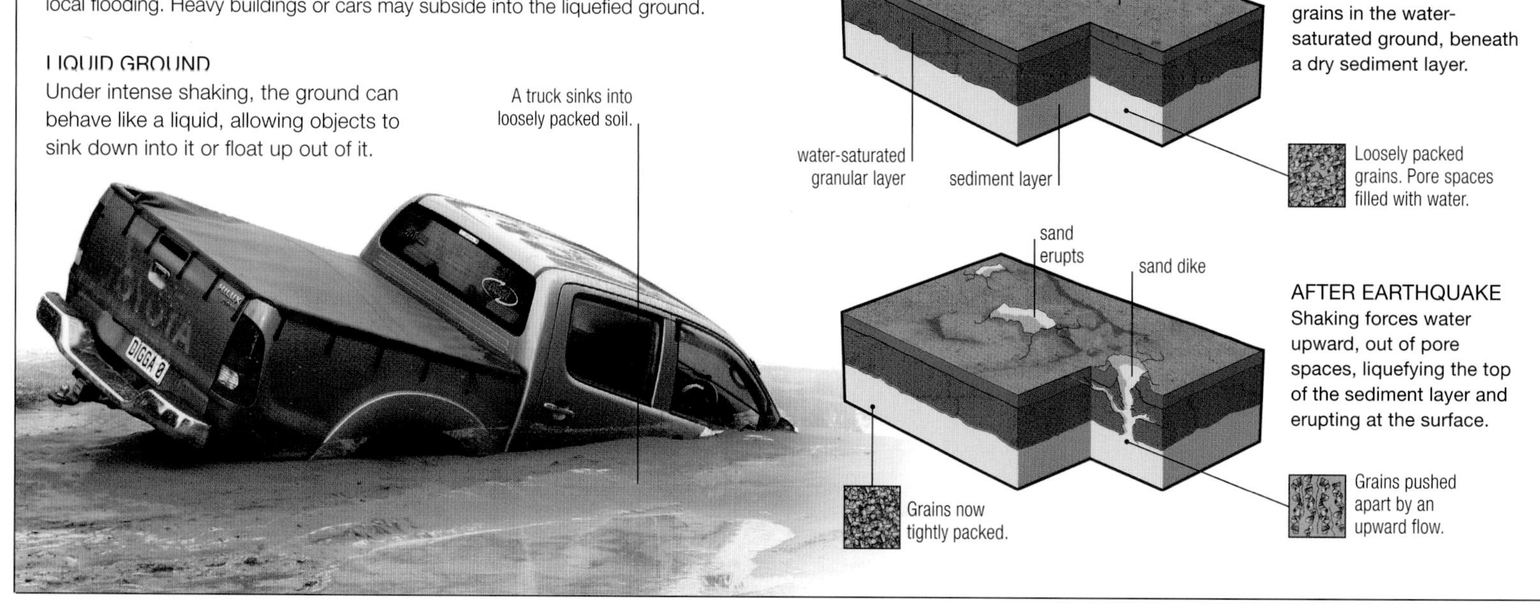

LIQUID GROUND
Under intense shaking, the ground can behave like a liquid, allowing objects to sink down into it or float up out of it.

BEFORE EARTHQUAKE
Water lies in pore spaces between loosely packed grains in the water-saturated ground, beneath a dry sediment layer.

AFTER EARTHQUAKE
Shaking forces water upward, out of pore spaces, liquefying the top of the sediment layer and erupting at the surface.

FUKUSHIMA NUCLEAR PLANT
Damage from the earthquake and tsunami in Japan in March 2011 caused fires to break out and reactors to overheat.

FIRES

Fires are a serious potential hazard in an earthquake-hit area. For example, when an earthquake severely shakes the ground beneath a city, electricity cables, gas pipes, and oil installations can be damaged or broken. Electrical arcing (a continuous discharge of electricity) or overheating often lead to explosions and fires that can rage through a city. Much of the loss of life during the San Francisco 1906 earthquake was due to uncontrollable fires that started this way. A new hazard has become apparent in the recent earthquake in Japan in 2011, where damage to electrical supplies during the quake and tsunami led to overheating and fires in the nuclear power plant at Fukushima. Partial core meltdown and hydrogen explosions led to leaks of radiation.

FLOODING

Landslides triggered by an earthquake may block rivers, causing widespread flooding. Debris shaken loose by the earthquake falls into rivers and blocks the flow. Water begins to pool behind the blockage, which is unstable, and can burst to flood the surrounding area. The 2008 Sichuan earthquake in China (see pp.220–21) triggered landslides that dammed several rivers. Massive amounts of water gathered behind the dams forming a number of quake lakes. The lakes threatened to flood cities already hit by the earthquake and they had to be carefully drained to prevent this. Man-made dams that have been damaged by an earthquake can also trigger flooding. If there is heavy rain after an earthquake, blocked drains may result in urban flooding.

DAMAGED CITY
This image shows the earthquake-hit city of Beichuan in China's southwestern province of Sichuan on May 27, 2008. Many buildings had been destroyed by the earthquake on May 12, 2008 and landslides in the mountainous region were triggered.

QUAKE LAKES
This image taken on June 10, 2008 shows how the area became flooded when rivers in the region were blocked by debris from landslides. A large body of water quickly grew behind these dams and started to flood the damaged city with a huge quake lake.

BAM 2003

On December 26, 2003, at about 5:30 in the morning, a magnitude-6.6 earthquake struck the ancient town of Bam, located on the old Silk Road in southeastern Iran. The ground shook with such acceleration that the historic citadel, the Arg-e Bam—which was one of the largest mud-brick constructions in the world, and at least 2,000 years old—was completely destroyed. The huge number of casualties, which included about 30,000 fatalities, with a similar amount injured, and about 100,000 people made homeless—was due to the roofs of many houses collapsing when most people were still in bed. Buildings as well as roads and infrastructure were damaged. Maximum intensities of shaking were felt in Bam and the nearby town of Baravat. The area around Bam is prone to earthquakes and the epicenter of this earthquake was close to a recognized fault line—the Bam fault, plus there were small local ruptures of the ground surface. However, detailed studies later showed that the main fault movements occurred at depth on a nearby unrecognized fault, involving right-lateral and reverse displacements of the rocks. The Bam earthquake was the result of the forces that have built up because of the northward movement of the Arabian Plate, which is moving at a rate of about 1.2in (3cm) per year, and colliding with the Eurasian Plate. This has resulted in the rise of the Zagros Mountains, as well as movement of blocks of crust along strike-slip faults in southeastern Iran.

DECEMBER 26, 2003

Location	Bam, southeastern Iran
Type	Reverse strike-slip fault
Magnitude	6.6

THE APPROXIMATE NUMBER OF PEOPLE KILLED

BUILDINGS DESTROYED
Poor quality construction and materials meant that most properties could not withstand the strong shaking generated by the quake and many houses collapsed.

ANCIENT CITY BEFORE
The citadel of Bam, an ancient fortified area in the north of the modern city of Bam, was constructed 2,000 years ago from mud. It had been well preserved until 2003, when the earthquake struck.

ANCIENT CITY AFTER
This photograph shows the devastation of the Bam citadel following the magnitude-6.6 earthquake in December 2003. Shaking destroyed the mud walls and arches, burying much of the area under debris.

BAM FLATTENED
This is an aerial view of the city of Bam after the 2003 earthquake. The ringed area shows the destroyed Arg-e Bam, or ancient citadel. Much of modern Bam was badly damaged as well.

"I LOST MY WIFE IN THIS EARTHQUAKE... I AM VERY SAD BUT I THINK BAM'S PEOPLE NEED INTERNATIONAL HELP."

ASGHAR GHASEMI, TEHRAN, IRAN

EARTHQUAKE ZONE

The city of Bam lies in an earthquake-prone area—there are major strike-slip faults in southeastern Iran, including the Bam Fault—which has lead to a long history of earthquakes, including four larger than magnitude-5.6 in the region to the northwest of Bam. However, prior to 2003 the city of Bam itself had no record of significant earthquake damage. Therefore, the massive destruction of the city and its surroundings caught the population unawares. The scale of the disaster prompted a change in international relations with Iran, and at least 44 countries, led by the United States, sent specialist teams to help with rescue and relief. The UN and International Red Cross launched an appeal for aid, raising tens of millions of dollars. The US sent five airlifts of supplies, including 1,146 tents, 4,448 kitchen sets, more than 10,000 blankets, and 75 tons of medical supplies.

TEMPORARY SHELTER
Four months after the earthquake, Iranian children were photographed standing in front of their temporary tent shelter. It was estimated that about three-quarters of all buildings in the area were destroyed, leaving 100,000 people homeless.

QUAKE-TRIGGERED LANDSLIDES

The violent tremors during even a moderate earthquake are enough to shake the ground loose and trigger huge landslides, especially where the ground is steep. In fact, geologists have long considered the scars of ancient landslides in earthquake-prone regions as tell-tale signs of past earthquakes.

UNSTABLE GROUND

Earth's surface is not flat, but has been molded by tectonic forces and sculpted by rivers and glaciers into valleys, hills, and mountains that have slopes. Their surface soil and rocks are held in a delicate balance of forces—they cling to the bedrock because of their weight, while gravity causes them to fall down. Because movements at faults are linked to earthquakes, the most mountainous regions—with the steepest surface slopes—are subjected to the most frequent and intense earthquakes. In the aftermath of a big earthquake, a landscape may become marked with the scars of landslides, particularly on the steepest slopes. These mark the regions where the shaking of the ground was most intense. The most obvious results are rockfalls, where boulders, already weakened by erosion and sitting loosely on the bedrock, tumble downhill. In addition, shaking of water-saturated soil can squeeze the water out, resulting in a very weak, liquid-like substratum—a process called liquefaction. Vast slumps of soil and rock slip and slide on this weakened layer. Finally, heavy rain after an earthquake may wash much of the loose ground downhill, sometimes in giant mudflows.

ROCK HAZARDS

The tops and bottoms of cliffs or steep ground are very unsafe places during an earthquake. Huge chunks of rock can detach themselves, gathering speed as they fall and crash into buildings and people below. A tumbling boulder during the 2011 earthquake in Christchurch, New Zealand, cut a path of destruction through gardens and houses in a suburb at the foot of some volcanic hills. In rugged and earthquake-prone terrain, geologists use rockfalls as indicators of past earthquakes in the region.

FALLING DANGER
Residents of Wajima, Japan, can be seen carrying their belongings past a huge boulder that fell from a cliff during an earthquake in 2007.

MASSIVE BLOCKADE
A 2010 landslide in northern Taiwan devastated a 980-ft (300-m) stretch of National Highway 3, depositing thousands of tons of rock.

BEICHUAN, CHINA
Evacuees carry their belongings near a landslide site in the Beichuan county of Sichuan Province, China, following an earthquake in May 2008.

MUDSLIDE IN GANSU
This satellite image shows a town in the Gansu province of China that was submerged after a rain-triggered mudslide flooded the town, burying many alive.

THE GANSU MUDSLIDE

In August 2010, a mudslide occurred in Zhouqu county, in the Gansu Province of China. It was caused by heavy rains falling on steep, weathered ground during the catastrophic 2010 floods in China. Although it was not an earthquake-triggered mudslide, it was typical of mudslides that can happen if heavy rains occur soon after an intense earthquake, washing over ground that has been shaken and weakened. During the 2010 floods, torrential rains in the mountainous region had created sludge-like conditions. Large amounts of mud and rock slid off the steep slopes resulting in a mudslide that buried one village completely and killed about 1,500 people.

LANDSLIDES

A natural geological process, landslides play an important role in sculpting the landscape. They are a form of what geologists call "slope failure." An earthquake acts as a trigger, starting a landslide in already unstable ground. Thus, all basic types of slope failure, observed by geologists and illustrated in these diagrams, can occur during an earthquake.

ROCKFALL
Loose or weakened rocks on steep ground or cliff faces will ultimately fall down, accumulating as a pile of scree.

EARTHFLOW
When the soil rests on a smooth rock surface, or is saturated by water, earthquakes easily trigger earthflows.

MUDFLOW
Heavy rain running off steep hillsides, especially after an earthquake, can create mudflows that move downhill.

LIVING WITH EARTHQUAKES

Most of the world's population live in cities located in earthquake-prone areas. Engineers have used their ingenuity to design new buildings and reinforce existing buildings to withstand the huge ground movements during an earthquake. There are also warning systems in place to pick up unusual seismic activity in these zones.

EARTHQUAKE RISK

One of the main long-term goals of scientists is to forecast the next "big one." Current understanding of the earthquake phenomenon suggests that this is many years away, if it is possible, because earthquakes are an uncertain process. Seismologists have tried unsuccessfully to study animal behavior, natural gas leaks, or sequences of small precursor earthquakes, which might indicate incipient ground motion before the main shock. More promising have been detailed studies of the past history of earthquakes and fault movement in a particular region, combined with measurements of the steady growth of rock strain. These provide insight into the likelihood of a large earthquake in the next few decades or centuries. To date, most attention has been focused on the major fault lines. However, many earthquakes occur on previously unrecognized faults, making prediction more difficult.

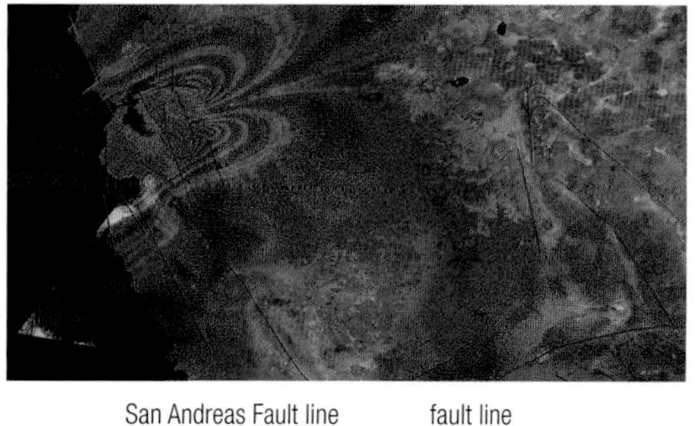

SAN FRANCISCO EARTHQUAKE RISK
This satellite image of California shows earthquake faults (red lines) and the San Andreas Fault (yellow line). The bands of color are synthetic aperture radar patterns, which indicate seismic deformations resulting from a model earthquake on the San Andreas Fault. The model has estimated that there is a 25 percent chance of an earthquake measuring magnitude-7 or above in the next 20 years on the San Andreas Fault.

JAPAN'S EARLY EARTHQUAKE WARNING SYSTEM

The Japanese Meteorological Agency has placed a network of sensitive instruments throughout Japan, designed to pick up the characteristic ground motions that occur in the first few seconds of a large earthquake. These motions trigger the Earthquake Early Warning System, linked to civil defense organizations and television and radio stations, which can then broadcast a coordinated plan of action.

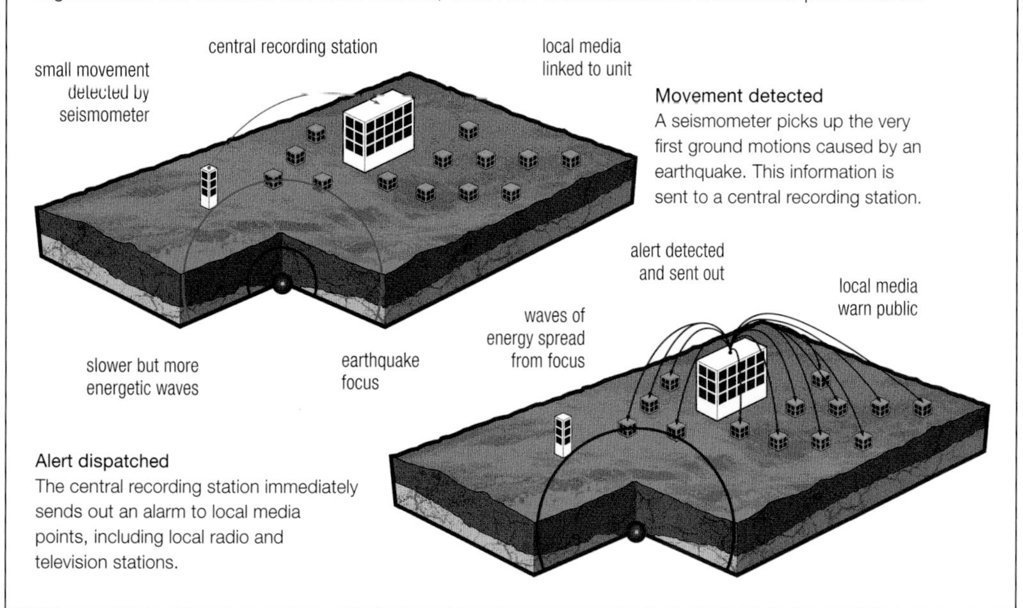

Movement detected
A seismometer picks up the very first ground motions caused by an earthquake. This information is sent to a central recording station.

Alert dispatched
The central recording station immediately sends out an alarm to local media points, including local radio and television stations.

TAIPEI 101
One of the world's tallest buildings at 1670ft (509m), the Taipai 101 relies on its mass dampers to counter high winds or earthquakes.

EARTHQUAKE-PROOF BUILDINGS

In reality there is no such thing as an earthquake-proof building, though simple strengthening will greatly reduce the damage from a moderate earthquake. Engineers have studied the effect of the complex ground movements during a big earthquake on buildings, bridges, and other major structures, in order to find ways to reduce the risk of collapse, or to allow the building to fall in such a way that the loss of life might be minimized by the provision of internal refuge pockets. One solution is to increase the strength of the internal structure and foundations, reinforcing them with steel or concrete bracing. The shaking of a building can also be reduced, either by isolating it with so-called base isolators, made of a soft metal such as lead, in the foundations, or by damping the vibration with heavy counterweights inside the building. Since the late 1960s, many countries have introduced design codes that specify the maximum ground movements a building should be able to withstand. These design codes have been updated over time, following studies of the effects of more recent earthquakes.

MASS DAMPER
A steel pendulum weighing 734 tons and costing $4 million acts as a mass damper. It is suspended from the 92nd to the 88th floor.

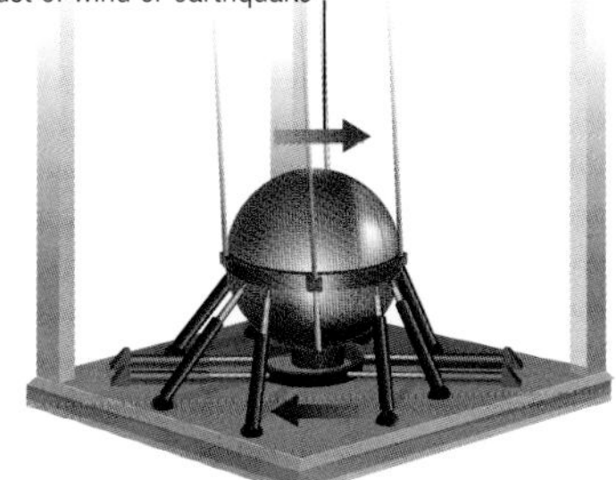

HOW THE MASS DAMPER WORKS
The mass damper is tuned to swing in a way that counteracts any swaying of the building caused by wind or earthquake shaking; in effect it acts as a gigantic shock absorber.

SEISMIC RETROFITTING

A large number of man-made structures in earthquake-prone regions were built when earthquake design codes were nonexistent or less stringent. Removal of obviously weak parts such as towers, protruberances, or parapets make a building safer on the outside. Adding cross braces or new structural walls, or damping and isolation systems, will prevent internal collapse when a structure experiences the maximum ground accelerations specified in the design code. Experience in recent large earthquakes in California, New Zealand, Taiwan, and Japan, show that retrofitting is largely successful.

EARTHQUAKE REINFORCEMENT
This building in Berkeley, California, near the Hayward Fault, has been encased in a cage of cross-braced steel struts to prevent it collapsing.

5

RESTLESS OCEANS

<< Wave power
Surf crashes over a coral reef platform in the Pacific Ocean.

HOW AN OCEAN ORIGINATES

From time to time throughout Earth's history, new oceans have formed through landmasses splitting up by a process called continental rifting. Other oceans have gradually shrunk and disappeared as a result of plate movements bringing continents together.

CONTINENTAL RIFTING

A continental rift is a region where the continental crust (the upper rocky layers of a continent) becomes thin, with sections collapsing downward to form a valley. This process is typically caused by heat flowing up from the interior at a hotspot (see pp.32–33) and stretching the crust. A continental rift is usually affected by volcanic activity as magma wells up and erupts at the surface. If rifting develops into pronounced spreading of the crust, it becomes a divergent plate boundary (see pp.28–29) and leads to the formation of an ocean, as the sea floods into the rifting region. The rift then turns into a mid-ocean ridge.

Rift
Split develops in continental crust

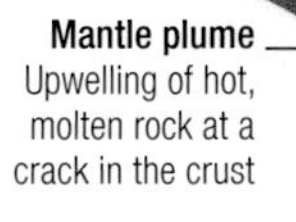

Mantle plume
Upwelling of hot, molten rock at a crack in the crust

1 SPLITS IN A CONTINENT
Upward flow of heat from Earth's interior can cause a region of continental crust to start rifting or splitting apart. It weakens and stretches the crust, causing it to thin. A three-branched split may develop in the crust.

Continental crust
Continues to thin, stretch out, and develop fissures

Active rift arm
Where thinning, widening, and volcanic activity continues

2 SPLITS WIDEN
Typically, two of the three fracture branches, called rift arms, begin to widen. Magma (molten rock) wells up and erupts at the surface along these active arms through volcanoes and fissures. The third arm, called a failed arm, develops no further.

THE OPENING OF THE ATLANTIC

A fairly recent example of a new ocean that developed through continental rifting is the Atlantic Ocean, which formed in three main stages. The first stage, about 180 million years ago, separated what is now North America from Africa and involved the opening of much of the central and northern parts of the Atlantic. Next, Africa and South America rifted apart, from about 130 million years ago, to create the South Atlantic. This is thought to have left a failed rift arm, the Benue Trough, in Nigeria. Beginning about 60 million years ago, the rift extended between what are now parts of northern Europe, on one side, and Greenland on the other, to form the northern Atlantic.

ISLE OF STAFFA
This small island off the west coast of Scotland, made of basaltic lava, is a relic of the rifting process that led to the opening of the most northerly part of the Atlantic. The basalt columns enclose a sea cave, Fingal's Cave.

THE SPLITTING OF AFRICA

Evidence suggests that the processes leading to the formation of a new ocean are already well advanced in northeastern Africa. At a triple junction centered on the Afar Triangle in Ethiopia, one active rift arm has already opened up the Red Sea, and a second arm—the East African Rift that extends southward from Ethiopia—is showing distinct signs of further thinning and spreading in its northernmost section. These signs include volcanic and earthquake activity, fissures appearing in the ground, and the spreading or extension of existing fissures. Researchers predict that as a result of the rifting in this region, a large part of the Horn of Africa—including Somalia, eastern Ethiopia, a part of Djibouti, and also possibly part of Kenya—will eventually separate off as a new island. A new branch of the Indian Ocean will form between this island and the rest of Africa. Some think this new ocean will have formed within about one million years.

DABBAHU FISSURE
In 2005, this 1,600-ft- (500-m-) long volcanic fissure opened up in the Afar region of Ethiopia, where the African continent is actively rifting apart. When it opened up, the fissure erupted spewing some light-colored ash, pumice, and large rocks.

3 FORMATION OF NEW OCEAN FLOOR

The active rift arms develop into a mid-ocean spreading ridge, where new ocean floor is continuously created and pushed away from the ridge. This process continues to widen the ocean and push apart the two blocks of continental crust.

CLOSING OF THE TETHYS OCEAN

As well as opening up and widening, oceans can also shrink and eventually disappear, or "close," as a result of plate movements. Between about 250 million and 15 million years ago, an ancient sea called the Tethys Ocean existed over an area extending roughly from present-day northeastern Africa to Indonesia. It gradually shrank, as plate movements pushed Africa and India northward. Most of its seafloor was pushed under Europe and Asia, though remnants still exist in locations such as the Black Sea. Other parts were lifted up and are now dry land, often containing interesting marine fossils.

Fossil in the Valley of the Whales
This fossil of a 49-ft- (15-m-) long whale that once inhabited the waters of the Tethys Ocean is one of many discovered in a small region of northeastern Egypt known as the Valley of the Whales.

THE OCEAN FLOOR

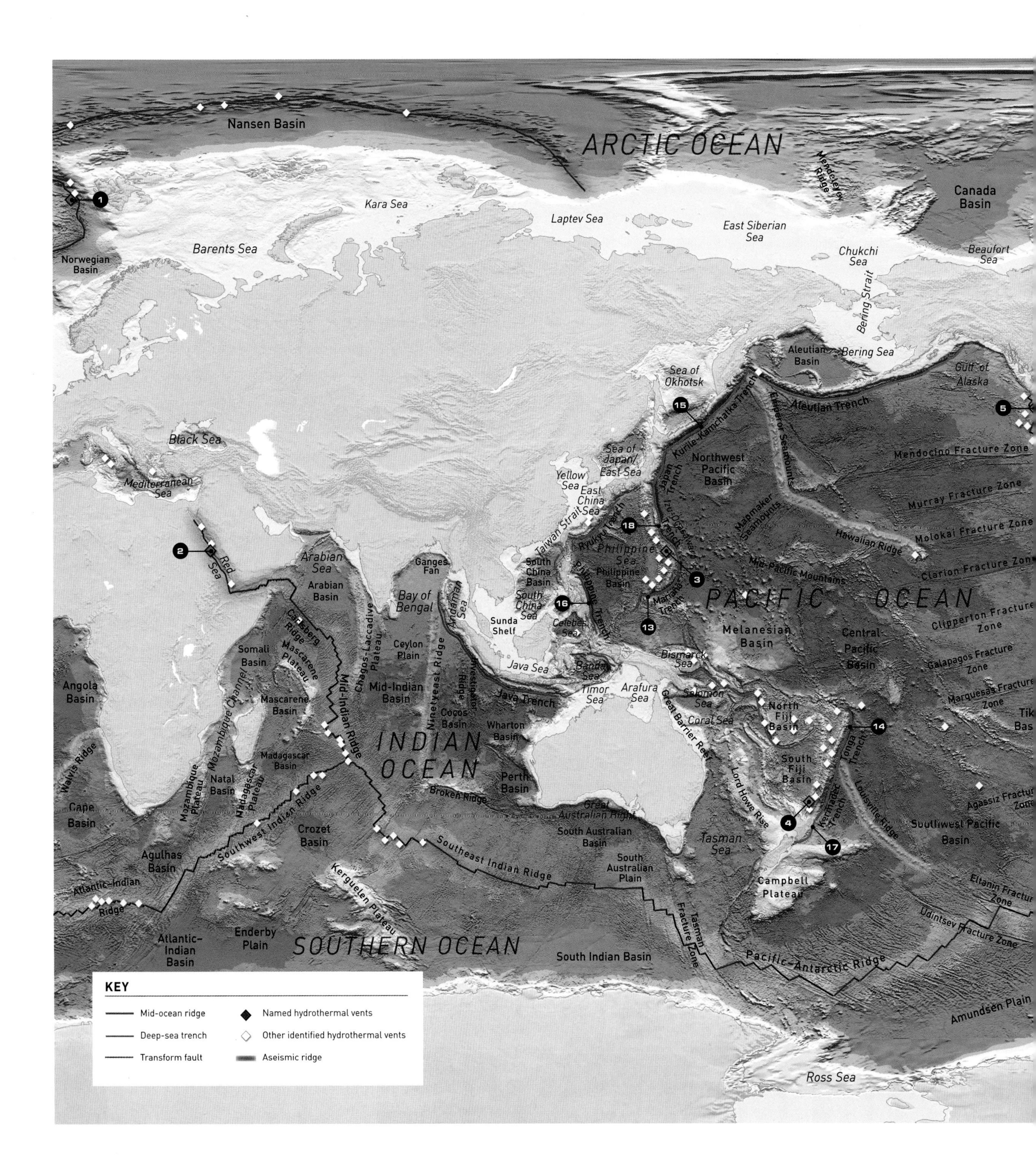

The ocean floor contains many features related to tectonic plates. Mid-ocean ridges are plate boundaries where new plate is made, and deep-sea trenches are where plate edges subduct or descend beneath neighboring plates. Aseismic ridges are lines of seamounts (see pp.258–59) caused by plate movements over hotspots (see pp.32–33).

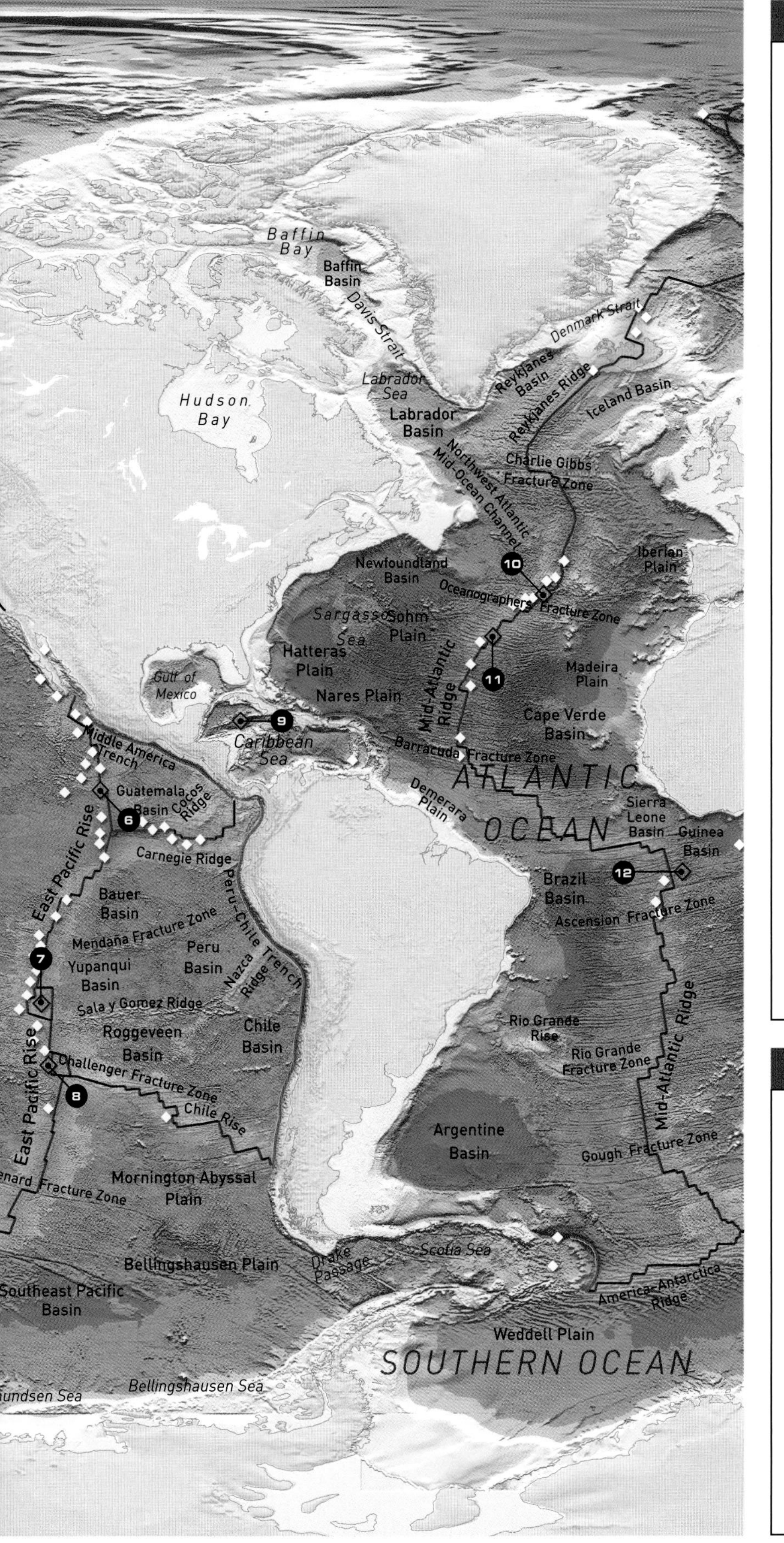

12 NAMED HYDROTHERMAL VENTS

Hydrothermal vents, which are plumes of heated water issuing from seafloor fissures, occur in some volcanically active regions of the seafloor, nearly all close to plate boundaries. Listed here are some named vents, although there are many more named and unnamed ones.

1 LOKI'S CASTLE

Location Mid-Atlantic Ridge
Depth 7,700ft (2,352m)
Features Black smokers

2 DISCOVERY DEEP

Location Red Sea
Depth 7,220ft (2,200m)
Features Warm brine pools

3 CHAMPAGNE VENT

Location NW Eifuku volcano, northern Mariana Arc, W Pacific
Depth 5,250ft (1,600m)
Features White smokers

4 BROTHERS

Location Kermadec Arc, SW Pacific
Depth 5,085ft (1,550m)
Features Black smokers

5 MAGIC MOUNTAIN

Location Explorer Ridge, NE Pacific
Depth 6,070ft (1,850m)
Features Black smokers

6 MEDUSA

Location E Pacific Rise
Depth 8,500ft (2,580m)
Features Black smokers

7 ANIMAL FARM

Location E Pacific Rise
Depth 8,700ft (2,660m)
Features Mussel beds

8 FRED'S FORTRESS

Location E Pacific Rise
Depth 7,650ft (2,330m)
Features Black smokers

9 PICCARD

Location Mid-Cayman Rise, Caribbean
Depth 16,400ft (5,000m)
Features Deepest known vent

10 LUCKY STRIKE

Location Mid-Atlantic Ridge, nr Azores
Depth 5,600ft (1,726m)
Features Black smokers

11 LOST CITY

Location Mid-Atlantic Ridge
Depth 2,460ft (750m)
Features Methane and hydrogen vents

12 MEPHISTO

Location Mid Atlantic near Ascension Island
Depth 9,997ft (3,047m)
Features Black smokers

THE OCEAN'S DEEPEST TRENCHES

13 MARIANA

Location W Pacific
Maximum depth 6.78 miles (10.91km)
Length 1,480 miles (2,550km)

14 TONGA

Location SW Pacific
Maximum depth 6.76 miles (10.88km)
Length 854 miles (1,375km)

15 KURIL-KAMCHATKA

Location NW Pacific
Maximum depth 6.55 miles (10.54km)
Length 1,240 miles (2,000km)

16 PHILIPPINE

Location W Pacific
Maximum depth 6.55 miles (10.54km)
Length 820 miles (1,320km)

17 KERMADEC

Location SW Pacific
Maximum depth 6.24 miles (10.05km)
Length 745 miles (1,200km)

18 IZU-OGASAWARA

Location W Pacific
Maximum depth 6.08 miles (9.78km)
Length 497 miles (800km)

SEAFLOOR TECTONICS

The floors of the oceans display many features relating to tectonic plates—the huge slabs of Earth's outer shell that gradually move around on its surface. Two of the most important are mid-ocean ridges, where new plate material is created, and deep-sea trenches, where it is destroyed.

MID-OCEAN RIDGES

Linked series of divergent plate boundaries in the oceans (see pp.244–45) are called mid-ocean ridges. These ridges rise above the general level of the seafloor like mountain ranges with gentle slopes, and are places where new oceanic lithosphere (oceanic crust) is created from magma welling up from Earth's interior. As new seafloor is created, it moves away from the ridge in a process called seafloor spreading. Mid-ocean ridges fall into two main types—fast- and slow-spreading—depending on the rate at which they create new lithosphere. Many moderate and small earthquakes occur at mid-ocean ridges, which are also the main site for hydrothermal vents, or subsea smokers, (see pp.250–51).

narrow axial cleft

oceanic lithosphere

upwelling mantle

FAST-SPREADING RIDGE
Common in the Pacific, a fast-spreading ridge has a relatively smooth surface and a narrow cleft running down the middle. The spreading rate is 2–6in (6–16cm) per year.

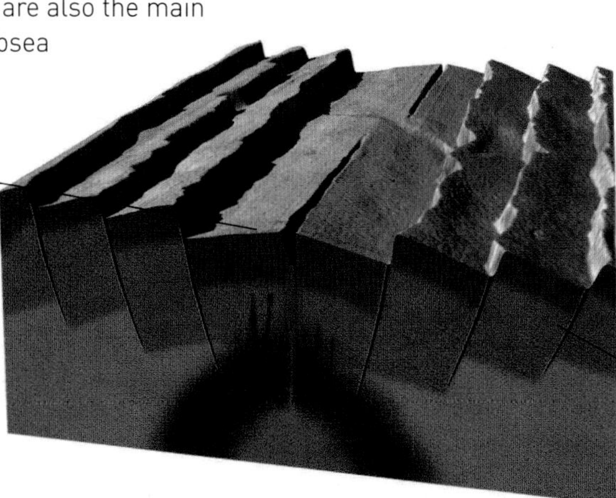

SLOW-SPREADING RIDGE
A common feature in oceans except the Pacific, this type of ridge has a wide central rift valley, and a rugged terrain. The spreading rate is typically 0.7–2in (2–5cm) per year.

SEAFLOOR SPREADING

The basic consequence of plate movements, seafloor spreading is extremely important for all types of tectonic activity, such as the formation of mountains and earthquakes. In the 1960s, American scientist Harry Hess first proposed the idea of seafloor spreading. It was confirmed by measuring the magnetic patterns in the ocean crust. This discovery was crucial to the development of plate tectonic theory.

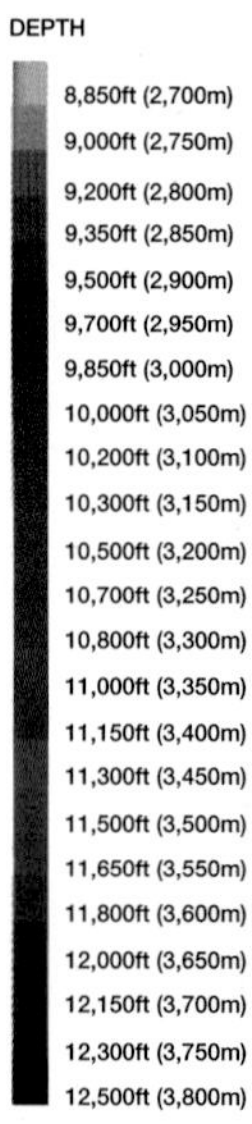

PACIFIC SEAFLOOR
This false-colored 3-D sonar map shows part of the East Pacific Rise, a major mid-ocean ridge in the Pacific. The seafloor on either side of the ridge is gradually drawing away from it.

DEEP-SEA TRENCHES

Oceanic or deep-sea trenches are asymmetric V-shaped valleys in the ocean floor. They occur where the edge of one plate, consisting of oceanic lithosphere, pushes down or subducts beneath a neighboring plate, which may also be oceanic lithosphere (at an ocean–ocean convergent boundary), or alternatively, continental lithosphere (at an ocean–continent boundary). They are places where plates are slowly destroyed. These trenches are typically 3 to 6 miles (5 to 10km) deep. The deepest spot (which is also the lowest point on Earth's surface) occurs in the Challenger Deep in the Mariana Trench. This has a depth of 6.77 miles (10.91km). It is pitch black at the bottom of trenches and the water is just above freezing. Nevertheless, some unusual animals live there. These trenches are also commonly affected by earthquakes, which occur when the strain that builds up as one plate grinds beneath another is suddenly released.

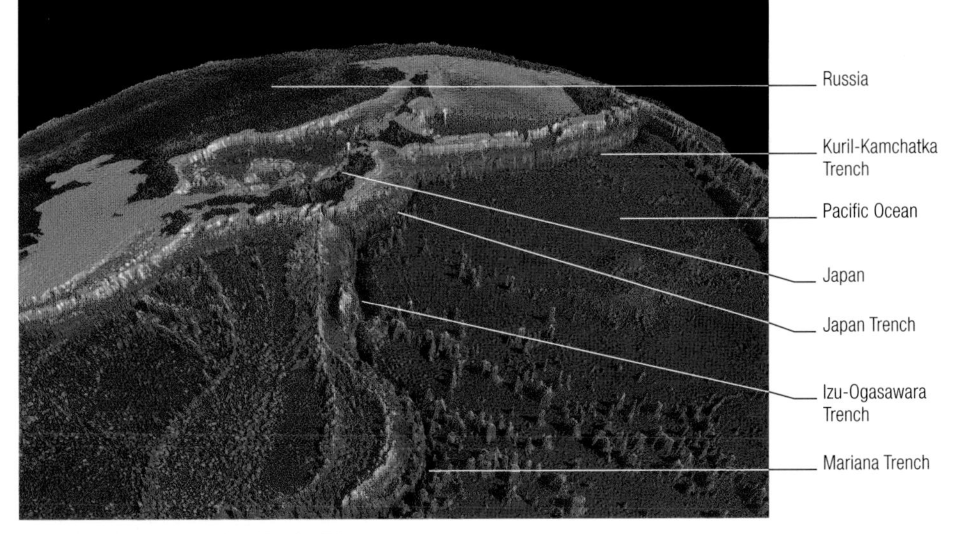

NORTHWEST PACIFIC TRENCHES
The floor of the northwestern Pacific features an array of deep-sea trenches. These mark places where the edge of the Pacific Plate is pushed below various neighboring plates.

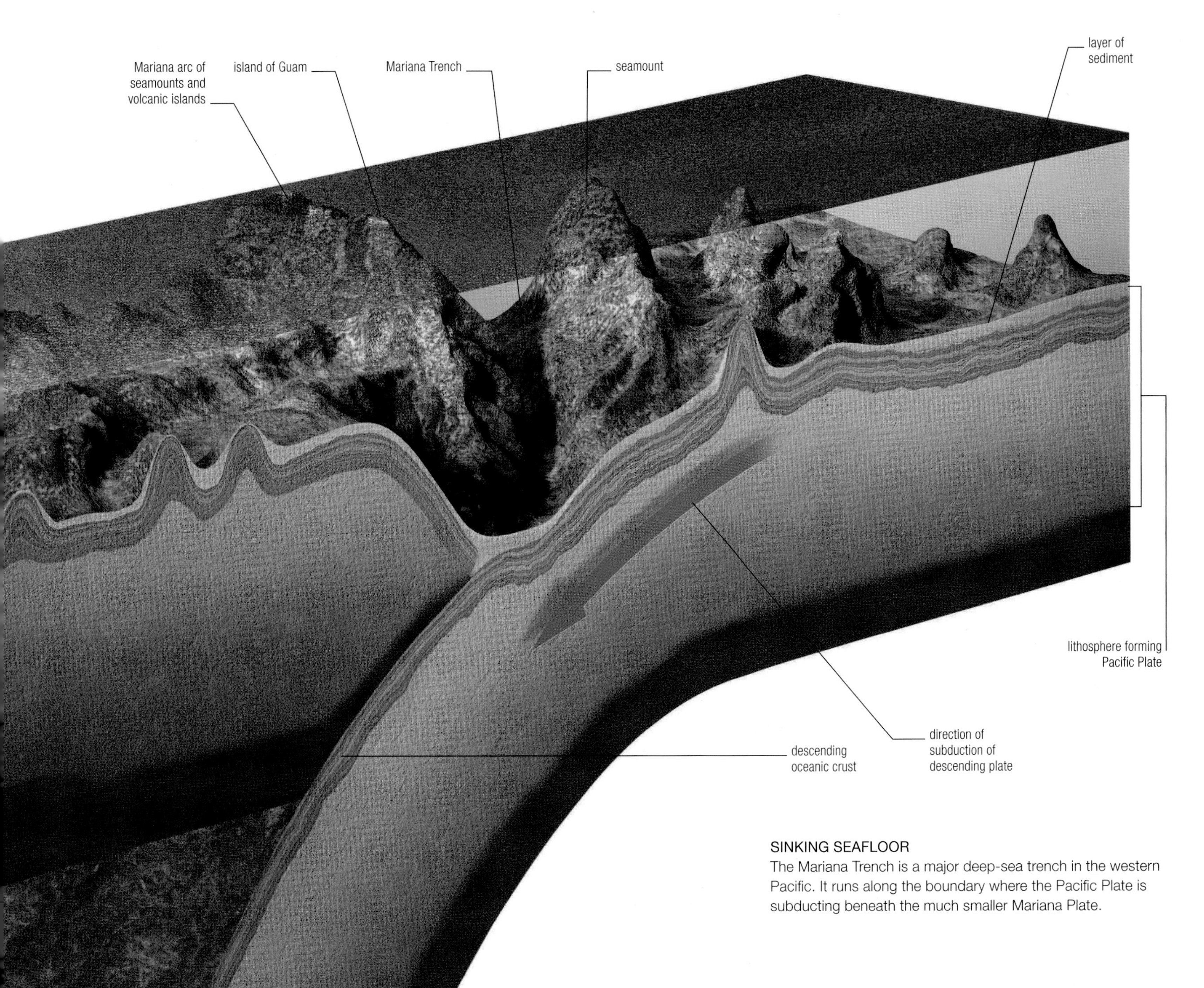

SINKING SEAFLOOR
The Mariana Trench is a major deep-sea trench in the western Pacific. It runs along the boundary where the Pacific Plate is subducting beneath the much smaller Mariana Plate.

CRACK BETWEEN PLATES
The crest of the Mid-Atlantic Ridge—a plate boundary—runs through Iceland, where it is covered by a lake in one section. A diver hovers over the ridge, which is marked by a crack in the lake bed. On the left side of the crack is the North American Plate and on the right, the Eurasian Plate.

SUBSEA SMOKERS

At numerous spots in the oceans, plumes of extremely hot, mineral-rich water gush out of fissures and odd-looking chimneys sticking out of the seafloor. These submarine counterparts of geysers are commonly called black (or white) "smokers" on account of the dark or light minerals that precipitate out of them.

VENT FORMATION AND LOCATION

Subsea smokers, or hydrothermal vents, are created when seawater seeps down through cracks in the seafloor, becomes heated by hot rocks or magma, and then escapes back into the cold water above. As the plume of water cools, minerals precipitate from the water to form solid particles, thus giving the appearance of "smoke." The first vents to be discovered, in 1977, were black smokers at the bottom of the Pacific, near the Galápagos Islands. Since then, hundreds more have been found, mostly around mid-ocean ridges or other seafloor regions where magma lies fairly close to the surface. Many have been given colorful names such as Magic Mountain and Loki's Castle. The average depth of these vents is about 6,900ft (2,100m) beneath the sea surface.

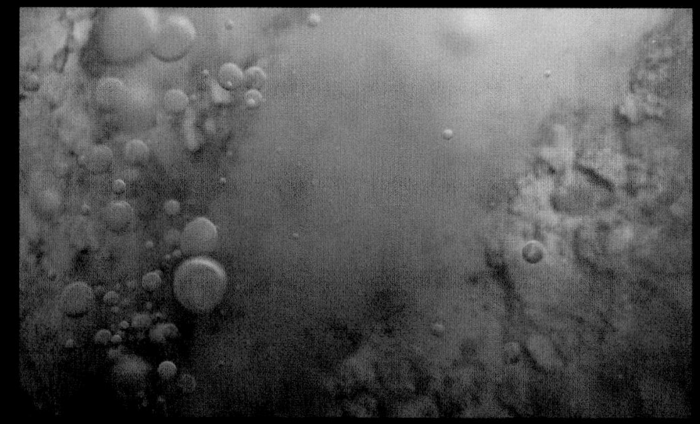

WHITE SMOKER
Bubbles of liquid carbon dioxide erupt from Champagne Vent, a white smoker. Champagne Vent is part of a submarine volcano called NW Eifuku, which lies within a volcanic arc south of Japan. Water escaping from white smokers is cooler than at black smokers.

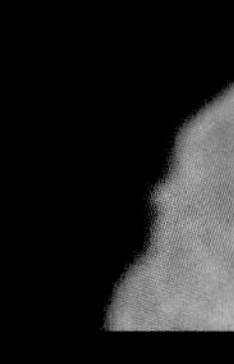

HOW A SUBSEA SMOKER WORKS
Seawater penetrating a few miles into the seafloor is heated by hot rocks or magma, dissolving minerals from surrounding rocks. As the hot water escapes back into the sea, it cools and the dissolved minerals precipitate out.

Black plume
Occurs where dark-colored minerals, such as iron sulphide, precipitate

White plume
Occurs where light-colored minerals, such as anhydrite, a form of calcium sulphate, precipitate

Heat from magma or hot rocks
Warms the percolating seawater to 660–750°F (350–400°C); high pressure stops the water from boiling, instead making it "superheated"

Mineral deposits
These build up from various precipitated minerals and include chimney-shaped cones that can be tens of feet high

Cold seawater
Has a temperature of around 36°F (2°C) and seeps through cracks

Conduit
Carries superheated water to surface

Superheated water
Dissolves minerals as it moves through cracks in rocks

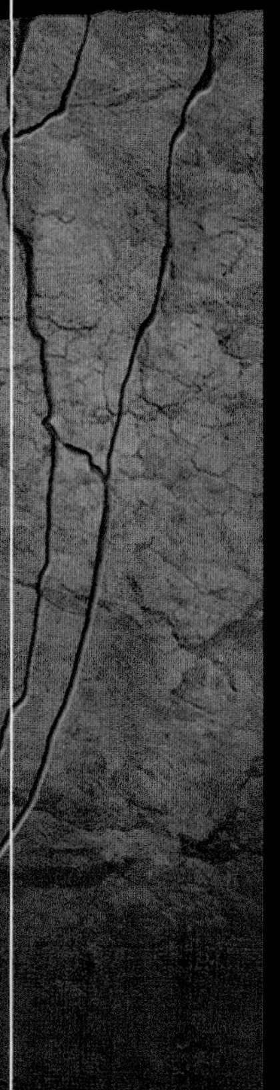

BLACK SMOKER
The water coming out of this black smoker, located near a mid-oceanic ridge, is scaldingly hot, around 750°F (400°C). It has about the same acidity as vinegar due to the chemicals it contains.

VENT COMMUNITIES

The heat and dissolved minerals around vents support some extraordinary animal communities, often including shrimp, crabs, mussels, and tube worms. Remarkably, in the absence of plants (which need light) as a food source, vent communities thrive on specialized microbes that obtain energy from oxidizing or reducing chemicals in the vent fluids.

TUBE WORMS
A bed of tube worms cover the base of a black smoker in the Pacific. These worms can be up to 8ft (2.4m) long and as thick as a human arm.

WORLD'S DEEPEST-KNOWN VENTS

CAYMAN VENT
Dark, superheated water gushes out of the top of a two-story-high smoker chimney at one of the deep vents in the Cayman Trough.

AUTOSUB6000

In April 2010, a British expedition filmed the world's deepest-known hydrothermal vents in the Cayman Trough, a 3-mile- (5-km-) deep depression in the Caribbean Sea. The expedition revealed the vents to be 2,625ft (800m) deeper than any previously known. The discovery followed up on work by scientists from the Woods Hole Oceanographic Institution in the United States, who had already detected the vents by sensing chemicals typical of vent plumes in the overlying seawater. The British, using a robot submarine called Autosub6000 equipped with sonar and chemical sensors, further pinpointed the vent locations. They then launched a deep-towed camera called HyBIS to film the vents, revealing slender mineral chimneys on the sea floor spewing out water hot enough to melt lead. Scientists plan to return to the vents in future to study the chemicals in the vent plumes and the range of marine life there.

SUBMARINE VOLCANOES

About 80 percent of the world's volcanic activity occurs in the oceans. New volcanoes are continually formed near mid-ocean ridges, over hotspots, and within volcanic island arcs. These volcanoes grow upward, a few of them eventually reaching the surface to produce some highly dramatic eruptions.

UNDERSEA ERUPTIONS

Volcanoes erupting on the deep sea floor have a difficult task to grow because of the enormous water pressure bearing down on them. Nevertheless, thousands do so by extruding vast amounts of lava at a high rate. Much of this underwater volcanic activity is, of course, invisible, though not undetectable, from the sea surface. Using submersibles, scientists have observed a few subsea volcanoes erupting at depth, including a volcano called NW Rota-1, whose main active vent lies 1,640ft (500m) underwater in the western Pacific.

When the tip of a submarine volcano reaches the sea surface, it can produce some spectacular eruptions. Such events occur several times every century. The eruption of Iriomote-jima in 1924 was one of the largest-ever eruptions in Japan; there were eruptions in 1939 and 1974 of a submarine volcano called Kick'em Jenny off Grenada in the Caribbean; and a submarine volcano produced the island of Surtsey in 1963 (see pp.256–57).

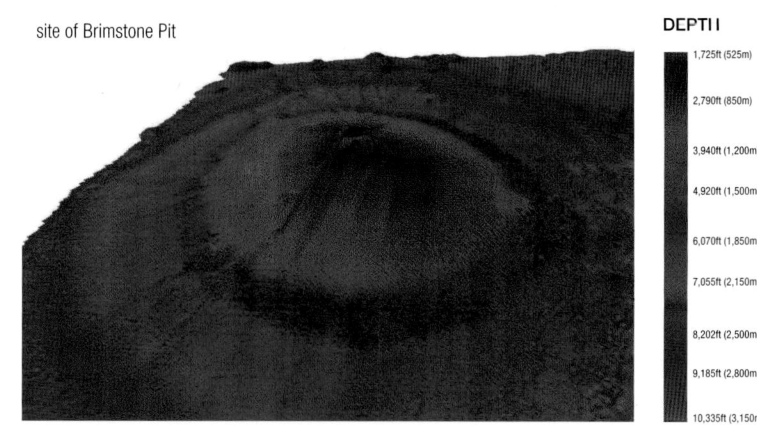

NW ROTA-1
This 3-D sonar map shows the submarine volcano NW Rota-1 in the Mariana Island arc of the western Pacific, along with its erupting vent—the Brimstone Pit.

VOLCANO AT HUNGA HA'APAI
In March 2009, a subsea volcano erupted near the island of Hunga Ha'apai, Tonga, in the southwest Pacific. The explosion of ash visible here reaches about 330ft (100m) above the sea's surface.

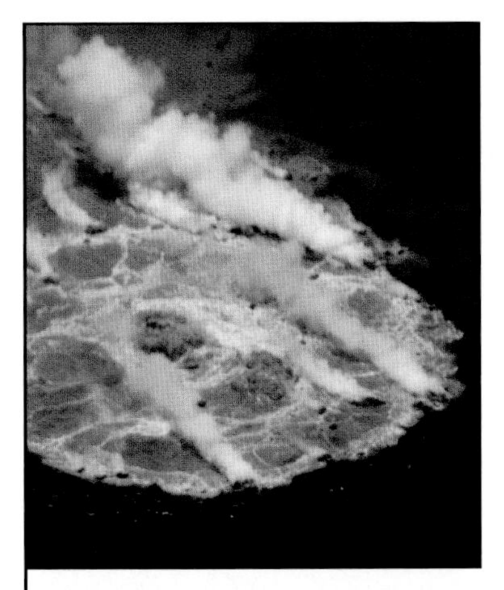

FUKUTOKU-OKANOBA

Location Near Minami Iwojima Island, Bonin Islands of Japan, western Pacific

Summit depth 0–46ft (0–14m)

This volcano has erupted at or near the sea surface many times in the past, most recently in 2010. Eruptions discolor the nearby seawater, produce steam and ash plumes, and sometimes, as in 2005 (above), create rafts of hot pumice.

LO'IHI

Location Southeast of Hawaii (Big Island), central Pacific

Summit depth 3,179ft (969m)

Lo'ihi lies on the flank of Mauna Loa, a huge shield volcano that forms much of the Big Island of Hawaii. Extensively studied by submersibles, this submarine volcano has been growing for about 400,000 years over the hotspot that formed all the Hawaiian Islands.

MAHANGETANG

Location Near Siau Island, off North Sulawesi, Indonesia

Summit depth 26ft (8m)

Located in a region of Indonesia that is highly active volcanically, Mahangetang rises some 1,300ft (400m) from the sea floor. Its surface is dotted with several volcanic vents or fumaroles that continuously release gases as a mass of tiny bubbles.

WEST MATA

Location Lau Basin, near Tonga, southwestern Pacific

Summit depth 3,852ft (1,174m)

Located not far from the Tonga Trench, this volcano was first discovered in November 2008, while it was vigorously erupting. A return expedition in 2009 discovered tubes of pillow lava, with glowing bands of magma (above) extruding from the volcano's vent.

EPHEMERAL ISLANDS

When the summit of a submarine volcano reaches the surface, it sometimes erupts enough material for an island to form. Usually the island lasts for only a short time before it is eroded away by wave action. Some submarine volcanoes repeatedly produce such ephemeral islands.

OCCURRENCE

The birth of a new volcanic island, invariably accompanied by huge explosions and clouds of steam and ash, is an astonishing sight, but the island that forms will not necessarily persist for long. The chances of it lasting for even a few years depend largely on the type of material erupted—particularly on whether it is mainly ash and cinders, or molten lava. It also depends on the persistence of the eruptions and on the strength of the waves and currents in the area of eruption. For a long-lived island to form, a series of vigorous eruptions is needed over many years. It also helps if a large proportion of the volcano's magma is erupted as molten lava, because as this congeals it can bind together loose rock, ash, and cinders into a tough, wave-resistant material. More often, however, eruptions continue for only a few weeks or months, and the island that forms consists mainly of just ash, cinders, and fragments of pumice. Generally, these islands are eroded away within a fairly short space of time—from a few months to a few decades—a mere "blink of an eye" in geologic terms.

AN EXPLOSIVE BIRTH
On January 20, 1986, the underwater volcano Fukutoku-Okanoba in Japan's Bonin Islands erupted at the sea surface. Within days, it built a large island, but the eruption soon stopped. By March 8, the island had been eroded away.

RECURRENT ISLANDS

The summits of some submarine volcanoes sporadically appear and disappear over many decades, producing a series of short-lived islands. These islands, spaced apart in time rather than location, often carry a single name, although their various incarnations may look different. For example, in 1865, the British ship HMS *Falcon* reported a new island in Tonga, in the southwestern Pacific, naming it Falcon Island. A few years later, it had gone, but different-looking islands, also called Falcon Island, were reported in 1885, 1927, and 1933. Other recurrent islands have also been recorded in Tonga—submarine volcanoes at Home Reef and Metis Shoal have each produced a handful in the past 200 years. The volcano Kavachi in the Solomon Islands is the champion builder of such islands—it has produced more than 11 since 1950.

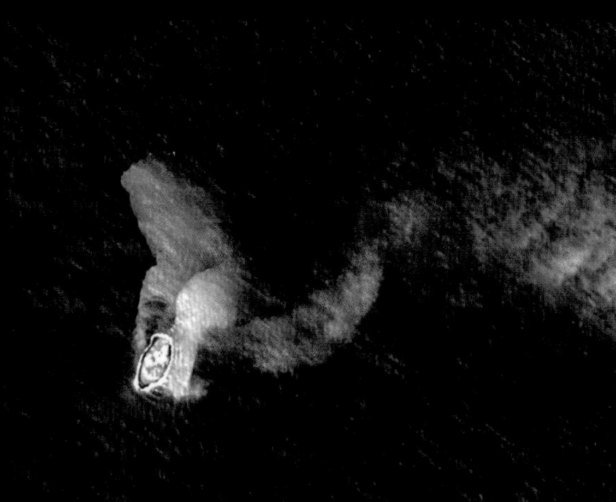

HOME REEF, TONGA
This satellite image shows an ephemeral island, about 2,625ft (800m) long, formed by the submarine volcano Home Reef in October 2006. By February 2007, most of the island had disappeared below the surface of the sea.

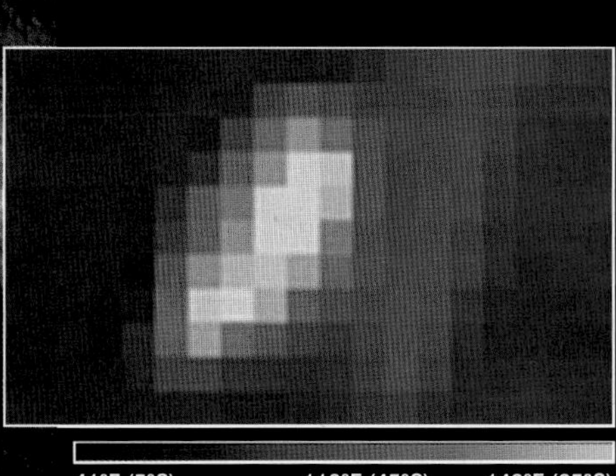

TEMPERATURE SCAN
A thermal image of the island formed by Home Reef showed that the surface temperature at its center was 149°F (65°C)—likely due to the presence of freshly erupted cinders and lava.

METIS SHOAL, TONGA
In June 1995, a new island appeared at Metis Shoal in Tonga. When this photograph was taken, it was about 140ft (43m) high, 920ft (280m) across, and had a temperature of 1,300 to 1,500°F (700 to 800°C). It was a lava dome (huge mound of hot lava) extruded by an undersea volcano, and was washed away soon afterward.

A MEDITERRANEAN EPHEMERAL ISLAND

The Mediterranean has its own submarine volcano, nicknamed Empedocles, which intermittently forms an ephemeral island southwest of Sicily, Italy. The island it produces is usually referred to as Ferdinandea but has also been called Julia and Graham Island. Its last appearance was in 1863 and before that in 1831, when a dispute arose between France, Britain, and the King of Naples as to who owned it. Before the issue could be resolved, the island stopped erupting and it soon sank beneath the waves.

1831 ERUPTION
Ferdinandea's 1831 eruption led rapidly to the development of an islet with a central crater spewing forth ash and lava. The eruption continued for six months.

KAVACHI, SOLOMON ISLANDS
A cloud-like formation appears after an eruption of the Kavachi submarine volcano. Every time Kavachi sticks its head above the water, it seems to do so in a slightly different place. This is because it has a broad summit with several eruptive vents.

SURTSEY 1963

In November 1963, a series of dramatic explosions at the sea surface to the south of Iceland attracted first the attention of local fishermen, then volcanologists, and finally the world media. Within weeks, the eruptions gave rise to a substantial new island—the most famous of the 20th century—which the Icelandic government named Surtsey, after Surtur, a fire god of Norse mythology. It continued to emit clouds of ash and fountains of lava, growing and transforming, until falling quiet in the summer of 1967. Since then, it has been gradually eroding, while acquiring a reputation as an outstanding laboratory for the study of biocolonization, the invasion of new land by life forms.

NOVEMBER 1963	
Location	Off the coast of Iceland
Volcano type	Submarine volcano
Eruption type	Surtseyan
Explosivity index	3

35,000,000,000

VOLUME OF MATERIAL ERUPTED IN CU FT (1 BILLION CU M), 1963–67

AN ISLAND IS BORN

The eruption that led to the formation of Surtsey is thought to have started just a few days before the island's appearance, from volcanic vents lying 430ft (130m) underwater. These vents are part of a volcanic system lying on the Vestmannaeyjar Ridge, an offshoot of the Mid-Atlantic Ridge (see p.246). The underwater build-up of lava proceeded rapidly, until eruptions became visible as clouds of steam and ash rising from the sea surface, in what is now known as a "Surtseyan" style of eruption. Within a day, the tip of the submarine volcano could be seen protruding above the surface.

KEY

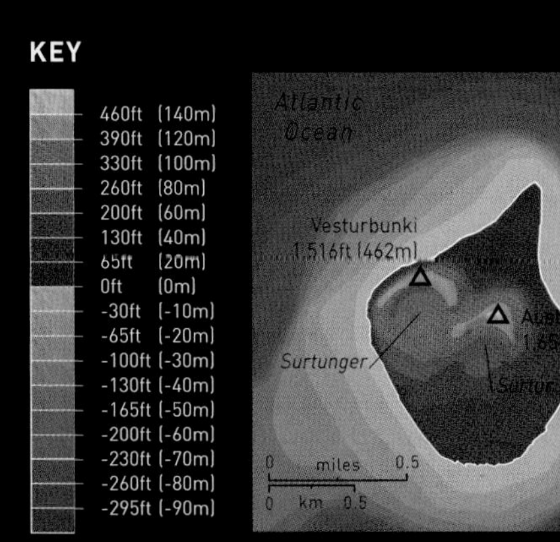

MAP OF SURTSEY On this map, the semicircular crater at the center, Surtur, marks Surtsey's original volcanic vent. The second crater, Surtungur, marks where a second vent opened up some three months into the eruptions.

> **"ON SURTSEY (THE) LANDSCAPE CREATED WAS SO VARIED... IT WAS ALMOST BEYOND BELIEF."**
>
> **SIGURDUR THORARINSSON**, ICELANDIC GEOLOGIST

STAGES IN THE ERUPTION

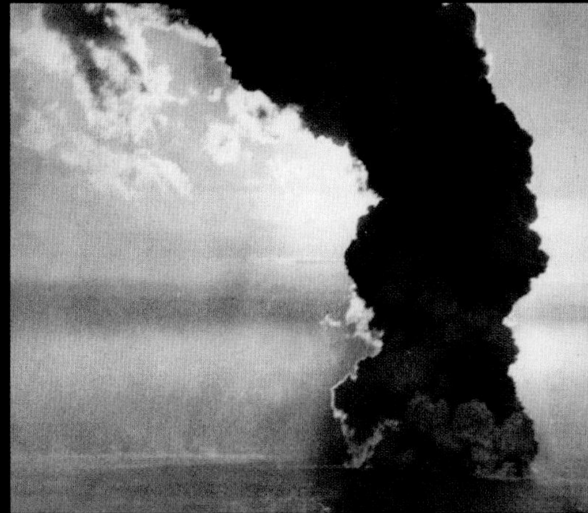

1 THE FIRST EXPLOSIONS On November 14, 1963, fishermen spotted a column of ash and steam rising from the sea surface about 21 miles (33km) off Iceland's south coast. While going toward it, they witnessed explosions generating a huge plume of dark ash.

2 EARLY STAGES For the next 10 days or so, the interaction between hot magma and seawater kept producing spectacular eruptions—a mixture of dark ash and steam—and an ash cloud that extended to a height of 6 miles (10km).

3 AN ISLAND FORMS By November 27, 1963, the amassed ash and cinders had formed a distinct, but partially formed, cone some 1,600ft (500m) in diameter. This had a wide crater enclosing the vent, from which ash and steam kept spewing.

4 LAVA FLOWS Over the next few years, lava flows occurred both from Surtsey's original vent and a second vent that had opened up. As this lava solidified, it built a protective cap over a large part of the island.

AERIAL VIEW
In this view of Surtsey, less than a year after its original eruption, steam is seen venting from Surtungur. Dark lava covers the southern part of the island, at the top here. Light-colored ash makes up most of the rest.

SURTSEY TAKES SHAPE

Initially, the eruptions focused on one main vent, building a roughly circular island with a crater at the center. Because of their vigor and the rate at which new material was added, the eruptions kept pace with the rate at which waves were eroding the island. In February 1964, a second vent opened up to the northwest of the first, building a second crater. Soon the island reached such a size that seawater could no longer easily reach the vents. The eruptions became less explosive and lava fountains became the main activity. As a result, a protective layer of hardened lava formed over much of the loose material underneath.

FUTURE OF SURTSEY

Surtsey attained its maximum surface area of 1.04 sq miles (2.7 sq km) in June 1967, when it stopped erupting. Since then it has gradually eroded, and by 2011 it was down to less than half its maximum size. Although the rate of erosion is likely to slow as the island shrinks to its core of hardened lava, Surtsey is expected to eventually disappear altogether, probably within a few centuries. For now, it is of great interest to biologists, who can study how life takes hold on a barren island. Lichens and mosses appeared first, and as birds began nesting, their droppings enriched the soil, helping plants to survive.

LIFE ON SURTSEY
This crater has been colonized by mosses and lichens, around 100 species of which have been recorded across Surtsey. There are also 30 species of higher plants and more than 80 bird species.

ATOLLS, SEAMOUNTS, AND GUYOTS

Many ocean regions that were once in areas of high volcanic activity, such as hotspots (see pp.32–33), have subsequently moved away from these areas because of plate movements. Three common types of oceanic features in these regions are atolls, seamounts, and guyots. Each was once either a volcanic island or a submarine volcano.

ATOLLS

An atoll is a ring of coral reefs or low-lying coral islands encircling a lagoon. Large numbers of atolls exist in the Pacific and Indian oceans. It was English naturalist Charles Darwin, in 1842, who first explained the way in which an atoll forms from a volcanic island. As a volcanic island ages, it gradually sinks because of the compression of the sea floor that it sits on. If the island is in an area where coral organisms thrive (mainly the tropics), then these organisms build a fringing coral reef around the island. As it sinks further, this turns into a barrier reef, separated from the sinking island by a lagoon. Finally, the island sinks out of sight altogether, leaving behind an atoll.

SEAMOUNTS

Defined as mountains that rise at least 3,280ft (1km) from the sea floor, seamounts are either flat topped—called guyots—(see right), or ordinary round or irregularly topped ones. Of the latter, most are the remains of submarine volcanoes that never grew large enough or strongly enough to form volcanic islands. After the volcanoes became extinct, they gradually subsided and were eroded away to form seamounts. Many seamounts are arranged in chains or elongated groups because they originated from linear volcanic fissures on the sea floor, or formed from a series of volcanoes that erupted sequentially at a single hotspot.

LIFE ON A SEAMOUNT

Often isolated in the deep ocean, the summits of seamounts and guyots provide habitats for a variety of marine life adapted to shallow water. Underwater currents are forced upward when they meet seamounts, bringing nutrient-rich water closer to the surface. The nutrients support the growth of plankton, which, in turn, provides food for many types of animals.

Pink sea fan
This huge sea fan—a type of soft coral—lies at the summit of a seamount in the Flores Sea, Indonesia. Sea fans catch plankton with their tiny tentacles as it drifts past them in the ocean.

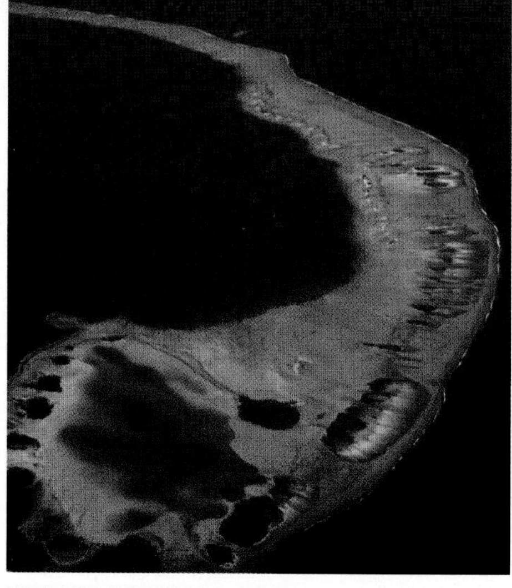

PART OF RANGIROA ATOLL
Located in French Polynesia, Rangiroa is one of the world's largest atolls. A part of its encircling ring of coral islands is visible here.

active submarine volcano

volcanic cone forms from solidified lava

corals grow underwater near shoreline, forming fringing reef

active volcano

SUBMARINE VOLCANOES

VOLCANIC ISLANDS

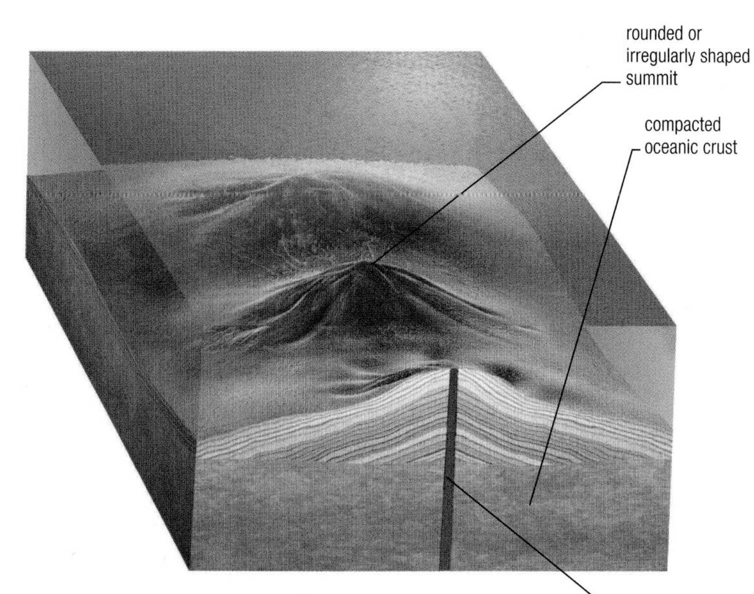

ORDINARY SEAMOUNTS
Tens of thousands of these seamounts exist on the floors of the five oceans. Most are extinct submarine volcanoes that never formed volcanic islands.

MARAKEI ATOLL
This relatively small Pacific atoll is in the Gilbert Islands, part of the Republic of Kiribati. Covering an area of 5 sq miles (13.5 sq km), it almost completely encloses its central lagoon.

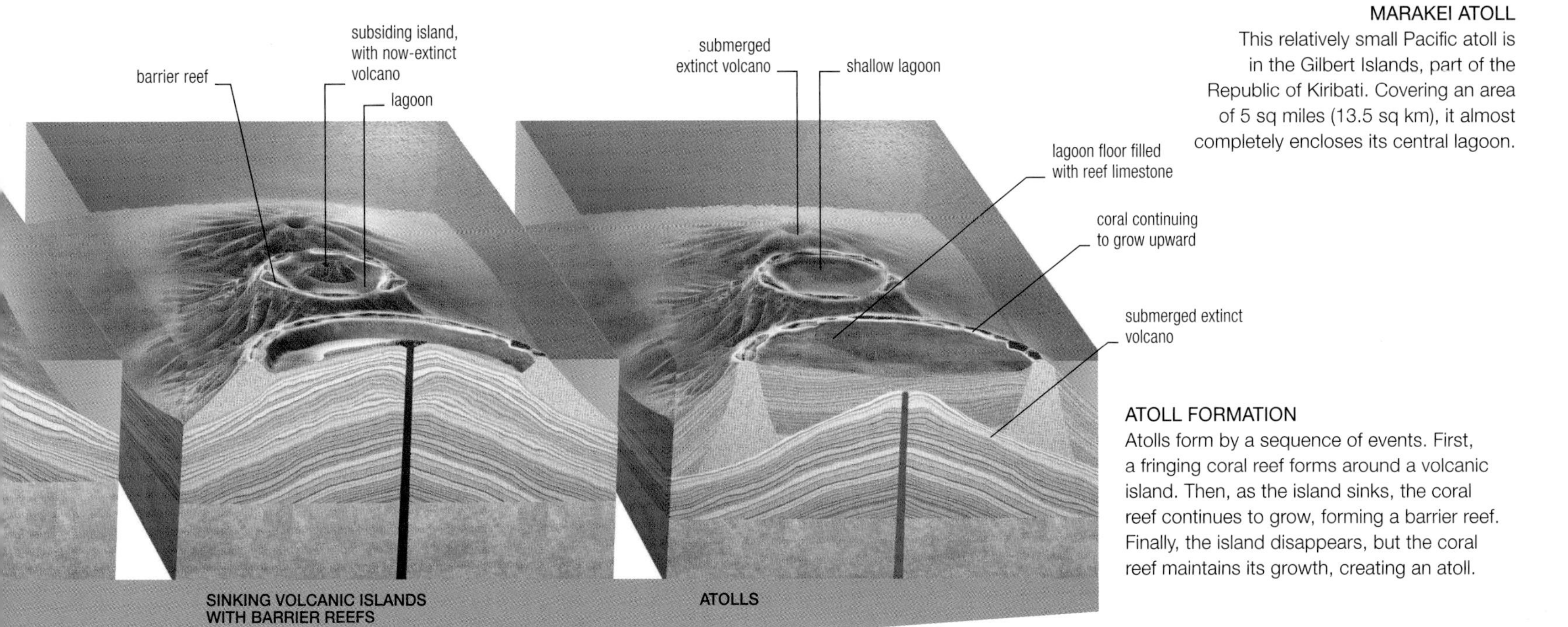

ATOLL FORMATION
Atolls form by a sequence of events. First, a fringing coral reef forms around a volcanic island. Then, as the island sinks, the coral reef continues to grow, forming a barrier reef. Finally, the island disappears, but the coral reef maintains its growth, creating an atoll.

GUYOTS

A guyot, or tablemount, is a flat-topped seamount with its top usually more than 660ft (200m) below sea level. Most are found in the Pacific Ocean. When a volcanic island sinks, coral reef formation may halt at some point or may never even start because the island exists in a region where coral-reef-forming organisms don't thrive. In this case a guyot may form. As the island sinks, its top is worn flat by wave action, wind, and atmospheric processes. The flat top is retained as the island continues to sink. Some guyots are 100 million years old or more, and the largest rise as high as 2.5 miles (4km) above the seafloor.

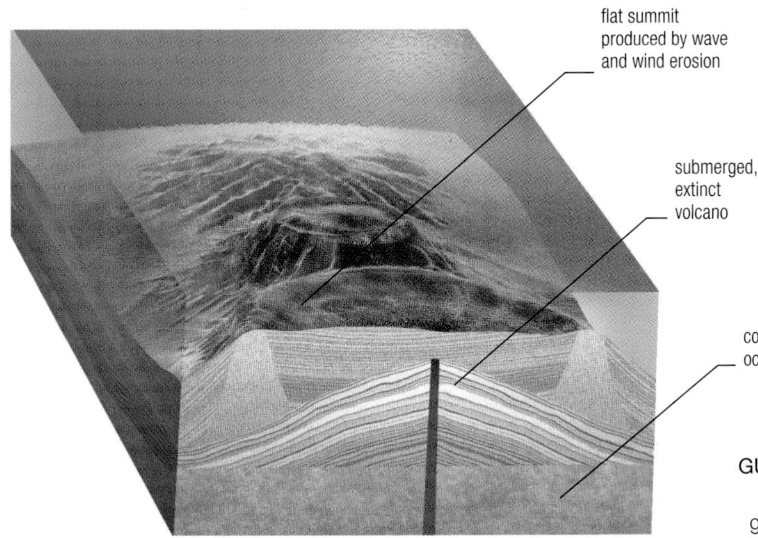

GUYOTS
The flat surface of a guyot, which can be up to 6 miles (10km) wide, is caused by wave action flattening down the top of a volcanic island as it sinks. The flat top and rim may contain remains of old coral reefs.

GUYOTS ON DRY LAND
This once-submerged guyot, along with others discovered in the Afar Depression of Ethiopia, is now exposed on land. It shows that this part of Africa was underwater for a period of time.

THRASHING THE SHORE

This wave was observed smashing into a lighthouse at Porthcawl in Wales, UK, in November 2009. Relative to the sea state at the time, it was a huge wave—other large waves were less than half the size—and thus could qualify as a rogue wave.

ROGUE WAVES AND EXTREME TIDES

Waves and tides are two types of natural ocean-surface phenomena that are not normally considered particularly violent but have their roguish sides. The more extreme variants of both can be highly dangerous or even destructive.

ROGUE WAVES

Large waves caused by storms at sea are always a hazard to shipping, but freak or rogue waves are something quite different. These are rare, unusually large waves that measure more than twice the average height of the larger waves observed in a sea area. Up to 98ft (30m) high, their existence has only been confirmed since 1995 through satellite imaging. Rogue waves are strongly suspected of having caused the loss of at least one ship at sea—the *München*, a German ship, in 1978—and severely damaging others. Their exact cause is not known, but rogue waves most probably result from a combination of strong winds and fast currents having a focusing effect that joins a number of ordinary-sized waves together. Rogue waves are quite different in nature from tsunamis (see pp.262–63).

BEACHED
In 2008, the ferry *Riverdance* was hit by a rogue wave in the Irish Sea, causing its cargo to shift. It listed heavily and could no longer be steered, ending up stranded on a beach.

TIDAL BORES

In coastal areas where a rising tide funnels into a narrowing river estuary, it can form a powerful wave called a tidal bore, which surges upstream. Some tidal bores are dangerous. The world's largest occurs at the mouth of the Qiántáng Jiang River, near Hangzhou in China. It can be up to 30ft (9m) high and is considered life-threatening.

Thundering by
Spectators regularly line up on the shores of the Qiántáng Jiang to watch its famous tidal bore. It roars by with a thunderous sound at speeds of up to 25mph (40kph).

STRONG TIDES

Tides are predictable vertical fluctuations in local sea level, caused by gravitational interactions between Earth, the Moon, and the Sun. They are actually brought about by horizontal flows of seawater called tidal currents. Around coasts, obstructions such as narrow channels and promontories can turn these currents into extremely strong flows, known as tidal races, that develop two or four times a day. These in themselves can be a hazard to boats and ships, but where the currents converge, meet underwater obstructions, or interact with strong winds, further dangers can result, such as whirlpools and rough sea areas called overfalls. Twice a month—when Earth, the Sun, and the Moon align—unusually intense tides called "spring" tides occur that can accentuate these hazards. Scotland's Corryvreckan whirlpool, for example, has caused many sinkings and emergencies over the years.

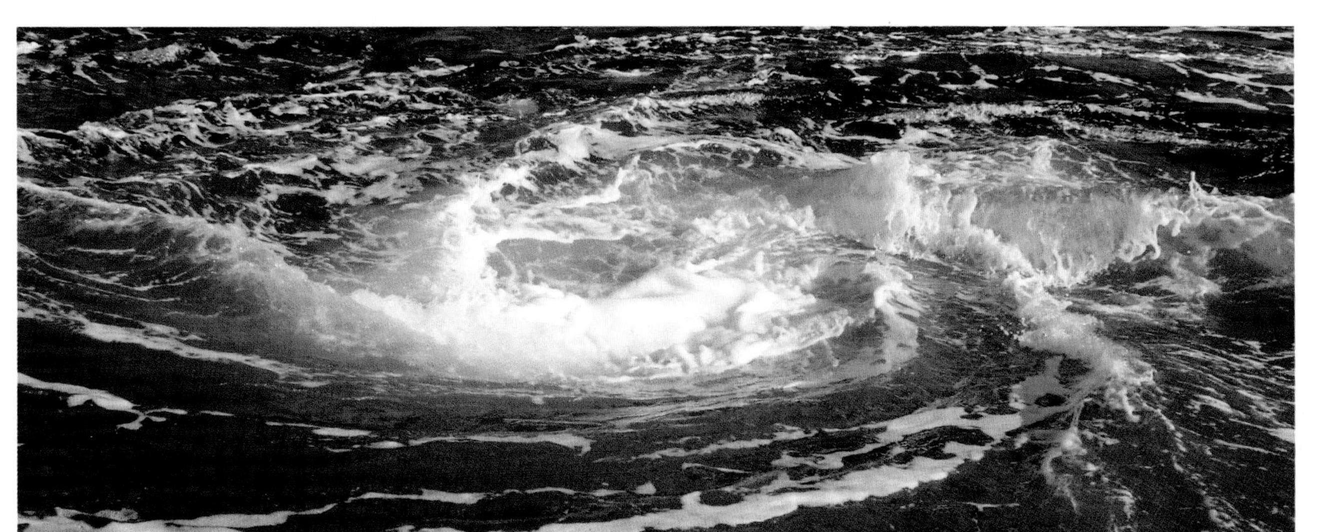

CORRYVRECKAN WHIRLPOOL
This whirlpool on the west coast of Scotland is caused by seawater rushing through a narrow strait between two islands twice a day and interacting with various unusual seabed features. It is locally regarded as violent and dangerous.

TSUNAMIS

A tsunami is a powerful pulse of energy that can propagate for a great distance on the ocean surface in the form of high-speed waves. Hardly noticeable in the open ocean, tsunami waves dramatically increase in size when they reach shallow water, and can cause a colossal amount of destruction as they surge onshore.

GENERATION AND PROPAGATION

Tsunamis are most often triggered by submarine earthquakes, which cause a rupture under a section of the sea floor. As a result, huge slabs of the sea floor are suddenly pushed upward along faults on convergent plate boundaries. This displaces water, leading to tsunamis. Other causes of tsunamis include volcanic eruptions that occur underwater or trigger landslides into the sea, sudden slumps of large amounts of sediments underwater, or a massive meteorite impacting an ocean. Tsunamis travel at speeds of 310 to 500mph (500 to 800kph), much faster than large, wind-generated ocean waves. They have longer wavelengths—the crest-to-crest distance of a tsunami wave is about 125 miles (200km). But in the open ocean, tsunami waves have a low amplitude of less than 3ft (1m). In fact, the term tsunami, which means "harbor wave," was coined by Japanese fishermen who returned to port, having noticed nothing unusual at sea, only to find their home harbors devastated.

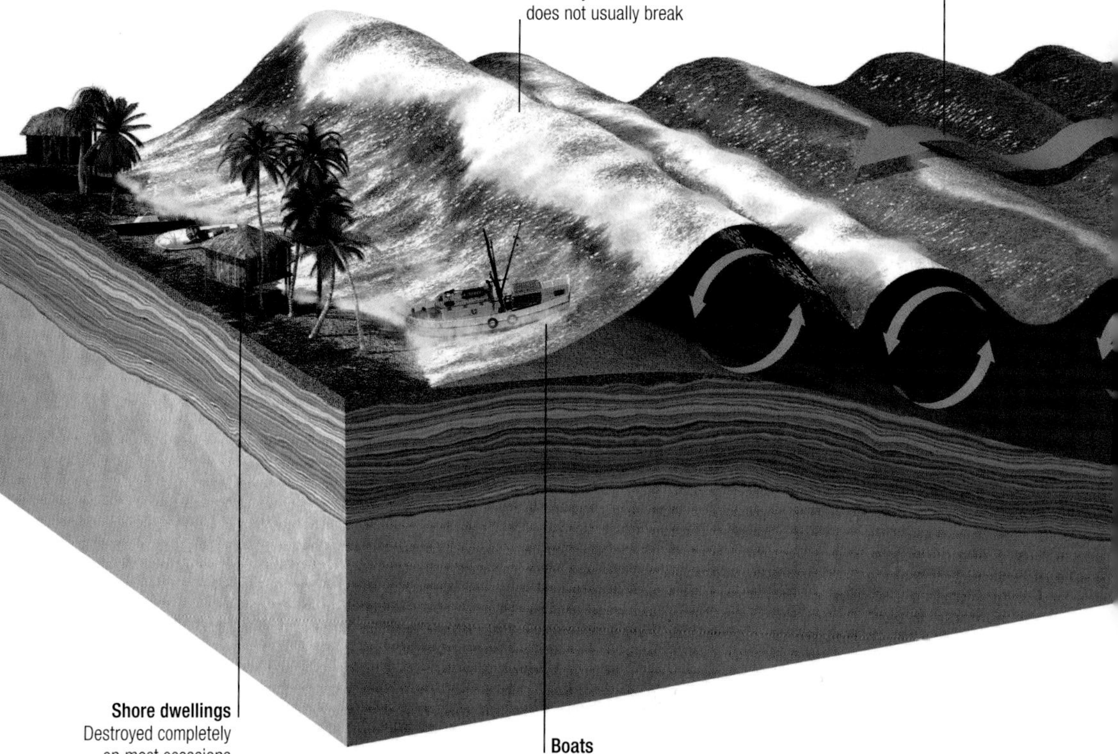

ARRIVAL ON SHORE

The initial signs of a tsunami about to hit shore vary, depending on whether a wave trough or crest arrives first. If a trough comes first, the sea recedes dramatically, whereas in the case of a crest, a large incoming wave appears. As it reaches shallow water, the wave slows to less than 50mph (80kph), but its height increases drastically, in extreme cases up to 100ft (30m). Usually, it does not break but surges forward powerfully. Onshore, there is a sudden onrush of water and a rapid rise in its level, as the water keeps moving inland. With a large tsunami, the surge of water can lift and smash boats, vehicles, houses, and people. After about 20 minutes, the water pulls back strongly, drawing vast amounts of debris, and even people, out to sea. This process may repeat several times over an hour or so.

ROUGH SEAS
Massive tsunami waves hit the northeast coast of Japan's largest island, Honshu, following the submarine earthquake of March 11, 2011.

Waves
High energy but low amplitude waves are generated and spread out

Seawater above rupture
Suddenly thrust upward

Rupture at fault
Causes sudden uplift of sea floor

Shock waves
Generated by earthquakes, these spread out from rupture

TSUNAMI FORMATION
Most tsunamis are caused by ruptures of large sections of rock under the sea floor. A rupture can cause a section of the sea floor to suddenly jerk upward by up to several feet. This displaces overlying seawater, generating tsunami waves.

DISTRIBUTION OF DART TSUNAMI BUOYS

DART (Deep Ocean Assessment and Reporting of Tsunamis) buoys play a critical role in tsunami warning systems. Nearly all the buoys are positioned close to ocean–ocean and ocean–continent convergent plate boundaries, where sea floor fractures and earthquakes that generate large tsunamis are most likely to occur. Currently, there are 41 buoys in the Pacific Ocean, seven in the Atlantic Ocean (including one in the Gulf of Mexico), and four in the Indian Ocean. Most of these buoys are owned by the US National Data Buoy Center (NDBC). Thailand and Indonesia have each installed a DART buoy in the Indian Ocean, and Australia, Russia, and Chile own the rest.

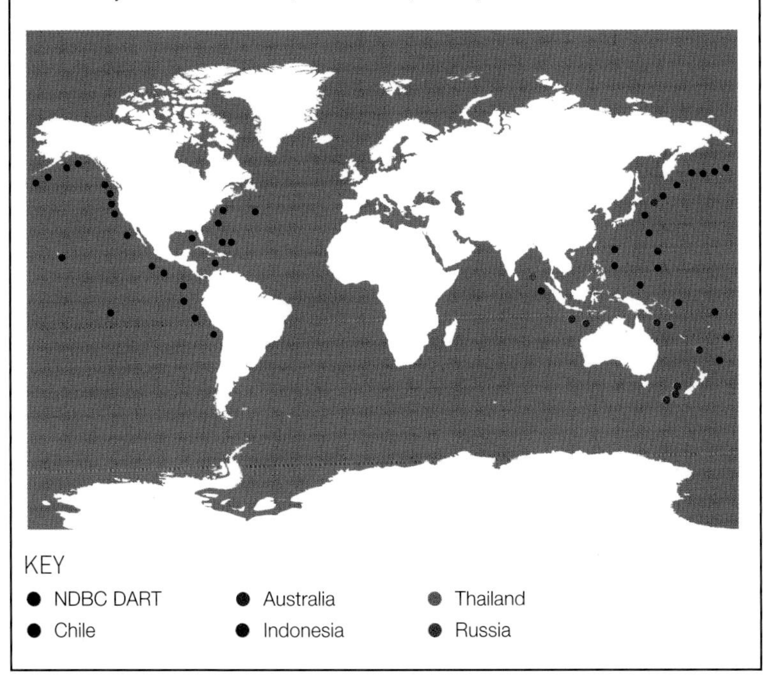

TSUNAMI ALERT SYSTEM

Large tsunamis are notorious for the destruction they cause. Recent catastrophes, in Japan in 2011 (see pp.268–69) and the Indian Ocean in 2004 (see pp.264–65), as well as significant disasters in Java, Samoa, and the Solomon Islands, have greatly increased awareness of tsunami hazards. Warning systems have been established to look for distinctive signs of tsunamis. These systems have three main components—a network of earthquake detecting stations for sensing earthquakes that might trigger tsunamis, the deployment of systems in the open ocean to detect actual tsunamis (see below), and a means of quickly disseminating alerts and warnings to affected populations. A well-developed tsunami alert system currently operates in the Pacific Ocean, where the warnings are issued by the Pacific Tsunami Warning Center. Warning systems have also recently started operating in the Atlantic and Indian oceans.

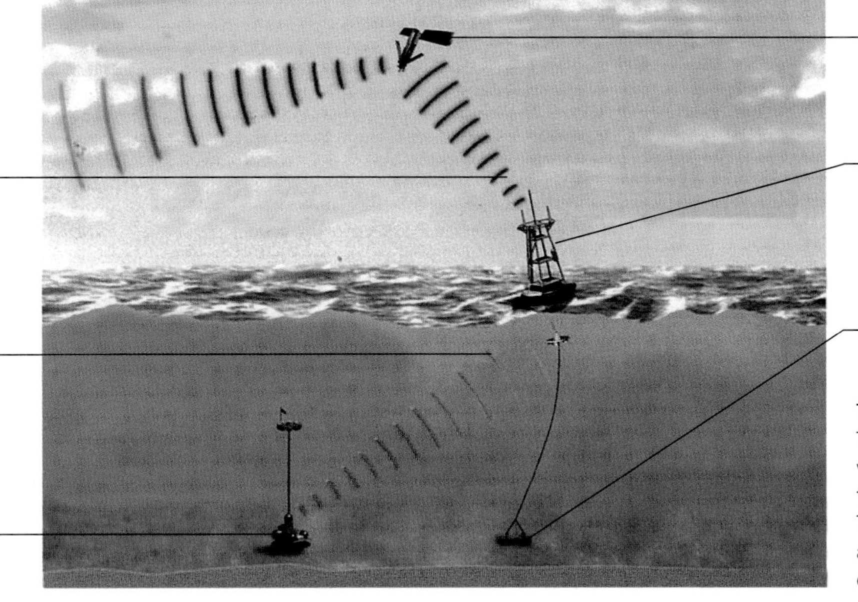

Warning data
Relayed to a satellite network

Sound waves
Used to transmit data to the surface buoy

Sensor
Set on the sea floor, it measures changes in pressure that may indicate passage of a tsunami wave at the sea surface

Satellite
Transmits data to tsunami warning centers

Surface buoy
Receives data from detector and converts it to a radio signal

Heavy weight
Used to anchor surface buoy at depths of up to 4 miles (6km)

TSUNAMI DETECTION SYSTEM
The buoy's basic function is to receive warning signals from the sensor and transmit these to a satellite network. The latest buoys can be put on special alert when a possible tsunami-triggering earthquake is detected.

INDIAN OCEAN TSUNAMI 2004

In December 2004, a devastating tsunami caused a huge number of deaths and cataclysmic destruction on coasts across a wide area of the Indian Ocean. One of the worst natural disasters of all time, the tsunami was caused by a colossal rupture along the Sunda Trench subduction zone. The accompanying earthquake ranks as one of the three most powerful ever recorded. The coastal regions and countries most affected were the northwestern coast of the island of Sumatra in Indonesia, Sri Lanka, parts of the coasts of Thailand and India, and the Maldives.

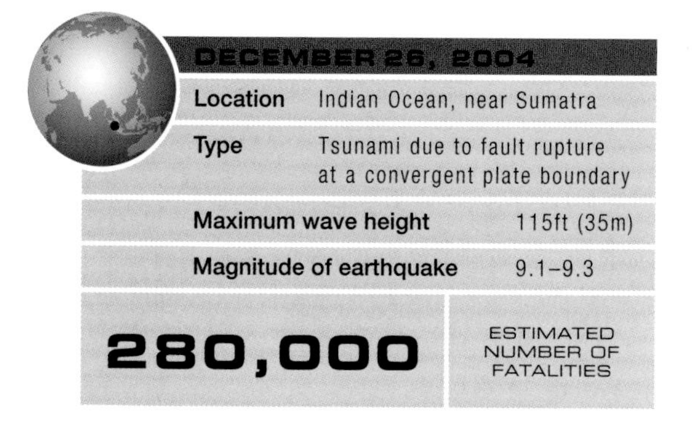

DECEMBER 26, 2004	
Location	Indian Ocean, near Sumatra
Type	Tsunami due to fault rupture at a convergent plate boundary
Maximum wave height	115ft (35m)
Magnitude of earthquake	9.1–9.3
280,000	ESTIMATED NUMBER OF FATALITIES

DASH FOR SHORE
The sea receded before the tsunami hit the shore in southern Thailand. This family on the beach at Hat Rai Lay near Krabi can be seen running away from the 16 to 32-ft- (5 to 10-m-) high waves. Remarkably, the whole family survived.

CAUSE

The tsunami was the result of a sudden fracture in Earth's crust along part of a fault where the India–Australia Plate pushes under the Burma part of the Eurasian Plate. The fracture started at a point about 19 miles (30km) beneath the seafloor off the northwestern coast of Sumatra and ran for about 990 miles (1,600km) in a northerly direction. As a result, the two plates moved about 50ft (15m) relative to each other. At the same time, long, narrow parts of the sea floor above the rupture were suddenly pushed upward. This movement triggered tsunami waves, which propagated to the east and west.

Time after earthquake (in hours)

KEY 0 1 2 3 4 5 6 7 8 9 10

INDIA
AFRICA
MALDIVES
THAILAND
Indian Ocean

SPREAD OF TSUNAMI WAVES
This map shows how the tsunami waves spread across the Indian Ocean from the rupture site on the seafloor. Most of the severely damaged coasts were hit within three hours of the earthquake. Coasts in Africa were not hit until more than six hours later.

KHAO LAK BEFORE TSUNAMI
This satellite image of a part of the coastline of Thailand was taken on January 13, 2003, almost two years before the disaster. This 1-mile- (2-km-) wide area is part of a peninsula near Khao Lak, one of the hardest hit regions in Thailand.

KHAO LAK AFTER TSUNAMI
A photograph of the same section of the coastline, taken three days after the tsunami struck, shows much of the vegetation lost or covered in mud, an area with buildings destroyed (on the left), and beaches denuded of sand.

THE TSUNAMI HITS

1 COASTLINE INUNDATED
The coast of northwestern Sumatra was the first to be hit by waves up to 115ft (35m) high. These destroyed buildings, stripped away vegetation, and flooded everything up to several miles inland.

2 MAMMOTH WAVES SWEEP IN
Waves hitting shores about an hour or so after the earthquake were 16 to 32ft (5 to 10m) high, as seen here at Penang in Malaysia. The waves surged forward over whatever obstructions they met.

3 INLAND FLOODING
In southeastern India, near Chennai and surrounding areas, the tsunami waves surged far inland within the first three hours, carrying everything they encountered with them.

4 MOUNTAINS OF DEBRIS
Smashed boats, dwellings and their contents, vehicles, bodies, and mud were left in a chaotic pile along thousands of miles of coastline, as seen here in the Galle district of Sri Lanka.

EFFECTS AND CASUALTIES

Because no warning system was in place for the Indian Ocean at the time, people on the affected coasts were taken largely by surprise. Overall, the tsunami consisted of a succession of waves, occurring in cycles with a period of more than 30 minutes. The biggest waves, which hit the Sumatran coastal town of Lhoknga, are estimated to have been 115ft (35m) in height. The tsunami resulted in many human casualties, principally in Indonesia, with more than 235,000 estimated deaths, followed by Sri Lanka (more than 35,000), India (more than 18,000), and Thailand (more than 8,000). Thousands of coastal buildings were destroyed and there was extensive damage to the environment, in particular to coral reefs.

“PEOPLE WERE SCREAMING AND KIDS WERE SCREAMING ALL OVER THE PLACE... AND AFTER A FEW MINUTES YOU DIDN’T HEAR THE KIDS ANY MORE...”

PETRA NEMCOVA, CZECH MODEL, AT KHAO LAK IN THAILAND WHEN THE TSUNAMI STRUCK

三王岩
浄土ヶ浜
シートピアなあど

TSUNAMI STRIKES JAPAN
On March 11, 2011 a tsunami wave crashed into the harbor of the port of Miyako in northeastern Japan. A huge mass of seawater spilt over the 10-ft- (3-m-) high harbor wall and cascaded into the street below. Some of the boats in the background followed soon afterward.

JAPANESE TSUNAMI 2011

On Friday March 11, 2011, the world looked on in horror as reports and video footage emerged of a devastating tsunami sweeping across a coastal region of Honshu, the main island of Japan. The irresistable surge of water overwhelmed and carried everything before it—including ships, houses, cars, and people. The water swept through towns and across fields, roads, and airports, leaving a vast area of the country flooded and littered with a horrific jumble of debris. Many people died or went missing; hundreds of thousands were displaced; and a huge number of buildings were destroyed, with an economic cost measuring in hundreds of billions of dollars. On top of that, damage to a nuclear power plant raised concerns over the leak of radioactive materials.

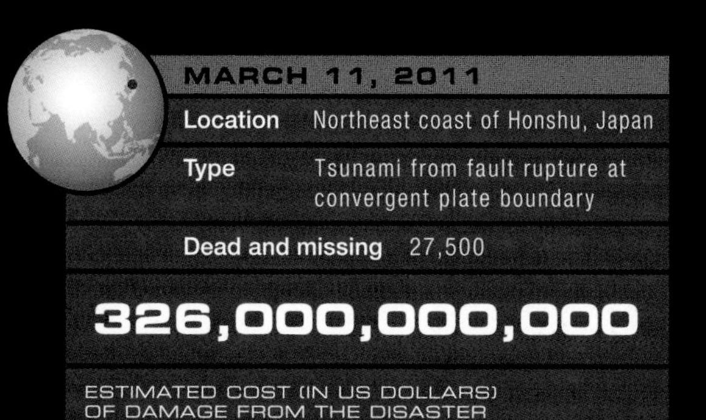

THE TSUNAMI

The tsunami was caused when a length of the existing fault under the seafloor off the northeast coast of Honshu ruptured suddenly. This rupture—about 300 miles (500km) in length—triggered a magnitude-9.0 earthquake, the most powerful ever to hit Japan. The seafloor on the leading edge of the Japan plate sprang upward when the rupture occurred, displacing the overlying seawater and triggering tsunami waves. Within minutes, these waves had reached the coast, washed over anti-tsunami seawalls, and surged inland for distances up to 6 miles (10km). Off the east coast of Japan, the waves were estimated to be 125ft (38m) high, but most were 10 to 40ft (3 to 12m) high. Within an hour or so, an area of around 180 sq miles (470 sq km) had been inundated, with a vast amount of destruction to buildings and infrastructure in addition to the heavy death toll.

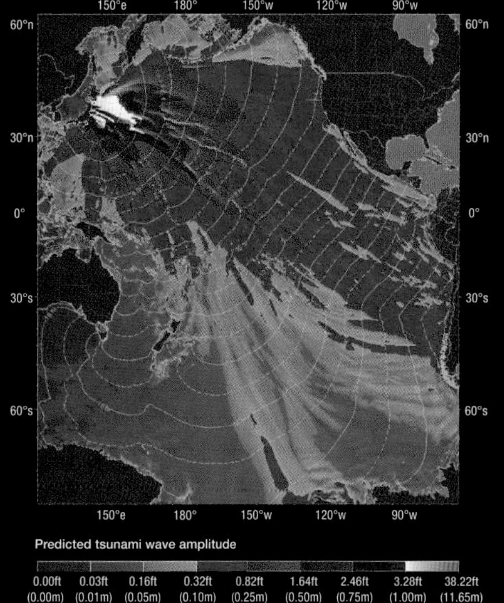

WAVE HEIGHT PREDICTION MAP
This map, generated minutes after the tsunami was first detected, predicted how the height of the waves would decrease as they moved across the Pacific.

DISASTER STRIKES

1 EARTHQUAKE HITS HONSHU
The earthquake severely damaged roads, as seen here in Urayasu City, near Tokyo. The massive tremors also led to the failure of a dam and set two oil refineries on fire. Two very large aftershocks (M-7.9 and M-7.7) followed soon after.

2 TSUNAMI SURGES
Within minutes, massive tsunami waves had reached a long section of the coast of Honshu, surging through harbors, ports, and towns, and across fields, roads, railways, and canals, as seen here at Iwanuma, Miyagi Prefecture.

3 FIRES, FLOTSAM, AND JETSAM
Houses and their contents were smashed or lifted up, aggregated into floating islands of dwellings and debris, then dumped on the landscape. A combination of fuel leaks and sparking from damaged electrical equipment caused fires.

WARNING SYSTEM

Japan is well equipped for responding to earthquakes and has an advanced tsunami warning system. A warning with details of which prefectures were likely to be hit hardest was put out within three minutes of the earthquake. Although it saved some people, on this occasion the waves were so big and arrived so quickly that the overall impact of the warning was limited.

WARNING SIGN
Signs such as this on Japan's tsunami-affected coasts indicate where people should run to—usually higher ground—in the event of a large earthquake or a tsunami.

"THE WAVE HIT ME AND PUSHED ME INTO A DOORWAY. I CLUNG ON TO THE DOOR TO SURVIVE."

HIROMI ONODERA, TSUNAMI SURVIVOR

4 EVACUATION SHELTERS
Relief operations were soon underway, with evacuees housed in improvised shelters such as the gymnasium shown here. But the Japanese government was faced with some huge challenges—300,000 displaced persons; fuel shortages and frequent power outages in the affected region; and difficulties in getting food, water, and medicines to the survivors.

6 EXTREME WEATHER

<< Lightning
Lightning strikes during a thunderstorm above the city of San Francisco.

WHAT IS WEATHER?

The term "weather" refers to the changes in the atmosphere we experience day to day. These are caused by the sun, air, and water and bring rain or sunshine, snow or drizzle, and wind or calm. The variety of our weather depends on our location and the season.

LAYERS OF THE ATMOSPHERE

The atmosphere can be divided into layers according to whether temperature decreases or increases with height. The lowest layer, the troposphere, is where the air cools with height and stretches from Earth's surface up to the chilliest point called the tropopause. The stratosphere is next, where temperature is first constant with height and then increases upward to the stratopause. The temperature increases because stratospheric ozone absorbs some of the solar radiation streaming down through the atmosphere, warming the air at great heights. The mesosphere is higher still, where temperature decreases with height up to the mesopause—the coldest part of the atmosphere. Above this is the thermosphere where temperature increases with height, rapidly beyond 55 miles (88km) because solar radiation is absorbed when it ionizes atmospheric gases such as oxygen (causing aurorae). The outer edge of the thermosphere blends gradually into space.

THE WEATHER LAYER

The troposphere is where all weather occurs. It is thinner over cooler areas of the globe and thicker where the air is warmest, meaning that the tallest clouds and heaviest rain can occur in the most strongly heated, tropical atmosphere. Air circulates up and down to produce the clouds and clear skies we see. Rising air cools while sinking air warms, so the weather layer becomes colder higher up. It contains the weather systems that produce gales, torrential rain, dust storms, hail, tornadoes, and virtually all the planet's pollution.

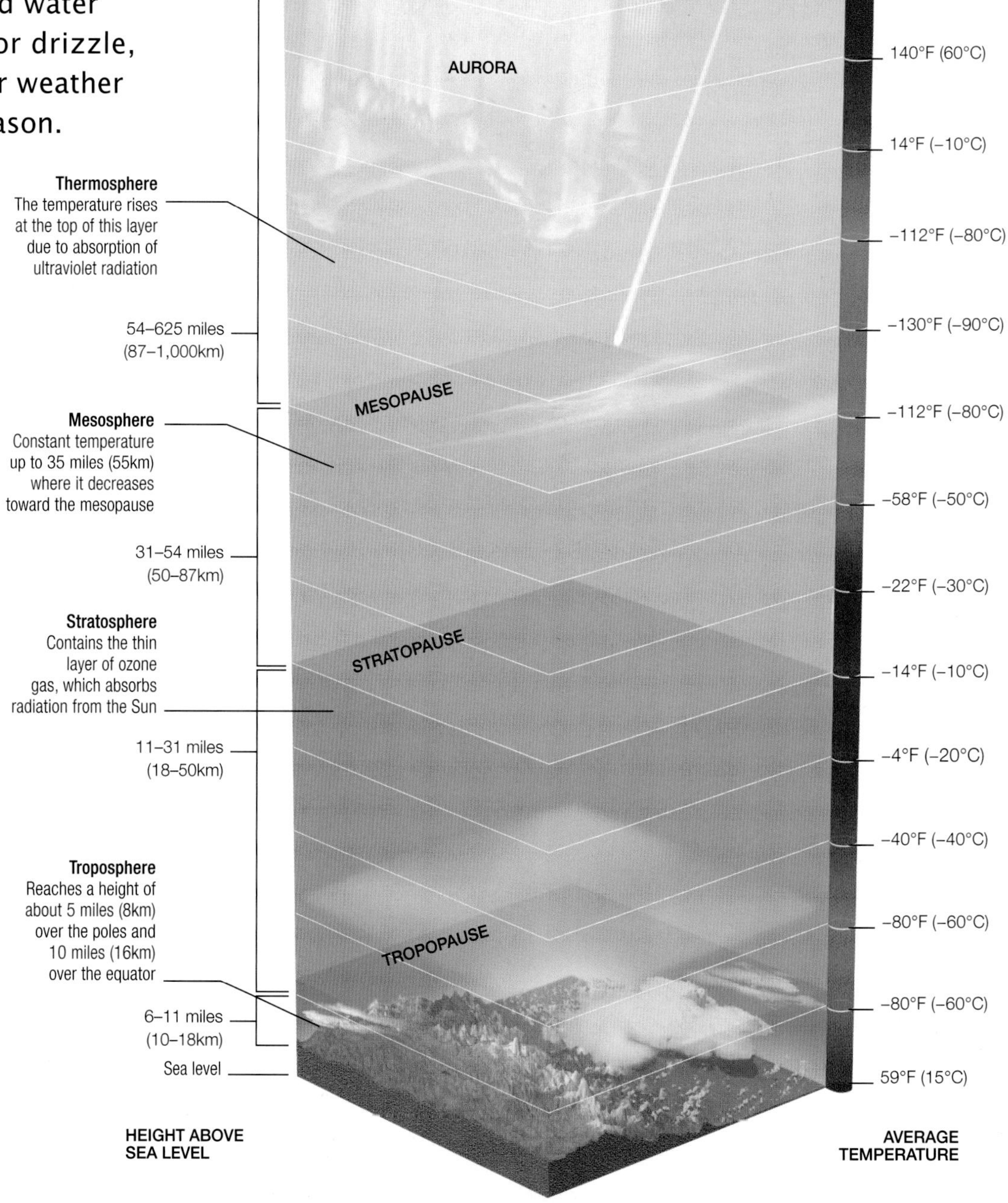

ATMOSPHERIC TEMPERATURE
Progression upward through the troposphere brings lower temperatures. In the stratosphere beyond, the temperature rises, then it falls again in the mesosphere and increases though the thermosphere.

EARTH'S CLIMATE ZONES

Weather may change daily but climate describes a long-term pattern of temperature and precipitation in a region. One way of defining Earth's climates is to use vegetation distribution to map the different zones. Climate and vegetation were linked by Russian climatologist Wladimir Köppen in 1936. He described climate zones using averages of temperature and precipitation and these values were used to establish seasonal variation and thresholds that are biologically significant, such as length of growing season and water availability. His classification ranges from tropical rainforests to arid deserts and the highest latitude polar regions where there is an absence of plant life. Temperate climate zones were further divided according to their proximity to seas and oceans. Mountain was a separate zone since elevation affects climate.

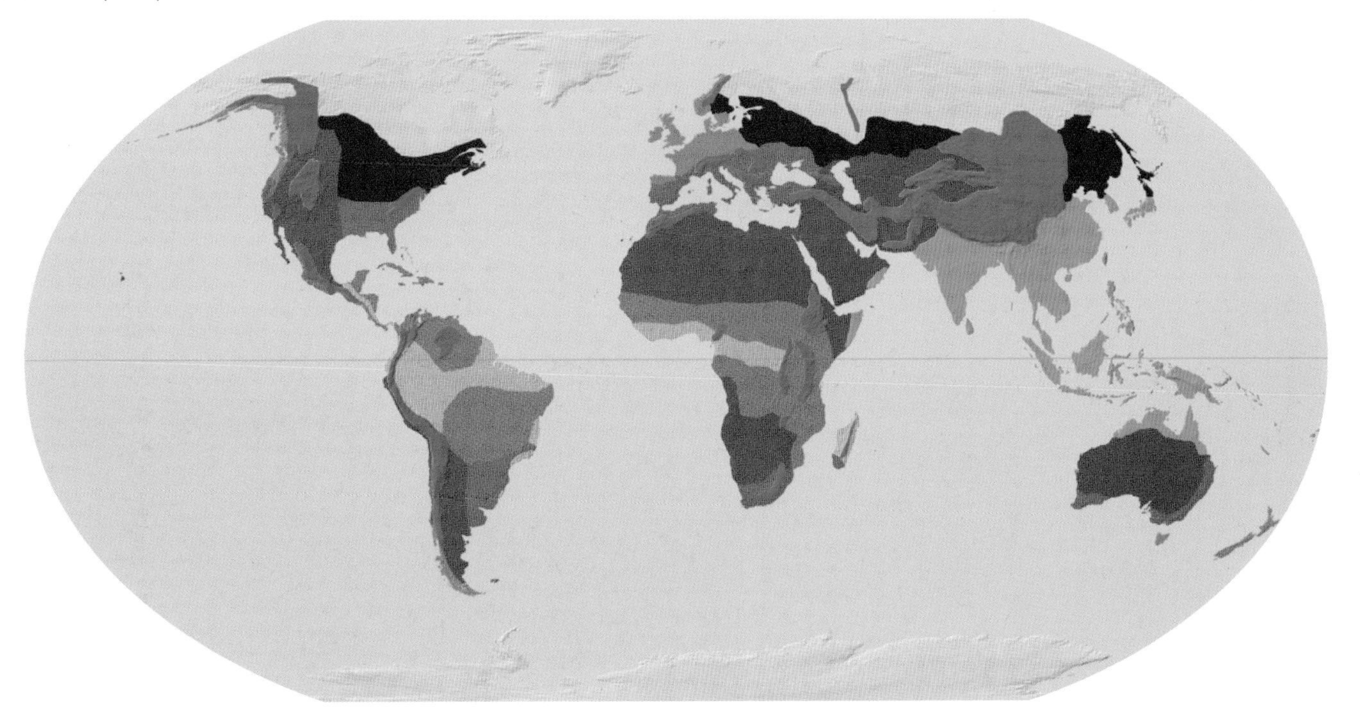

CLIMATE MAP
The map above shows nine climate zones named according to their temperature and precipitation, including seasonal variations.

KEY

- Hot climates with year-round rain
- Hot climates with monsoon rain
- Hot climates with seasonal rain
- Hot, dry climates
- Cool, temperate maritime climates
- Warm, temperate climates
- Cool, temperate continental climates
- Cold climates
- Mountain climates

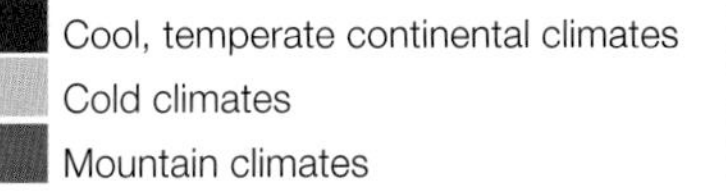

HOT, RAINY CLIMATES
Tropical rainforests lie in equatorial regions where the thunderstorms of the ITCZ (see p.277) supply plentiful rain.

HOT, DRY CLIMATES
Hot deserts are found where the air in the troposphere sinks, suppressing cloud formation and bringing clear skies.

WARM TEMPERATE CLIMATES
Temperate zones are typically green with a mild climate. They have wet summers or winters, or are wet all year.

COLD CLIMATES
Polar regions are arid. They include the massive continental glaciers of Antarctica as well as Greenland.

CHANGING WEATHER

Weather is the varying result of pressure systems (see pp.274–75) whose highs and lows can produce dry and calm conditions or cloudy, wet, and windy days. If the pressure systems are mobile, then there is great variability. In parts of the world where there are much more stable pressure features, there is less change in the weather. The middle latitudes tend to have a lot of variability from passing highs and lows while the world's hot deserts or subtropical oceans have much less variety, being dominated by persistent high pressure. Occasionally, there is violent weather, such as the gales of the middle latitude oceans or the tropical cyclones of the lower latitude seas.

SAME PLACE DIFFERENT WEATHER
One region well known for great variety in the weather is the middle latitudes. Traveling low and high pressure features can mean low cloud one day and constant sunshine the next in Paris, France.

GLOBAL PRESSURE

Day-to-day weather is the result of changes in high- and low-pressure systems. An area of high pressure occurs where a large mass of air descends through the atmosphere, a low-pressure zone forms where the air is rising. Air moving from highs to lows is felt as wind. Highs and lows form weather systems, which travel around the globe with variable winds.

GLOBAL PRESSURE

Barometric pressure is a very important weather measure because when it is recorded around the world simultaneously, and then mapped, it defines where the world's weather-producing highs and lows are. The value of pressure depends on the total mass of air above the barometer up to the upper edge of the atmosphere. High mass means high pressure, low mass means low pressure. So, mean sea-level pressure of 980 millibars means there is 11 tons of air above every square meter at sea level, while 1020 millibars (mb) means there is 11.4 tons. The movement of weather systems is monitored by analyzing a sequence of synoptic pressure maps, which show mean sea-level pressure. Maps reveal a number of weather systems, which can be interpreted and used for weather forecasting.

December to February

June to August

millibars

1040
1030
1025
1022.5
1020
1017.5
1015
1012.5
1010
1007.5
1005
1002.5
1000
995
990
985
980
975
970

GLOBAL PRESSURE MAPS
The intensity and location of pressure centers varies according to the seasons. The northern hemisphere winter (December to February average) experiences wet and windy weather from strong (deep) lows (cyclones or depressions)—these are shown as yellow on the map near Iceland and the Aleutian Islands. Across southern continents, highs (anticyclones) dominate. In contrast, the northern hemisphere summer (June to August average) has much weaker lows and winds, while the southern hemisphere has deeper lows across the Southern Ocean and highs dominate the continents.

AIR CIRCULATION

Across Earth's surface, winds spiral out from highs and are drawn toward the center of lows, and this flow of air—millions of tons of it—connects them. The air swirling into lows has to ascend, often producing widespread cloud and precipitation. The outward flowing wind of the high is supplied by air that is subsiding in depth, meaning that the ascent needed to produce tall (rain) clouds will be suppressed. Consequently, highs tend to be dominated by dry weather—although sometimes with a lot of low-level layered cloud. Higher up, air diverges above lows and converges above highs, maintaining the circulation pattern.

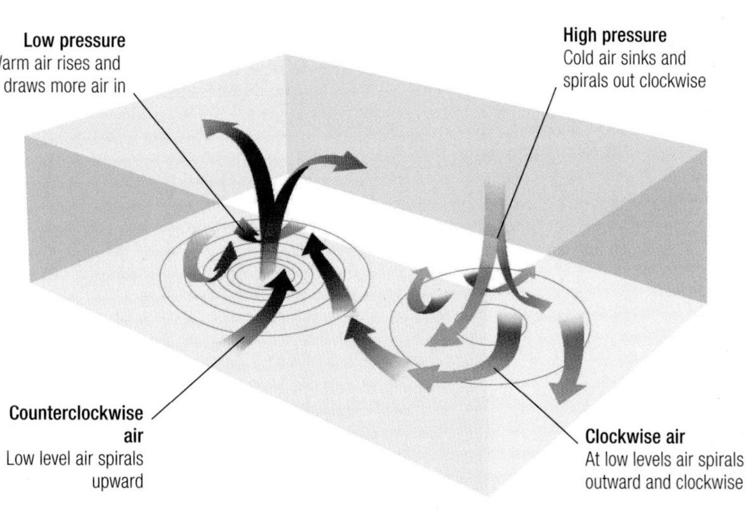

HIGH AND LOW PRESSURE
There are no set values of pressure that guarantee a feature is high or low, instead it is relative day to day. The features are defined by the highest and lowest pressures that occur on a given day. On one day a center of 1010 millibars could be described as a high, while a center of 1010 millibars the next day could be registered as a low.

LOW-PRESSURE SYSTEM
Lows (cyclones or depressions) are generally mobile, with the lowest pressure at their center. They are usually more compact than highs and hold a smaller mass of air. The inward-swirling air at low levels ascends to produce commonplace cloud and precipitation, while strong pressure gradients around the low can produce damagingly strong winds. Hurricanes and middle latitude depressions are examples of lows that produce this weather.

GLOBAL WEATHER CELLS

Taking an average of the global wind data through the troposphere shows how the atmosphere's circulation divides into overturning cells in each hemisphere. The biggest are the Hadley Cells, with their surface trade winds blowing into the east–west line of the thunderstorms of the Inter-Tropical Convergence Zone (ITCZ). Here the air ascends to great heights, flowing toward higher latitudes and sinking over subtropical latitudes. At the surface, this air heads either toward the equator as trade winds or as warm, humid air toward middle latitudes, where the Ferrel Cell dominates. Here the air ascends in frontal zones, spreading out aloft toward the subtropics and polar latitudes. The Polar Cell is smallest, involving a sinking motion over the highest latitudes and surface flow toward middle latitudes.

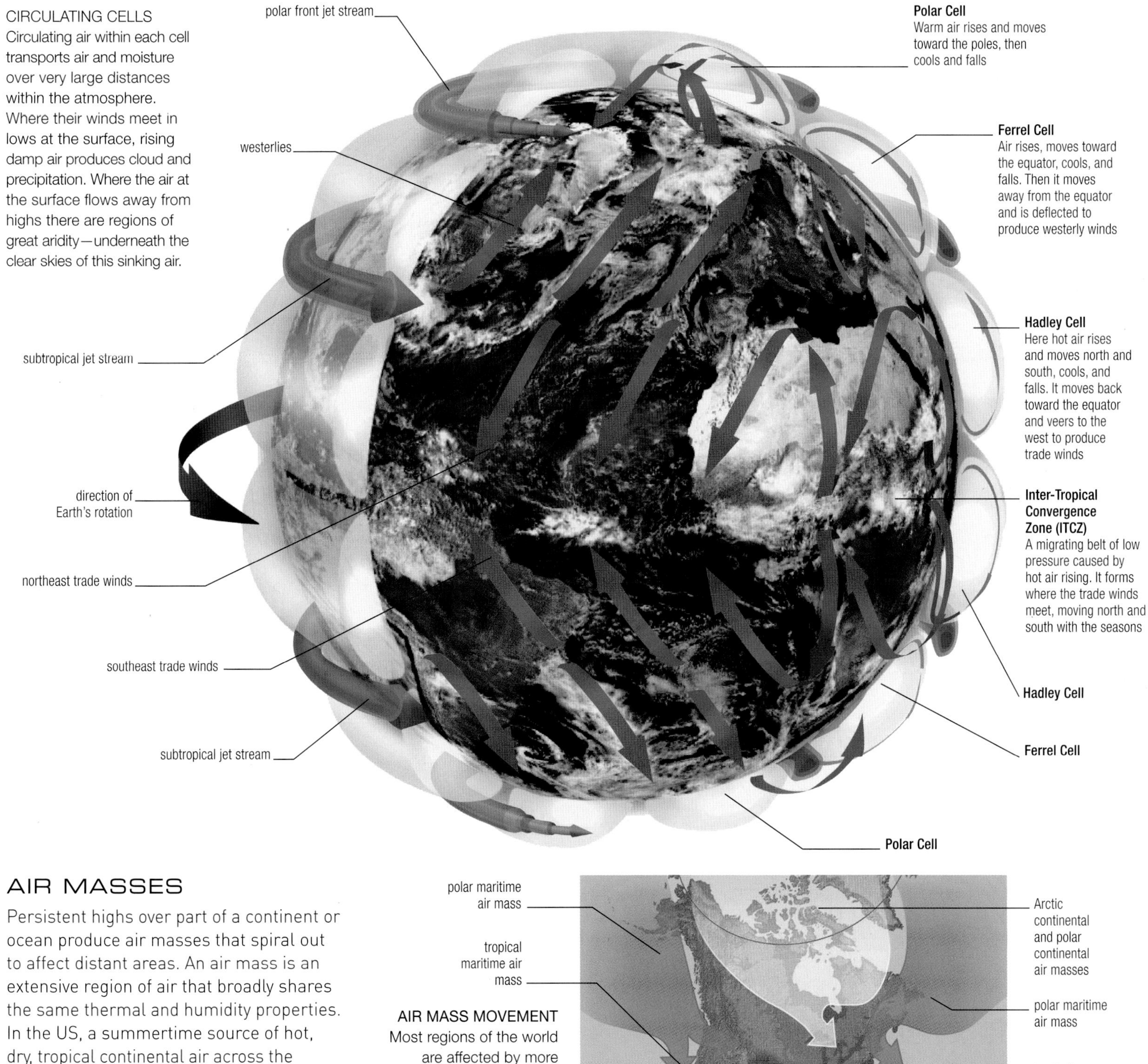

CIRCULATING CELLS
Circulating air within each cell transports air and moisture over very large distances within the atmosphere. Where their winds meet in lows at the surface, rising damp air produces cloud and precipitation. Where the air at the surface flows away from highs there are regions of great aridity—underneath the clear skies of this sinking air.

AIR MASSES

Persistent highs over part of a continent or ocean produce air masses that spiral out to affect distant areas. An air mass is an extensive region of air that broadly shares the same thermal and humidity properties. In the US, a summertime source of hot, dry, tropical continental air across the southwestern US and Mexico is critical in the development of the severe storms in the High Plains area. The US's weather is also influenced by cold, dry, polar continental air from Canada, polar maritime air from middle latitude oceans, and tropical maritime air from the lower latitude oceans.

polar maritime air mass
tropical maritime air mass
Arctic continental and polar continental air masses
polar maritime air mass
tropical continental air mass
tropical maritime air mass

AIR MASS MOVEMENT
Most regions of the world are affected by more than one air mass. For example, the US is influenced by five. The source of the air mass—whether it formed over land or ocean and from tropical or polar regions—determines its qualities.

WINDS AROUND THE WORLD

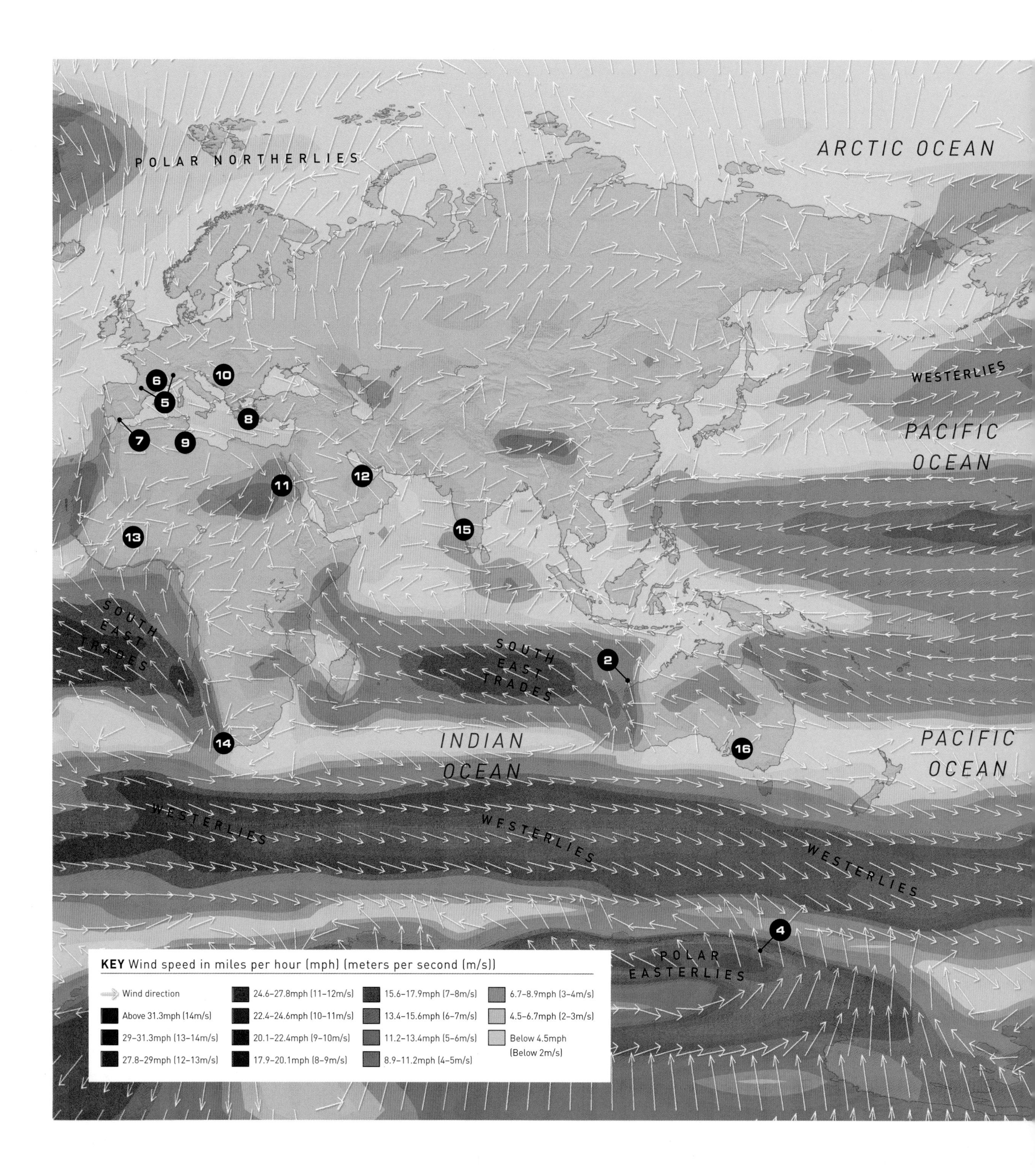

Winds move air around the world from areas of high pressure toward areas of low pressure. The world's windiest regions are along the tracks of the deep extratropical cyclones (see pp.302–03) in the mid-latitude oceans and in the trade wind zone. Strong pressure differences produce strong winds in individual weather systems, while regional winds occur in particular seasons.

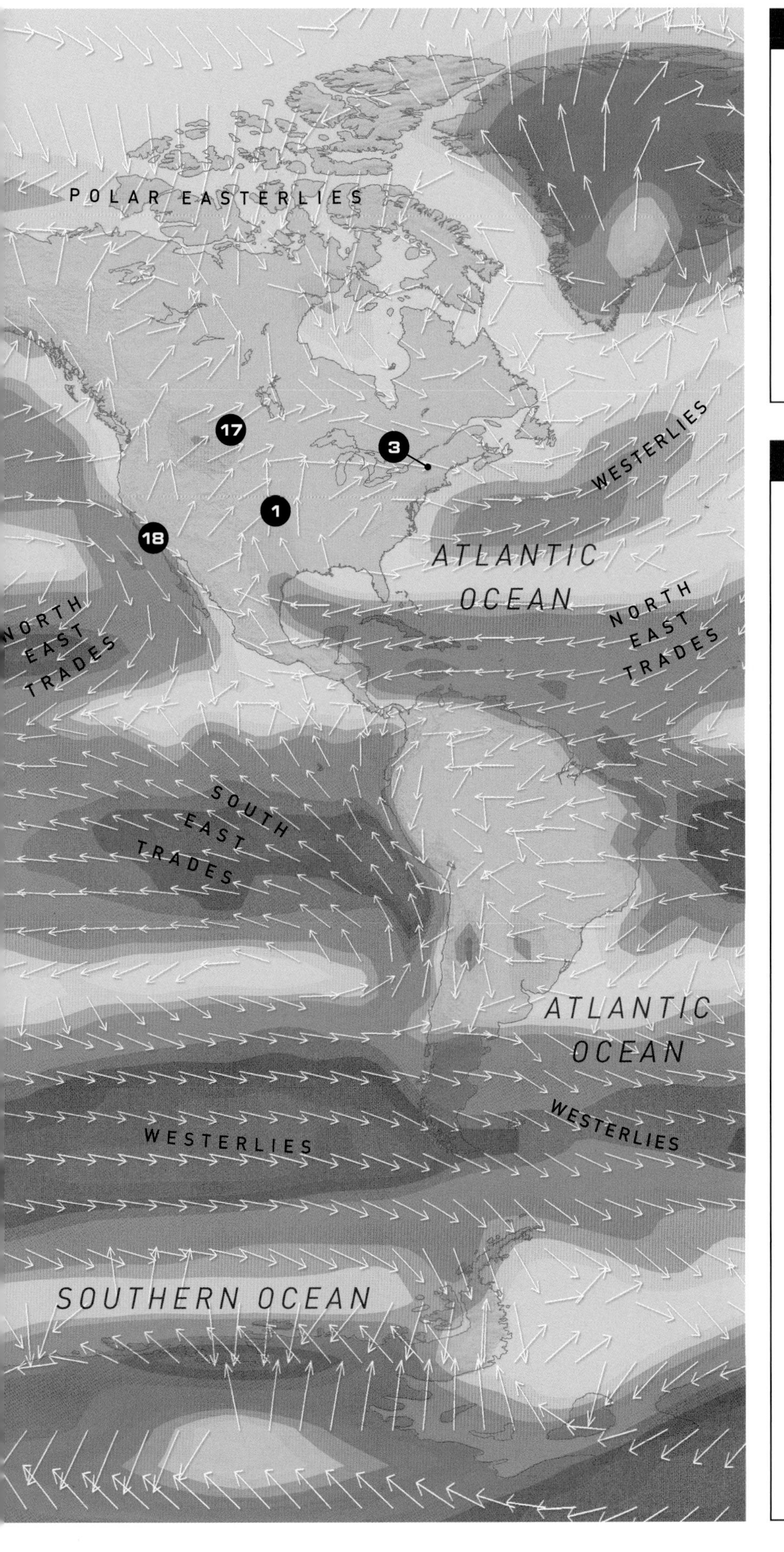

The highest windspeeds are recorded inside the funnel of a tornado, otherwise, winds that form around mountains and along the coast of Antarctica are extremely strong. The highest speeds usually occur during a short burst or gust of wind.

1 TORNADO FUNNEL

Country US
Speed 300mph (480kph)
When Frequent

2 BARROW ISLAND

Country Australia
Speed 253mph (408kph)
When April 10, 1996

3 MOUNT WASHINGTON

Country New Hampshire, US
Speed 231mph (371kph)
When April 12, 1934

4 COMMONWEALTH BAY

Country Antarctica
Speed 200mph (322kph)
When Common

Local winds affect areas of the continents at particular times of year. Regional winds are produced by factors such as topography that "squeezes" wind through valleys, and surface heating patterns that form pressure differences to affect flow.

5 TRAMONTANE

Area Pyrenees or Alps to Mediterranean
Type Cold and dry
When Winter

6 MISTRAL

Area Southern France
Type Cold, dry, northerly
When Mostly winter and spring

7 LEVANTE

Area Straits of Gibraltar
Type Strong easterlies
When Especially winter

8 MELTEMI

Area Aegean Sea
Type Strong, dry, northerly
When May–September

9 SIROCCO

Area North Africa and Mediterranean
Type Southerly, hot, dry, dusty
When Mostly spring and fall

10 BORA

Area Eastern Europe to Italy
Type Cold northwesterly
When Mostly winter

11 KHAMSIN

Area North Africa and Arabia
Type Hot, dry, dusty, southerly
When February–June

12 SHAMAL

Area Persian Gulf
Type Dry, northwesterly
When Mostly spring and fall

13 HARMATTAN

Area West Africa
Type Dry, dusty, northerly
When November–March

14 CAPE DOCTOR

Area South African coast
Type Dry, southeasterly
When Spring to late summer

15 ELEPHANTA

Area Malabar Coast, India
Type Strong south or southeasterly
When September–October

16 BRICKFIELDER

Area Southern Australia
Type Hot, dry, northerly
When Summer

17 CHINOOK

Area North America (from Rockies)
Type Warm, dry, westerly
When Mostly winter

18 SANTA ANA

Area California, US
Type Dry, offshore wind
When Late fall and winter

FRONTS

Fronts are important weather-related features that occur mainly across mid-latitudes and are associated with traveling depressions. Their boundaries are continually shifting, depending on where air masses of different temperatures and humidities meet. Many of the most heavily populated places in the world rely on the activity of fronts to provide rainfall. However, sometimes the rising moist air that flows over fronts produces widespread cloud and precipitation, some of which can lead to devastating floods. Warm fronts are more gently sloping than cold fronts, and their precipitation generally lasts longer. As depressions evolve, an occluded front lengthens as the faster cold front gradually scoops warm air up away from the surface, producing a narrow band of rain.

ROTATING FRONTS
This satellite image shows an approaching front. The wider band of cloud marks the cold front across the warm sector, with the occluded front swirling into the depression's low center. Speckled showers stream behind the cold front.

THE GULF STREAM
There are also fronts in the ocean. The Gulf Stream is a current that moves warmer water toward the poles. Warm water (in orange and yellow) flows from the Gulf of Mexico along the coast of the US before crossing the Atlantic toward northern Europe.

JET STREAMS

Jet streams are ribbonlike, fast-moving currents of air in the upper troposphere. The rapidly flowing airstream is formed by contrasts in temperature between tropical and polar air masses that occur right up through the troposphere; the greater the temperature difference, the faster the air flows through the jet stream. Swedish scientist Carl Rossby identified models to describe the motion of jet streams, now known as Rossby waves. Jet streams have a core where the air is flowing fastest, sometimes up to about 200 mph (322kph) in the winter. In mid-latitudes the "polar front" jet stream is a component of frontal depressions, which can develop into cyclones.

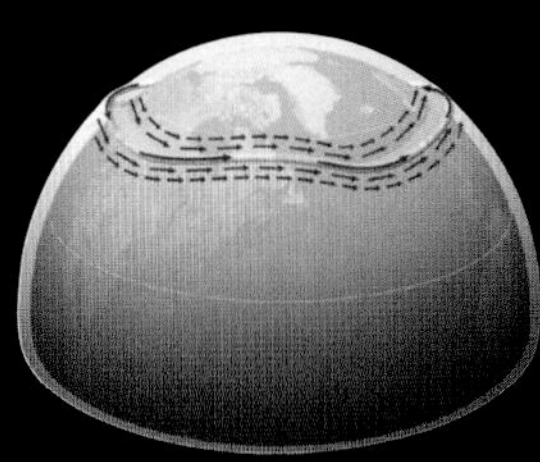

ZONAL ROSSBY WAVES
Rossby waves occur in the mid and upper troposphere and relate to surface weather. When the air blows from west to east, it it known as "zonal."

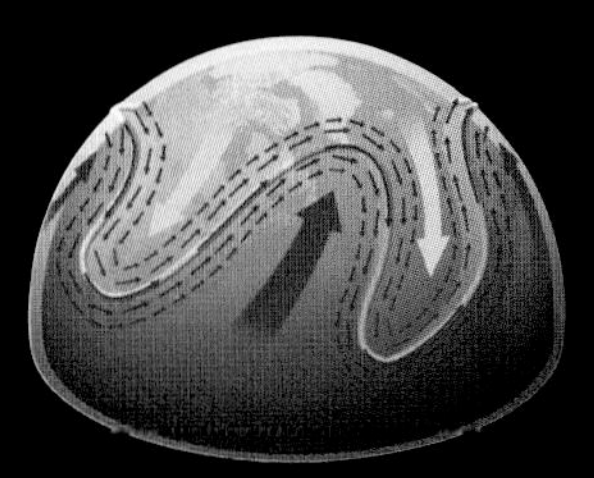

MERIDIONAL ROSSBY WAVES
Sometimes the air becomes very wavy or "merdional," with cool air flowing to lower latitudes and warm air moving to higher latitudes.

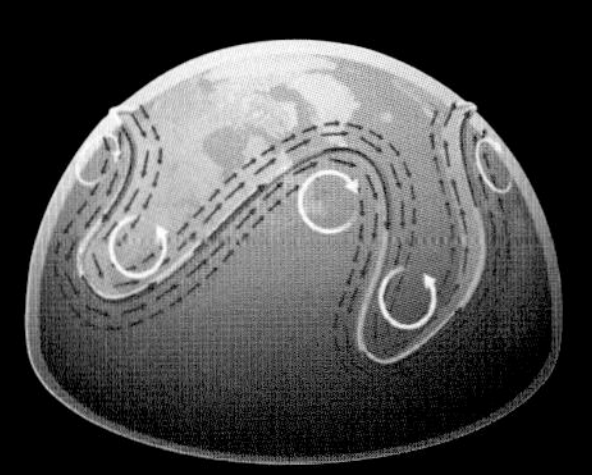

CUT-OFF ROSSBY WAVES
When the Rossby waves dig deep toward the lower latitudes, they can lead to swirling cold "cut-off" circulations that can bring showers to the subtropics.

SUBTROPICAL JET STREAM
The subtropical jet stream is a year-round feature of the upper troposphere that occurs at lower latitudes than the "polar front" jet stream. Here it is marked by a narrow strip of high cloud spanning the Red Sea.

WEATHER FRONTS

Norwegian meteorologists conceived the idea of weather fronts in the early decades of the 20th century as a "battleground" between air from different origins. Some modifications aside, this concept has stood the test of time. Fronts are significant features for weather analysts and forecasters because they symbolize the presence of large-scale, often thick, cloud and precipitation. This means that a classic sequence of clouds, combined with other measures such as a gradual change of wind direction, can herald the approach of a warm front. Major winter storms are usually linked to the presence of a warm front, which can create heavy snow and strong winds that can whip up blizzards. Cold fronts separate air masses, leading to dramatic falls in temperature. Sometimes they occur within the tropics when they penetrate into lower latitudes and bring rain to arid regions during the winter. They can also kick up dust storms in these areas.

Cold air mass High up are thin cirrostratus and altostratus clouds

Thick rain clouds Cirrus clouds are common ahead of the surface front

Surface front Warm air rises over the cool air, forming a boundary with a shallow slope

Rising air A mass of warm air rises toward the surface front

WARM FRONTS
This is the leading edge of the warm air in a frontal depression. The warm, damp air flowing over the front produces widespread cloud and rain, mainly ahead of the surface front.

AN APPROACHING WARM FRONT
A classic harbinger of a warm front is the gradual advance of the ice crystal cirrus clouds—the "mares' tails" cloud. Lower cumulus clouds sometimes occur ahead of the approaching warm front.

Cold front clouds Air from the warm sector is scooped up by advancing cold air

Heavy rain Rainband ahead of the front followed by clear, blue skies

Pressure build-up A boundary is formed when when cold air cuts under warm air

Warm sector The warm sector behind a warm front is followed by a sweep of much cooler and drier air

COLD FRONTS
This is the leading edge of cold air that streams in behind the frontal depression. The advancing cold air scoops up the warmer air ahead of it to produce cloud and rain.

COLD FRONTS AND RAIN SHOWERS
Behind a cold front, the chilly air may be warmed from below to generate cumulus clouds, some of which produce showers. This can happen over the sea in winter and over the land in spring.

Rising air Warm air is forced upward by the cold air of the cold front

Fast-moving air Cold air mass moves faster than warm air

Pushing ahead Cold air pushes under the warm air

Thick clouds A narrow band of thick cloud can produce heavy rain

OCCLUDED FRONTS
The cold front moves faster than the warm one, which means that gradually the warm sector air between them is lifted above the surface. This is known as occlusion, and generally produces a narrow band of rain.

RAIN AND SNOW WITHIN AN OCCLUDED FRONT
Occluded fronts have a relatively narrow band of low-layer cloud from which precipitation, such as rain or snow, will fall. The passage of an occlusion is not marked by much of a temperature change.

PRECIPITATION

Water and ice that falls from Earth's atmospshere is known as precipitation. Some parts of the world enjoy moderate amounts of rainfall, while continuously drought-stricken areas often need more. Excessive precipitation can lead to devastating floods.

RAIN AND SNOW

Rain and snow are the most common types of precipitation. In both forms they can range from widespread and prolonged, like those produced by tropical and mid-latitude cyclones, to localized and intense, such as those related to thunderstorms. Although most parts of the world experience nonthreatening amounts of precipitation, some extreme weather patterns involving rain or snow can lead to serious problems. Widespread flooding can occur with the passage of mid-latitude frontal depressions, and the most intense rain falls from very deep tropical cumulonimbus clouds. In the middle latitudes, the heaviest rain is from frontal depressions that have a strong, persistent supply of very damp, tropical maritime air. Large falls of snow, on the other hand, are most commonly associated with mid-latitude frontal lows supplied with relatively mild air that is rich in water vapor. Very cold air does not contain enough water to be the source of heavy snow.

RAIN WORLDWIDE

Highest average annual rainfall total:
460in (11,680mm) at
Mount Wai-ale-ale, Hawaii

Highest rainfall in one year:
1,042in (26,461mm) at
Cherrapunji, India, from
August 1, 1860 to July 31, 1861

Highest rainfall in one calendar month:
366in (9,300mm) at
Cherrapunji, India, in July 1861

Highest rainfall in 24 hours:
72in (1,825mm) at
Fac Fac, La Réunion Island,
Indian Ocean

Highest rainfall in 12 hours:
53in (1,350mm) at
Belouve, La Réunion Island,
Indian Ocean

GLOBAL PRECIPITATION
This map shows the level of precipitation around the world. The heaviest rainfall is concentrated in tropical regions.

Polar regions
Like deserts, these arid areas experience very little precipitation

Inter-Tropical Convergence Zone
The convergence of moist air into this region and mid-latitude lows produces the wettest weather

Chile
The town of Iquique receives heavy rain less than five times a century. Nearby, the Atacama Desert is one of the driest places in the world

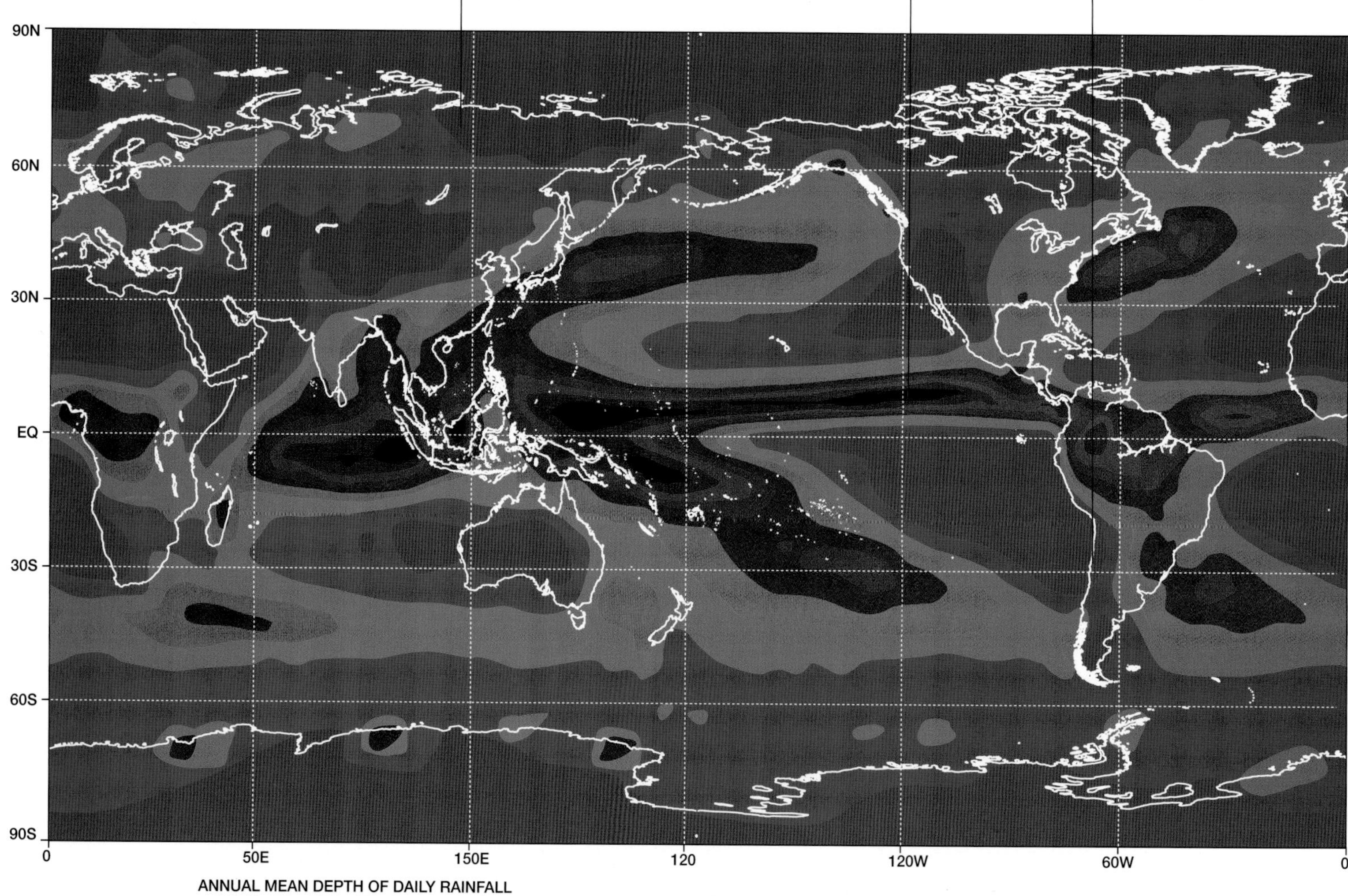

ANNUAL MEAN DEPTH OF DAILY RAINFALL

0.04in (1mm) 0.08in (2mm) 0.16in (4mm) 0.24in (6mm) 0.4in (10mm) 0.6in (15mm) 0.8in (20mm) 1in (25mm) 1.2in (30mm)

THE WATER CYCLE

The water cycle is a truly global phenomenon. Precipitation is just one part of it, and the amount and type of precipitation depends on what happens at various points during the water cycle. For example, the process of evaporation is contingent on sea temperature and moisture content in the soil and vegetation; so, warmer water and saturated soil will have a greater potential to create more precipitation. Similarly, a stronger wind may lead to more evaporation before the water condenses into clouds, potentially leading to greater levels of precipitation. Further along the cycle, some of the clouds will produce precipitation over land, some will fall on to the ocean. When rain falls on to land, some of it will evaporate again, but some will boost levels of soil moisture, as the rain soaks into the ground to supply rivers. Precipitation in the form of snow may last through winter or even longer to enhance the fluvial part of the cycle as spring snowmelt.

Snow precipitation Water returns to the land as snow

Rain precipitation Water returns to the land as rain

Transporting water Clouds carry water toward land

Plants Vegetation and plants lose water by transpiration

Lakes Water evaporates from lakes

Cloud formation Water evaporates from the sea, and condenses to form clouds

Seeping precipitation Precipitation is soaked into the ground and flows into the sea

Water returns to sea Water is carried downhill and returns to the sea via streams and rivers

Water storage Water accumulates in seas and oceans

TYPES OF PRECIPITATION

Precipitation takes many forms, including rain, drizzle, snow, and hail, and falls with varying intensity and duration. Stronger updrafts occur in cumulus clouds, while weaker ones form stratus clouds. Outside the tropics, most rain starts as snow high up in the cloud and melts as it descends.

DRIZZLE

These liquid droplets are around 0.02in (0.5mm) in diameter and fall from shallow layer clouds. They are seldom a risk unless they reach the surface as supercooled droplets that freeze on impact with any surface, leading to slippery conditions.

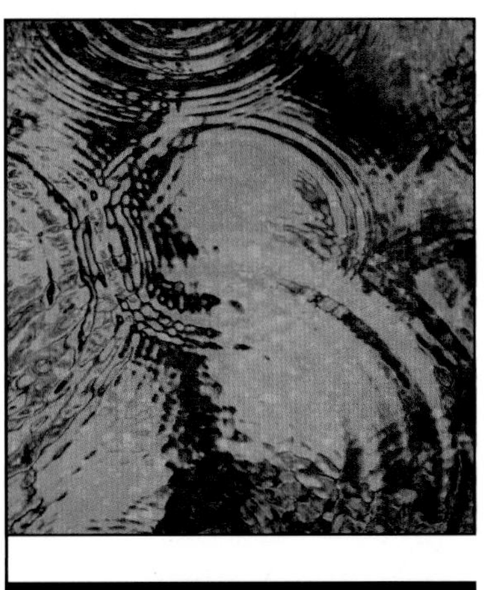

RAIN

The drops from this liquid precipitation are usually between 0.02–0.2in (0.5–6mm) in diameter. The heaviest rain falls from tall cumulonimbus clouds or very thick layer clouds. The most intense rain occurs in the tropics from thunderclouds.

FREEZING RAIN

This dangerous form of precipitation occurs when liquid drops are supercooled. They will freeze once they make contact with any surface, such as roads or sidewalks, leading to hazardous conditions, especially for motorists and pedestrians.

SNOW

Snow forms when tiny ice crystals in clouds stick together to become snowflakes. Deep snow and blowing snow are both very dangerous, causing serious and prolonged disruption. Frontal depressions are the most common source of widespread snow.

MAMMATUS CLOUDS

These bulbous, drooping clouds usually form on the underside of cumulonimbus anvils and commonly indicate thundery weather. They form in saturated, sinking air that is cooler than its surroundings. Here their unusual shape is dramatically illuminated by a Sun that is low in the sky.

EL NIÑO AND LA NIÑA

Variations in normal weather patterns, such as El Niño and La Niña, bring about some vast and extreme weather events. These involve unusual sea temperatures, changes to ocean currents, and effects on pressure systems. Together these phenomena are part of a complex global weather pattern known as El Niño–Southern Oscillation (ENSO).

EL NIÑO

The Spanish term "El Niño" originated in Peru where, at the end of the year, the normally cool coastal waters of the Peru current are replaced by warm water for a few weeks. Because this occurs around Christmas, locals called it El Niño—meaning "little boy" or "Christ child." However, the equatorial Pacific regularly experiences a much more extensive warming at irregular periods of between two and seven years, and this significant large-scale event is also known as El Niño. It has an impact on seasonal weather patterns across an area larger than the tropics. These seasonal anomalies range from increased droughts to substantially wetter, stormy conditions. Its origin is related to a see-saw effect known as the Southern Oscillation, which causes a sea-level pressure difference between a high in the southeast Pacific and a low in Indonesia. The pressure gradient between them drives the trade winds. When the difference is small, the trades weaken or even reverse into westerlies along part of the equator. This sets warm water drifting eastward across the Pacific.

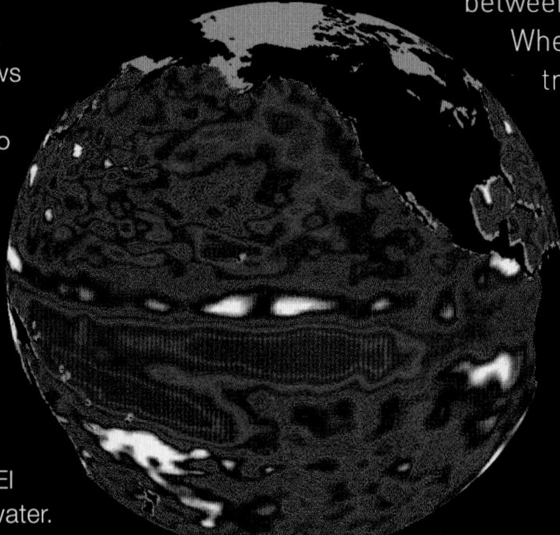

RAISED SEA LEVELS
This satellite image shows the height of the ocean surface during an El Niño event on July 11, 1998. As the surface warms it expands, so height is an indicator of the temperature of the ocean. The red-and-white band across the equator indicates a sea level about 3in (8cm) above normal, marking El Niño's migrating warm water.

RECORD WAVES IN HAWAII
Storms in the Pacific Ocean become more violent and more frequent when an El Niño event is taking place. Huge waves reaching 35ft (10m) have been recorded on the island of Oahu, Hawaii.

FOREST FIRE
Areas such as Borneo and the Amazon Basin become more susceptible to forest fires due to the increased sinking, drying motion of the atmospheric circulation during an El Niño. Here a fire rages in the Amazon rainforest during the El Niño of 1998.

NORMAL PATTERN
In a normal year trade winds blow from a high across the southeast Pacific toward a low over Indonesia. The southeasterly winds drive an equatorial current that leads to very warm water around Indonesia, which sparks frequent rain showers. Conditions are dry in the east over equatorial South America.

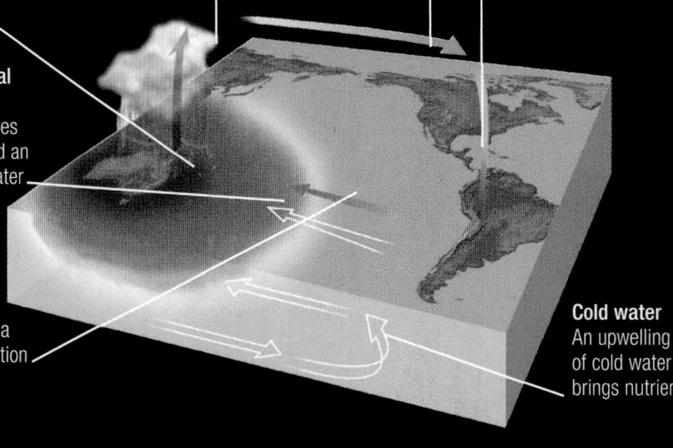

EL NIÑO PATTERN
El Niño events begin with a weakening of the normal pressure difference between the southeast Pacific high and Indonesian low and therefore of the trade winds too. This leads to the warm water area slowly migrating east along the equator, as far as South America. The unusually warm water sparks unseasonal showers, bringing serious flooding.

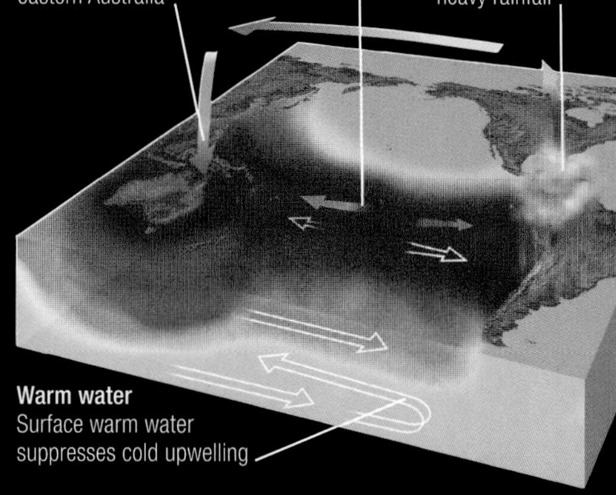

NORTH-ATLANTIC OSCILLATION

A large-scale pressure see-saw, the North-Atlantic Oscillation (NAO) affects winter weather in the northern hemisphere. The two major pressure centers involved are the Iceland low and the Azores high, with southwesterly winds usually blowing between them, bringing mild air to Europe. A large pressure difference between them means stronger winds and a small difference brings much weaker winds. A change in this pressure difference from year to year or over longer periods affects the nature of winters across much of Europe. Meteorologists use the difference in monthly mean pressure between the Azores and Iceland to study this phenomenon, linking it to the strength and range of winter weather systems over Europe.

Strong high pressure
This brings a strong flow of damp mild air across to northern Europe

Deeper low pressure
The depression tracks mainly across northern Europe, leaving the Mediterraean dry

POSITIVE NAO
A large pressure difference, known as a positive NAO brings generally mild and wet or snowy winters across northern Europe, with mainly dry conditions and water shortages over the Mediterranean region. The southeastern US tends to enjoy milder and wetter winters at this time.

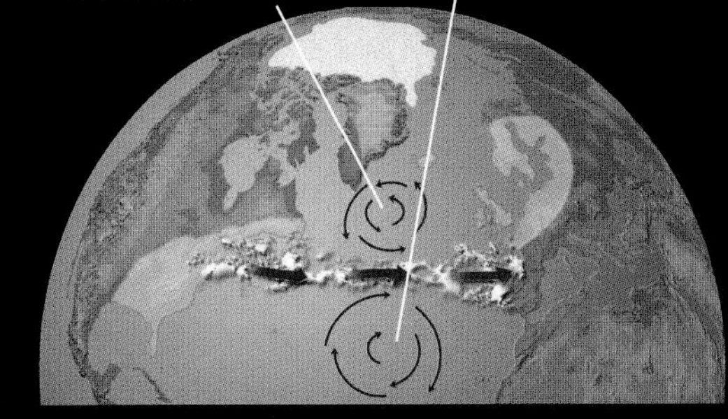

NEGATIVE NAO
A small difference in the pressure values is associated with a quite different pattern of European winter weather. In this situation, there are mainly cool, dry conditions over northern Europe and much more disturbed weather across the Mediterranean. Cooler winters tend to occur in the eastern US.

LA NIÑA

La Niña (Spanish for "little girl") develops with a similar frequency to El Niño though in reverse. Also like El Niño, it can trigger extreme weather patterns. It is believed, for example, that a recent La Niña event could have provided the conditions for the development of Cyclone Yasi that struck Queensland, Australia, in 2011. It is also known to be related to changes in the winter jet stream flow across the Pacific and North America. There are typically wetter than average winters in the Pacific Northwest and warmer, drier conditions in the southern US.

BEACH EROSION
The southern branch of the jet stream during periods of La Niña drives frontal depressions toward the US Pacific Northwest, pounding the coast. This can lead to significant coastal erosion such as here at Olympic National Park, Washington.

Low pressure
System of low pressure is located further west

Trade winds
Southeast trade winds are stronger

High pressure
Falling air brings dry conditions and high pressure

LA NIÑA PATTERN
La Niña, the antithesis of El Niño, occurs when the trade winds are stronger than average, enhancing the upwelling of cool water from the depths off the equatorial South American coast and transporting it westward across the ocean. The area of warm water and rainy weather is then confined to the western Pacific.

QUEENSLAND FLOODS 2010

December 2010 was the wettest month on record in Queensland, Australia. Heavy rainfall brought by Cyclone Tasha and weather conditions caused by La Niña (see pp.284–85) saturated the Brisbane River and other river basins. The town of Rockhampton—about 310 miles (483km) north of Brisbane—experienced more than 16in (400mm) of rain, massively higher than the average December rainfall of 4.3in (108.5mm). As rivers burst their banks, flooding became widespread across the state. Thousands of people were evacuated from towns and cities, roads were cut off, and homes were submerged in muddy water. The flooding continued into 2011, with 20,000 houses inundated in the city of Brisbane on January 13. Torrential downpours caused massive flash floods that resulted in fatalities. The cost of the damage to homes and infrastructure is estimated to be $10 billion. To make matters worse, Cyclone Yasi arrived in early February, weaving a path of destruction along the north Queensland coast. Wind speeds were estimated at more than 120mph (200kph), and in the city of Townsville alone, roofs were torn off buildings and sent hurtling down the streets.

1 FLASH FLOODS
Flash floodwaters cover a street in Toowoomba, Queensland, on January 10, 2011. Tragically, this event led to significant loss of life in the community.

DEC 2010–JAN 2011	
Location	Queensland, Australia
Type	Severe flooding
Fatalities	35

350,000

THE NUMBER OF PEOPLE DISPLACED

THE DELUGE
Rockhampton was severely hit. What began with massive rainfall in late December, turned into an ongoing disaster. This view from January 12 shows the flood's persistence in the city.

WATER LEVELS
This false-color satellite image shows the extensive inundation within Rockhampton. The unaffected areas are in red, and the flooded areas are in brown.

2 SEARCH FOR MISSING PEOPLE
An army helicopter lands by the remains of a home that was washed away by floodwater in Grantham, Lockyer Valley, on January 15. The military played a critical role in the rescue operation.

3 MAJOR DAMAGE
Further flash floods on January 11 in Toowoomba produced massive damage to property. The estimated 23-ft- (7-m-) high torrent piled up many vehicles like toys.

4 PUBLIC RESPONSE
Volunteers remove flood debris from houses in the Brisbane suburb of Fairfield on January 16. More than 35 of Brisbane's suburbs were severely affected.

“WHAT WE HAVE HERE IN QUEENSLAND IS A VERY GRIM AND DESPERATE SITUATION.”

ANNA BLIGH, PREMIER OF QUEENSLAND

MONSOONS

Monsoon comes from the Arabic "mausam", meaning season. Some regions of the world experience a seasonal reversal of wind direction and attendant changes from wet to dry seasons. Those affecting south Asia and West Africa are the best known.

WHAT CAUSES MONSOONS?

Seasonal heating patterns over continents lead to a change from high pressure in winter to low pressure in summer. Over south Asia, for example, the weather during the dry season (December–February) is dominated by air that flows in from the high pressure system in Siberia (also called the Siberian anticyclone). This air subsides over the Himalayas to the north, and northeasterly or northerly winds blow across India and adjacent countries at this time. During the wet summer season (June–September), low pressure develops over the India–Pakistan border region, drawing in warm, moist air across the tropical ocean to the south.

low pressure over Thar Desert

Inter-Tropical Convergence Zone

warm and wet southwest winds

high pressure over Indian Ocean

SUMMER RAIN
A low pressure centered over the Thar Desert between India and Pakistan combines with high pressure over the Indian Ocean to draw moist air across south Asia.

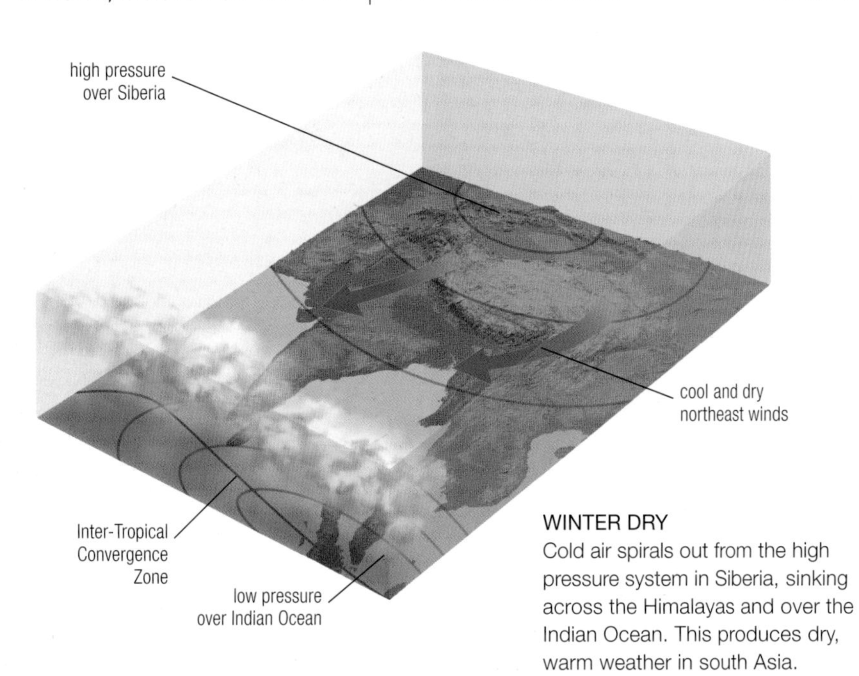

WINTER DRY
Cold air spirals out from the high pressure system in Siberia, sinking across the Himalayas and over the Indian Ocean. This produces dry, warm weather in south Asia.

4in (100mm) 8in (200mm) 12in (300mm) 16in (400mm)

HEAVY RAINFALL
Estimates of rainfall can be made from satellite observations. Heavy monsoon rains fall across parts of eastern Asia, stretching from central China through Korea, into Japan.

BRAHMAPUTRA BEFORE THE MONSOON
The Brahmaputra River is normally swollen during the wet summer monsoon season. In August 2007, there was some flooding, but nothing like what was to follow.

BRAHMAPUTRA AFTER THE MONSOON
Massive rainfall in September 2007 led to extensive flooding of the river, which had an impact on Bhutan, Bangladesh, and northeastern India. It was the third and worst flood of the season.

EFFECTS OF MONSOONS

Monsoon rains are occasionally disruptive in southern Asia. If excessive and prolonged, they lead to floods, resulting in loss of life and diseases, necessitating massive rescue and relief efforts. If below average, they can cause shortage of crops for the inhabitants of the region, requiring governments to offer welfare programs, particularly in rural areas. Monsoon rains are a substantial factor in the economy of India, where agriculture accounts for around two-thirds of the national income. High rainfall improves rice, cotton, and wheat yields, and replenishes reservoirs and groundwater, thus improving irrigation and aiding hydroelectric power generation.

IMPACT ON AGRICULTURE
Monsoons are essential to farmers, like the ones seen working here in the paddy fields of Myanmar. High temperatures and plentiful rain result in greater productivity.

MONSOON FLOODS
Torrential rain can cause severe disruption. In 2010, serious urban flooding occurred in Jammu, India, during the summer monsoon. Inundation of agricultural land can lead to significant crop damage in some years.

PAKISTAN FLOODS 2010

Unusually severe flooding in the Indus Valley in Pakistan followed the monsoons in June 2010. The origin of the floods was partly the result of the usual monsoon depression that moved into the region from the Bay of Bengal and partly related to a very unusual pattern of jet stream flow in the upper troposphere. In most years during the summer the flow of the jet stream is in a westerly direction to the north of the Himalayas. However, this year it meandered significantly so that it headed north across western Russia, causing an anomalous heat and dryness in the region, known as a "blocking high." Downstream it headed south to cause a trough, or low-pressure area, leading to persistent massive downpours. Unusually heavy rain fell in the Bay of Bengal throughout June, July, and August, as much as 11in (274mm) in a 24-hour period in Peshawar. The steep-sided upland regions in the headwaters of the Indus Basin suffered devastating flash flooding and related mudslides. The flood spread southward down the valley as a wave, with the heavy rains finally subsiding in September. Vast areas of highly productive land were swamped for long periods, destroying critically important crops such as rice and cotton. The floods eventually caused the deaths of more than 1,500 people and displaced over 6 million people.

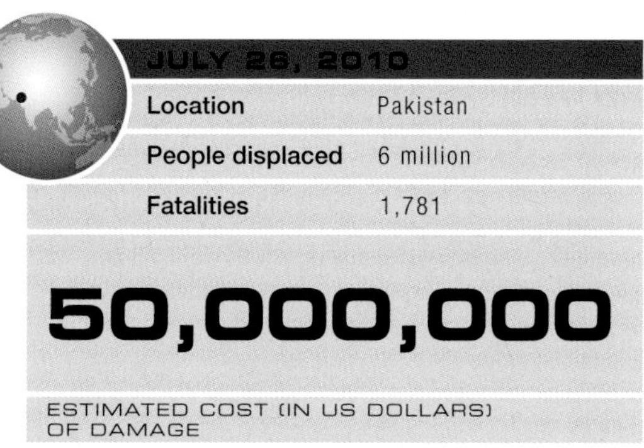

JULY 26, 2010	
Location	Pakistan
People displaced	6 million
Fatalities	1,781

50,000,000

ESTIMATED COST (IN US DOLLARS) OF DAMAGE

SWOLLEN RIVERS AND MUDSLIDES
The Swat Valley in Pakistan was very heavily damaged by flash flooding and devastating mudslides. As many as 34 of the region's 42 bridges were destroyed, along with a substantial number of roads. A lack of fresh water led to an outbreak of cholera.

WIDESPREAD FLOODING
The flooding in this area and on this scale was unprecedented in living memory. Millions of people were affected, escaping with what little they could carry. Gradually, the floods swept southward down the Indus Valley.

EFFECTS OF THE FLOODS

The enormity of the flooding, with one-fifth of the Indus River basin underwater, meant the event was the worst in at least 80 years. Substantial damage to domestic and public buildings, with more than 500,000 homes destroyed and many river bridges washed away, meant significant rebuilding and the rehousing of six million homeless individuals. Some 10,000 electricity transmission lines and transformers were destroyed along with extensive swathes of land on which rice, cotton, sugar cane, and tobacco were grown—at least 7.9 million acres (3.2 million hectares) of some of Pakistan's prime agricultural land was deluged. On August 18, the government announced that 15.4 million people had been directly affected by the disaster. Flooding also overwhelmed sewerage and water treatment plants causing a lack of clean drinking water, which led to disease.

FLOODED AREAS
This image, acquired by the Landsat 5 satellite on August 9, 2010, shows flooding near Kashmor, Pakistan. The second wave of flooding on the Indus was to affect this region a few days later.

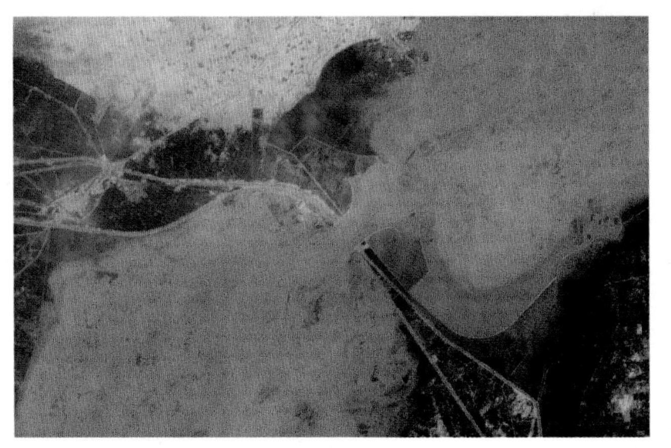

SEVERE INUNDATION
The region on August 12, 2010 shows serious, widespread flooding. River discharge though the gap in the center of the image reached 82,000 cu ft (25,000 cu m) a second.

> “I AM IN A VILLAGE WHICH WAS ENTIRELY SWEPT AWAY. I AM SURVIVING... BY RAILROAD TRACKS ON HIGH GROUND...”
>
> **ALEEM MAQBOOL**, BBC NEWS, NORTHWEST PAKISTAN

TROPICAL CYCLONES

Also known as hurricanes and typhoons, tropical cyclones form and develop across specific low-latitude ocean areas. They can cause widespread damage upon making landfall, generating storm surges and battering homes.

DEVELOPMENT

A tropical cyclone develops where the upper ocean layer is warmer than 79°F (26°C). It also requires the presence of tall thunderclouds and high humidity that prevents surrounding dry air from being drawn into the growing clouds. Usually, the first sign of a possible tropical cyclone is a "cloud cluster," which only grows if the direction and speed of the wind near the sea surface is similar to that in the upper levels. Otherwise this nascent system will "blow apart." Once the growing cloud cluster is fully formed, the spiraling, counterclockwise inflow over the ocean feeds strong updrafts (vertical movement of air) into the tall thunderclouds. The spiral motion is caused by the Coriolis effect, in which winds moving in a northerly and southerly direction are deflected due to Earth's rotation. On top of the tropical cyclone, the flow of air is outward, away from the center (the eye of the cyclone). If this upper outflow strengthens, it can lead to deepening of the low-pressure eye and stronger winds.

Atlantic hurricanes commonly develop off the coast of West Africa, sometimes near the Cape Verde Islands. Some that strike the Caribbean and US start life as westbound disturbances across West Africa. Tropical cyclones never form within about five degrees of the equator as organized circular motion does not occur there.

Cirrus clouds
Lie over cumulonimbus that forms bulk of clouds

Dry, warm air
Descends in the eye, an area of calm at the center of the cyclone

Sea surface
Rises at the center of the cyclone

Moist, warm air
Spirals upward in the eyewall, the area of strongest winds and rain

STAGES OF A CYCLONE

1 TRIGGERING A CYCLONE
The early stage of a cyclone is represented by the merger of a fairly scattered cluster of thunderclouds into a coherent cloud mass, with an indication of rotation around a low-pressure center.

2 A SWIRLING MASS
As the system intensifies, the circulation becomes more obvious with an indication of inward-flowing bands of cumulus clouds. The flow at this stage is more clearly associated with a center of circulation.

3 MATURE STAGE
The full-blown intense tropical cyclone is marked by a roughly circular cloud pattern and a distinct eye, with spiral rain bands. The cirrus clouds flowing out around the eye present a smooth texture.

High-level winds
Spiral outward from the center of the cyclone

Spiral rain bands
Extend for hundreds of miles from the cyclone center

Rain bands
Rising air in these forms dense clouds that cause heavy rainfall

CYCLONE ANATOMY
A strong tropical cyclone has an area of sinking air at the center of circulation. The eye is circular, and may range from 2 miles (3km) to 230 miles (370km) in diameter.

The western tropical Pacific is the warmest tropical ocean, and has the most number of cyclones, called typhoons in this region. Most tropical cyclones occur in the northern hemisphere, with 70 percent occurring from June to October. The southern hemisphere experiences cyclones from January to March. Cyclones are absent from the southeast Pacific and south Atlantic where cooler waters dominate.

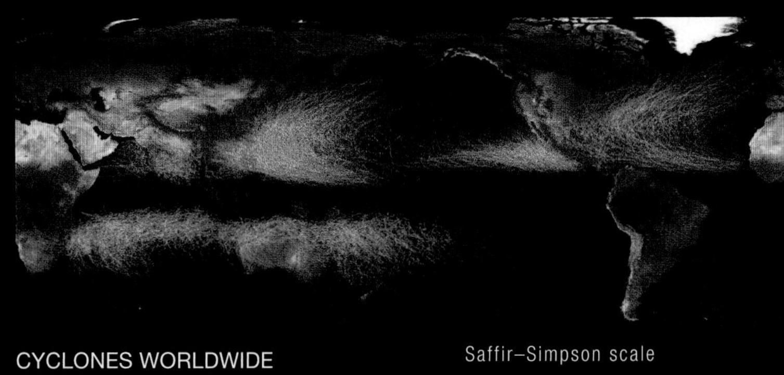

CYCLONES WORLDWIDE
Tropical cyclone tracks from 1985 to 2005 are shown here. These created havoc in tropical as well as mid-latitude regions.

MEASURING A HURRICANE

Named after a meteorologist, Bob Saffir, and an engineer, Herbert Simpson, the Saffir–Simpson Scale is a simple, useful scale of hurricane intensity. It is based on the nature of the damage caused, related to maximum sustained wind at a height of 33ft (10m) during a one-minute period. Intensity increases from 1 to 5, with a description of the damage and storm surge height above normal tidal levels. The scale is used and understood widely in hurricane-prone areas of the US. It is defined such that a category increase by one means a roughly fourfold increase in damage. A hurricane will probably weaken by one category after 0.6 miles (1km) of landfall.

CATEGORY 1

Wind speed 74–94mph (120–152kph)

Surge height 5ft (1.5m)

Example Hurricane Stan (2005)

Trees, shrubs, and unanchored mobile homes are damaged in category-1 hurricanes. Some damage also occurs to piers due to minor coastal flooding.

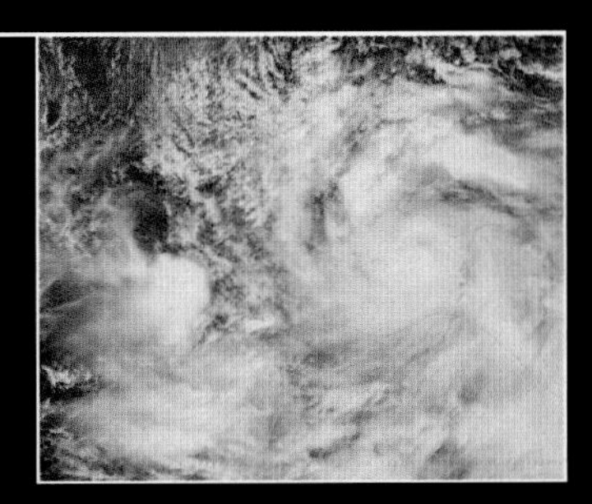

CATEGORY 2

Wind speed 95–109mph (153–176kph)

Surge height 6–8ft (2–2.5m)

Example Hurricane Nora (2003)

This category of hurricanes causes substantial damage to exposed mobile homes and piers. Trees are blown down and building roofs are also damaged.

CATEGORY 3

Wind speed 110–130mph (177–210kph)

Surge height 8–13ft (2.5–4m)

Example Hurricane Helene (2006)

Small buildings sustain structural damage and mobile homes are destroyed in category-3 hurricanes. Coastal flooding occurs and large trees are blown down.

CATEGORY 4

Wind speed 131–155mph (211–250kph)

Surge height 13–18ft (4–5.5m)

Example Hurricane Dennis (2005)

These hurricanes cause widespread damage to property, including windows, doors, and roofs, with flooding reaching up to 6 miles (10km) inland.

CATEGORY 5

Wind speed > 155mph (> 250kph)

Surge height > 18ft (> 5.5m)

Example Typhoon Jangmi (2008)

Buildings are severely damaged and lower floors sustain extensive damage if the cyclone is about 15ft (4.5m) above sea level within 1,640ft (500m) of shore.

STUDYING HURRICANES

Geosynchronous weather satellites are invaluable to forecasters, who keep an eye on the progress and development of tropical cyclones, also known as hurricanes in the Atlantic. These satellites provide high-quality images of evolving hurricanes across the tropical North Atlantic every half hour. Forecasters analyze the cloud patterns in these images to estimate a hurricane's intensity (surface wind speed). This is an indirect method of monitoring and predicting hurricanes. Their intensity is also measured directly by specialized aircraft, such as the C130 Hercules of the US Air Force. These aircraft carry out routine flights right across the eye of threatening storms, to measure temperature, pressure, and other variables from which sea-level wind speed can be calculated.

Predicting hurricanes is undertaken by weather services in some nations as an integral feature of their global numerical model forecast. The model predicts the direction and speed of the hurricane, its central pressure, and even the amount of rainfall that is likely to occur. It also forecasts the height and extent of the storm surge.

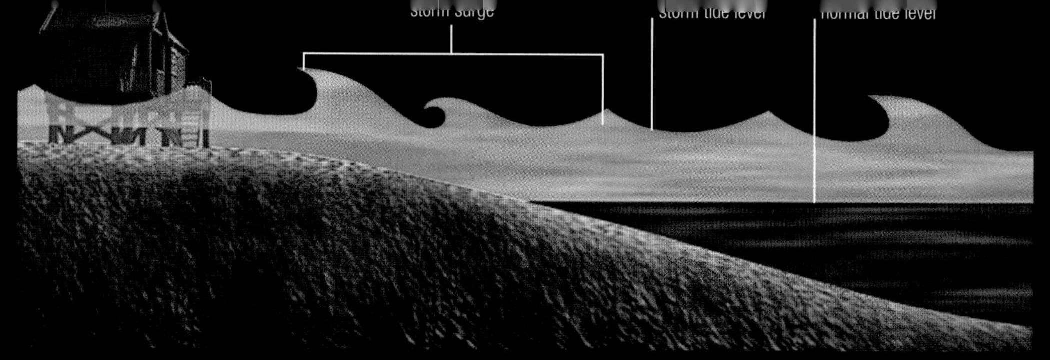

STORM SURGE

As a tropical cyclone moves across the ocean, the low pressure in its eye (see pp.292–93) domes the sea surface like a low, broad "hill" of water that moves along with the hurricane. When the cyclone hits land, the water in this dome surges over the coast in what is known as a storm surge. These amplify in shallow coastal waters—in extreme cases to a height of about 20–23ft (6–7m) above normal tidal levels. In the northern hemisphere, the eye track, coupled with strong onshore winds to its right, mean the surge is highest on the forward right quarter of a tropical cyclone approaching a coast line. In the southern hemisphere, the forward left quarter experiences the worst surge.

Storm surges can erode large sections of the coastline, cause flooding, and wash boats inland. Water movements associated with them can also damage coral reefs, especially those near the surface.

DANGER LEVEL
The most extensive flooding from a storm surge occurs if it happens to coincide with spring high tide—exceptionally high tides caused by the alignment of the Sun, Moon, and Earth.

HURRICANE ISABEL
A view of the prominent eye of 2003's Hurricane Isabel is seen here along with the "bubbling" cloud of massive thunderstorm tops embedded in one of its spiral rain bands, to the left of the eye.

CHASING HURRICANES

The ability to make direct observations of hurricanes that bear down on the Caribbean and United States is invaluable to forecasters. The 53rd Weather Reconnaissance Squadron of the US Air Force has a base at Biloxi, Mississippi, from where specialized aircraft such as C130 Hercules fly directly through all threatening hurricanes. They drop packages called dropsondes that descend in the storm, sending back details of the temperature, pressure, humidity, and wind. This data helps in predicting and monitoring the intensity of the storm.

Hurricane hunters
A dropsonde operator is seen in a C130 Hercules, part of the "Hurricane Hunters" team of the US Air Force. He is boosting the forecast quality by launching a dropsonde during Hurricane Floyd.

Storm chaser
The US National Oceanic and Atmospheric Administration (NOAA) monitors Atlantic hurricanes using specially equipped aircraft, such as this Lockheed WP-3D Orion.

DAMAGE CAUSED

Hurricanes cause damage in various ways. The dynamic pressure of the high winds can cause massive destruction, while torrential downpours often lead to serious flooding and mudslides. These worsen along low-lying coasts due to massively high, wind-driven ocean waves and the inland penetration of storm surges. The level of destruction forms the basis of the Saffir–Simpson scale (see p.293), which defines "significant" hurricanes as those between category 3 to 5. In the past, structures and vessels at sea were susceptible to damage, but today's forecasts and warnings have made this rare. However, essentially fixed features like oil and gas rigs can suffer extensively from passing hurricanes.

Category-5 tropical cyclones, such as Katrina (see pp.298–99) that passed through the Gulf of Mexico in the North Atlantic season of 2005, cause extensive damage to roofs of homes and industrial buildings, while some small buildings can be blown away. The magnitude of flooding from rain depends on the speed of progress of the cyclone—slower speeds result in greater flooding.

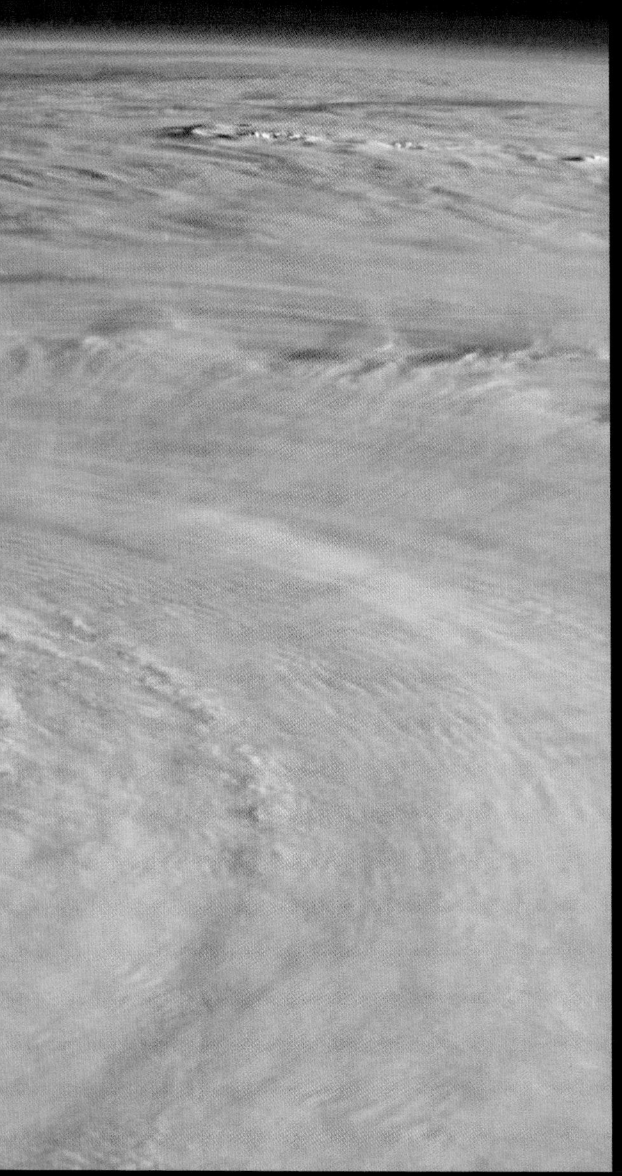

AFTER THE STORM
Mobile homes are not substantial enough to withstand a hurricane's fury. Unanchored ones are destroyed even by relatively weak category-1 storms.

SUBMERGED
Hurricane Ivan's storm surge swamped parts of the coastal city of Pensacola in Florida, in 2004. The very extensive low-lying coastline of the southern and southeastern US means such flooding is a perennial threat.

NARGIS LOOMS

A few days before landfall, Nargis exposed its power in the form of a well-developed eye and spiraling clouds. Forecasters often use satellite images such as this to study these features, which indicate a storm's severity.

CYCLONE NARGIS 2008

Considered to be the worst natural disaster to strike Myanmar, Nargis was the deadliest named cyclone ever recorded in the northern part of the Indian Ocean Basin. It made landfall on May 5, 2008, swamping the Irrawaddy Delta region some distance inland with a combination of a 12-ft- (3.6-m-) high storm surge and persistent torrential rain. The estimated 130-mph (210-kph) winds put Nargis into category 3 or 4 on the Saffir-Simpson Scale (see p.293). The shape of the Bay of Bengal region and the shallow waters in its northern part can amplify storm surges as they run across the bay. This, coupled with the extensive, heavily populated flat lands, leads to massive inundation. Although a cyclone is frequently monitored by satellite, getting warnings to the inhabitants can be very difficult. Slow evacuation in the region exacerbated the problem in 2008.

MAY 2008

Location	Myanmar
Type	Tropical cyclone
Damage	US $10 billion

138,000

THE NUMBER OF PEOPLE KILLED

NARGIS HITS HARD

1 HOVERING OVER MYANMAR
This satellite image shows the floodwaters over Myanmar's coasts. When Nargis hovered over the Bay of Bengal, it was already termed a severe cyclone. Here, the water is blue–black, vegetation is bright green, bare ground is tan, and clouds are white or light blue.

2 TORRENTIAL RAIN
Intense and prolonged rain resulted in widespread flooding and mudslides at different places. Here, victims await relief in the town of Dedaye, where rainwater ponds—their usual source of fresh water—were made useless by saltwater inundation.

3 MYANMAR DEVASTATED
Many villages were flooded and some communities suffered near total destruction of their homes. Some 2.4 million people were affected by hunger and homelessness. Inundation and infrastructure damage meant rescue and relief workers took days to reach victims.

THE PATH OF THE CYCLONE
The track of Nargis is based on frequent satellite images. It intensified into a Category-4 hurricane while heading for Myanmar. The Saffir-Simpson scale explains a hurricane's stages.

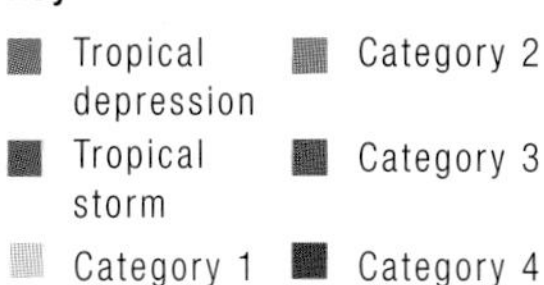

"OUR BIGGEST FEAR IS THAT THE AFTERMATH COULD BE MORE LETHAL THAN THE STORM ITSELF."

CARYL STERN, PRESIDENT AND CEO, US FUND FOR UNICEF, 2008

THE AFTERMATH

The death toll following Nargis was massive, due to drowning and the diseases that commonly occur after widespread flooding. Cholera and dysentery were rampant, as a result of poor sanitation and rotting bodies. Malaria and dengue fever also spread. Bogalay, in southern Myanmar, lost 365 of its 369 houses, while the destruction of one of the two oil refineries in the country led to a critical shortage of gasoline. Local and international aid workers delivered large quantities of rice, rehydration solution, and tarpaulin for shelter.

Makeshift shelter
Families affected by the cyclone take shelter from the rain in temporary accommodation along a road in the Shwepoukkan area of Yangon.

HURRICANE KATRINA 2005

The extremely active 2005 hurricane season in the North Atlantic saw 26 named storms, of which three attained category-5 status, including Hurricane Katrina. It was the costliest cyclone system ever and one of the five deadliest hurricanes to hit the US. It reached its most ferocious intensity after crossing a large patch of unusually warm water in the Gulf of Mexico. By midday on August 28—less than a day before landfall—winds peaked at 150 knots (170mph/280kph). The storm battered the coastline, and in New Orleans high waves and torrential rain caused widespread flooding. Thousands were stranded on rooftops for days before emergency services could reach them. In all, Katrina caused an estimated $90 billion worth of damage. The impact on New Orleans has been long-lasting. Many of the city's damaged houses still stand unrepaired and a significant number of its citizens never returned.

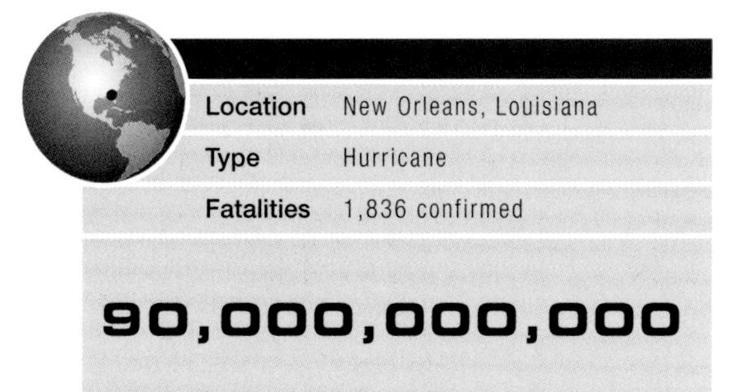

Location	New Orleans, Louisiana
Type	Hurricane
Fatalities	1,836 confirmed

90,000,000,000

ECONOMIC DAMAGE (IN US DOLLARS)

> “WE'RE GETTING REPORTS AND CALLS THAT (ARE) BREAKING MY HEART, FROM PEOPLE SAYING, 'THE WATER IS UP TO MY NECK. I DON'T THINK I CAN HOLD OUT'... AND THAT'S HAPPENING AS WE SPEAK.”
>
> **CLARENCE RAY NAGIN (JR.)**, MAYOR OF NEW ORLEANS, SEPTEMBER 2005

KATRINA STRIKES

1 DEPRESSION ADVANCES
While crossing over the Bahamas, a tropical depression intensified into a tropical storm. Typically, this is the stage when tropical cyclones are named. Hurricane Katrina was named on August 24, 2005.

2 FLORIDA GETS HIT
Katrina attained category-1 hurricane status with 70-knot (80-mph/130-kph) winds while clipping across southern Florida. It tracked over the Everglades wetlands and produced up to 14in (35cm) of rain locally.

INTENSE FLOODING

Katrina's massive storm surge, made worse by extremely high wind-driven waves, hit the Mississippi coast so violently that entire coastal settlements were destroyed. The inland penetration of the surge waters, combined with freshwater flooding from torrential rain, led to swollen rivers that overspilled the levees. A mandatory evacuation of New Orleans was ordered, but thousands of people had to seek shelter in the city's Louisiana Superdome stadium. The consequences of the widespread deep flooding were dire, with some 1,300 fatalities in Louisiana, the worst-affected state.

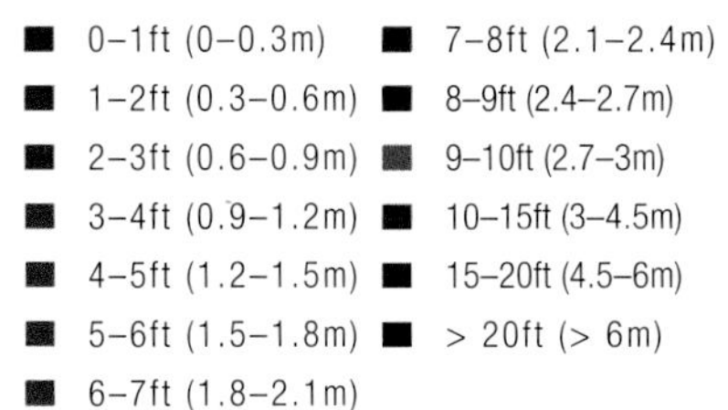

SCALE OF FLOODING
This false-color satellite image of the city of New Orleans shows the extent and depth of floodwaters in the city. Waters receded from the city six weeks later.

3 LEVEE SYSTEM BREAKS
Storm surge waters that traveled inland from the state's coast exceeded the height of the levee system along the Mississippi River in 53 places. A massive 19-mile- (30-km-) stretch of the coast suffered a surge up to 26ft (8.5m) above normal tidal levels.

4 NEW ORLEANS FLOODED
Approximately 80 percent of the city was flooded up to a depth of 23ft (7m). It was not until 43 days later that the Army Corps of Engineers declared that the floodwaters were entirely gone, after much misery for the city's inhabitants.

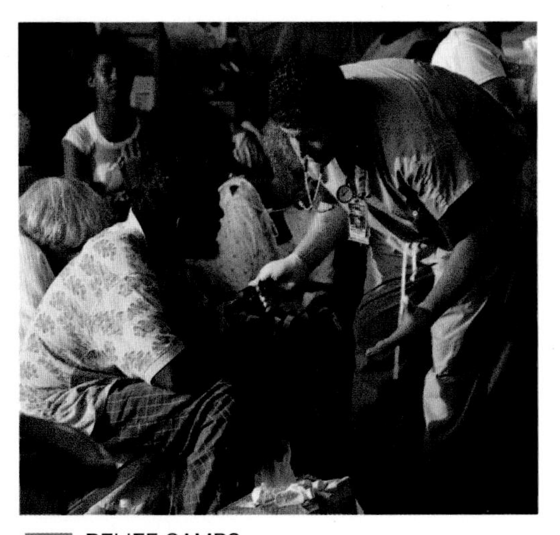

5 RELIEF CAMPS
Hundreds of thousands of New Orleans citizens became homeless. National and international voluntary and governmental organizations responded quickly, offering support. Many people have not returned to their home city, where unemployment has risen dramatically.

FLOODS AND FIRES
The impact of Katrina's storm surge combined with torrential rainfall produced extensive inundation in New Orleans, Louisiana. Floodwaters poured into the city from the Gulf of Mexico and the local river system. Then ruptured gas supplies and high winds caused the spread of serious fires.

EXTRATROPICAL CYCLONES

Extratropical cyclones are common phenomena across middle and higher latitudes. Occurring all year round, they are generally wetter and windier in fall and winter. Occasionally, they cause flooding and wind damage, but they are also the main source of water and wind power in these regions.

FORMATION AND DEVELOPMENT

Extratropical cyclones play an important role in transporting heat and moisture from the subtropics into higher latitudes. They form where cool, dry air intermingles with warm, humid air, creating a low pressure area, or depression. This happens primarily along the polar front that separates such air masses. The formation of the depression also depends on pulses of strong flow in the upper tropospheric jet stream (see pp.278–79). These conditions are more common in the western North Atlantic Ocean off the eastern seaboard of the US, and the mid-latitude North Pacific Ocean off the coast of east Asia. Once created, these frontal cyclones tend to track toward Europe, or western Canada and Alaska, carrying cloud and precipitation along with strong winds and rough seas. These cyclones typically last three to five days, gradually occluding (combining cold air with warm air) as they track across oceans and continents. At the end of their life cycle, the warm and cold air masses reach the same temperature, causing precipitation in a single region along the occluded front. Occasionally, extratropical cyclones can create havoc, resulting in rough conditions at sea, and threatening large-scale damage to the areas they run across on land.

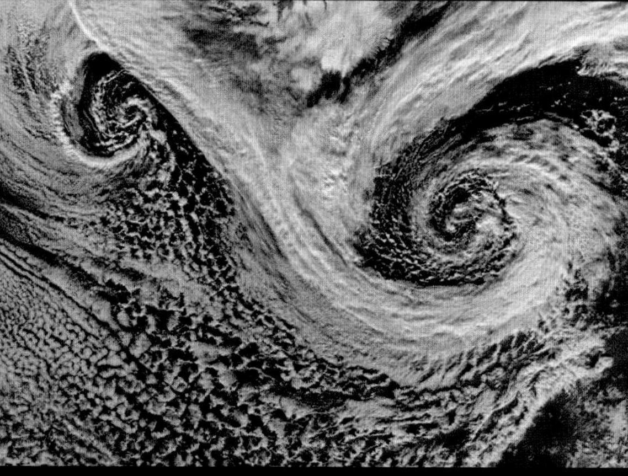

EXTRATROPICAL CYCLONE OVER ICELAND
This satellite view shows two cyclones to the south of Iceland in November 2006. Swirls of clouds mark the two depressions connected by frontal cloud bands.

NORTH SEA FLOODS
Huge waves at Sutton-on-Sea, England, were caused by a storm surge in the North Sea, on January 31, 1953. The Netherlands and Belgium were severely flooded and more than 2,300 people lost their lives, on land and at sea.

STING JETS

Severe mid-latitude cyclones can cause a lot of damage. However, meteorologists have now learnt that the most intense gusts are localized, so the worst damage occurs on quite a small scale. Explosive cyclones have certain characteristic features, including a relatively small "sting jet." This is known to start out about 3 miles (5km) up within a cloudhead—a thick layer of cloud that partly wraps around the low pressure center. The sting is roughly 3,280ft (1,000m) deep, plunging down to the surface in the span of a few hours, and bringing gusts of up to 100mph (160kmph) over an area some 30 miles (50km) wide. The evaporating rain and snow cool the descending air, giving it a further push. The sting lies in the "tail" of the low pressure system, and reaches the surface to the southwest of the pressure center.

FAMILY OF CYCLONES

Extratropical cyclones sometimes occur as a connected sequence, dominating the weather of entire ocean basins. Here, related low pressure systems affect the weather from eastern Russia to the western US, and the North Pacific in between.

THE PERFECT STORM 1991

A combination of unusual weather conditions in October 1991 led to the formation of an extratropical hurricane that caused extensive damage in the eastern US. It caused significant coastal flooding and beach erosion along much of the eastern seaboard, and a number of fatalities, both on land and at sea. The US National Weather Service coined the term "Perfect Storm" for this extraordinary event.

The storm developed when a low pressure system began forming on a cold front that moved off North America. By the middle of the day on October 28, it lay a couple of hundred miles off Nova Scotia, Canada, forming a classic "Nor'easter"—cyclonic storm with winds that blew from a northeasterly direction along the east coast of North America. As it blew hard across parts of the northeastern US and the Canadian Maritime Provinces, a hurricane called Grace, not far to the south, supplied moisture to the developing storm. By early morning on October 30, the Perfect Storm had reached its peak 370 miles (595km) south of Halifax, Nova Scotia. Waves reached heights of up to 40ft (12m) at this time.

The storm drifted across the warm Gulf Stream waters, reaching the status of a subtropical low around midday on October 31. The winds around the low pressure system's inner core gradually strengthened, turning the storm into a fully-fledged hurricane, the last of the 1991 season.

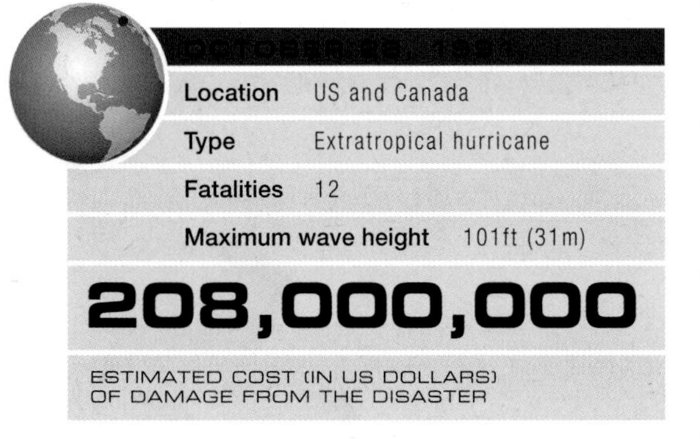

OCTOBER 28, 1991

Location	US and Canada
Type	Extratropical hurricane
Fatalities	12
Maximum wave height	101ft (31m)

208,000,000

ESTIMATED COST (IN US DOLLARS) OF DAMAGE FROM THE DISASTER

"... AN AWESOME EXAMPLE OF NATURE TAKING ADVANTAGE OF EVERYTHING SHE'S GOT AVAILABLE."

DAVID VALLEE, NATIONAL WEATHER SERVICE FORECASTER WHO TRACKED THE STORM, 2000

EVOLUTION OF THE STORM

1 LOW PRESSURE DEEPENS
On October 29, the extratropical cyclone lay off the shore of New England and Nova Scotia, as a classic "Nor'easter." Its circulation intensified on interaction with Hurricane Grace to the south, which fed humid, tropical air into the mid-latitude depression.

2 FORCE WEAKENS
The storm declined somewhat by October 31, but continued to wreak havoc at sea and along the US East Coast. Heavy surf, large swell, and high tides swamped many coastal areas. Extreme winds led to a state of emergency in some Massachusetts counties.

3 LANDFALL
By November 1, a tropical storm had developed, with the typical spiral banding intensifying into a Category-1 hurricane over the Atlantic. The storm finally began to weaken again, making landfall near Halifax on November 2.

PEAK FORM
The extratropical depression reached its peak in the early morning on October 30, 1991. Its central pressure fell to 972mb (millibars), with severe storm force winds reaching 60 knots (69mph/111kph).

IN FICTION

Director Wolfgang Petersen's movie, *The Perfect Storm* (2000), was based on Sebastian Junger's novel about the tragic, true-life story of the crew of the trawler, *Andrea Gail*, who were caught in the 1991 storm. The horrendous weather conditions made escape virtually impossible. Oddly, the crew did not activate the vessel's emergency position indicator that could have aided their rescue. The indicator was found washed ashore, switched off, on Sable Island, located 190 miles (300km) southeast of mainland Nova Scotia, on November 5, 1991.

Tossed on the ocean
This scene from the movie depicts the swordfishing boat, *Andrea Gail*, braving the storm. The boat is thought to have sunk sometime after midnight on October 28.

STORM TRACK
The unusual, sinuous track of the storm shows how the cyclone changed form (see Saffir–Simpson scale, p.293) over the period of a few days. Twisting and turning, it drifted over warmer waters, creating havoc at sea and on land.

Key
- Tropical depression
- Tropical storm
- Category 1

SNOWSTORMS AND BLIZZARDS

Snowstorms can wreak havoc, particularly in densely populated areas. Snow crystals form when water vapor in the cloud turns into ice. Combined with winds, a snowfall can create blizzards and drifts.

FORMATION

The most widespread snowfalls are produced by frontal depressions (see pp.278–79). Smaller-scale polar low pressure systems that occasionally form across the higher latitude oceans can also dump deep, but generally less extensive, snow. The heaviest snowfalls occur with fronts over which subzero, but not really cold, air flows. This humid air glides over the fronts to form thick, extensive cloud, which often generates prolonged and deep snowfall. Blizzards can occur if, for example, the depression is windy, blowing the falling snow around. They are also generated by wind whipping up settled snow from the surface.

SNOW PILE
Prolonged cold weather with recurring snowstorms can result in several feet of snow piling up. This buried car is being dug out of snow in Denver, Colorado.

IMPACT

Snowfalls on a large scale are defined as "slight" (up to 0.2 in or 5mm per hour), "moderate" (0.2in–0.5in or 5–40mm per hour), and "heavy" (more than 0.5in or 40mm per hour). Snowstorms can cause serious disruption to travel, as roads, highways, railtracks, and runways get blocked. The weight of the snow can damage power lines, leading to outages for days. In areas such as the southern states of the US, the weight of significant snowfalls can result in the collapse of roofs of buildings not designed to withstand such unusual events. Isolated snowfalls can usually be dealt with effectively by employing snowplows. However, if the storm is heavy and prolonged, the thick snow can settle quickly on recently cleared surfaces. Blizzards can reduce visibility drastically. Under certain circumstances, overcast and snow-covered surfaces can result in a "whiteout." Poor visibility makes it very difficult to distinguish between landmarks and to recognize reference points. Serious traffic accidents can sometimes occur in this highly disorientating weather condition. Snowstorms in mountainous regions can lead to avalanches in susceptible areas (see pp.308–09), and induce the risk of spring flooding during spells of rapid temperature increase.

COPING WITH SNOW
People in areas that are prone to snowstorms are often well prepared. Snowplows clear the roads and vehicles are fitted with snow tires. This snowplow is clearing a road in Oklahoma City in January 2010, after the area had been hit by an unusually heavy snowstorm.

WHITEOUT
A lone figure struggles through dense snow in the Scottish Borders, UK. The near-whiteout conditions illustrate the obvious danger to pedestrians, with visibility reduced to just a few feet.

STORM OF THE CENTURY, 1993

This major storm impacted 26 states in the US, and a large part of eastern Canada. Blizzards across the region caused damage to power lines that led to outages affecting more than 10 million people. Heavy snowfall occured in northern Florida, with up to 4in (10cm) recorded across the Panhandle. Certain parts of Appalachia received snowfalls of up to 3ft (1m), with drifting reaching up to 35ft (11m) in some places. Unprepared for the storm, some southern regions suffered a complete shutdown for three days. Thunder snow (thunderstorms with snow) was reported from Texas to Pennsylvania. This was considered to be partly responsible for the estimated 60,000 lightning strikes over the three days of the event.

Extent of storm
This curling sweep of frontal cloud affected a vast section of the eastern US. It is estimated that 40 percent of the country's population was affected by the storm.

GALTÜR AVALANCHE 1999

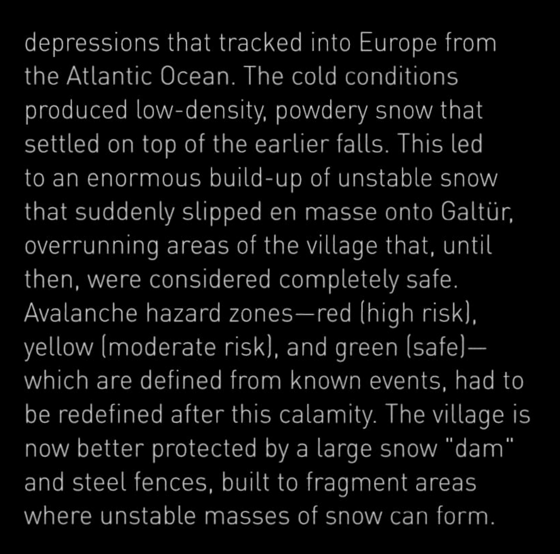

On the morning of February 23, 1999, a deadly avalanche hit the village of Galtür, Austria, in the central eastern Alps. Thirty-one people died in what came to be considered the worst avalanche in the Alps in 40 years.

The disaster occurred as a result of the unusual weather conditions that prevailed in the region that year. In January, mild weather conditions produced a "melt crust" on the upper surface of the snow lying in the region. This crust would melt during the day and freeze at night, leaving a smooth, slippery surface across the snowfields. February was cold, with record snowfalls of up to 13ft (4m), drifting in with stormy winds on the depressions that tracked into Europe from the Atlantic Ocean. The cold conditions produced low-density, powdery snow that settled on top of the earlier falls. This led to an enormous build-up of unstable snow that suddenly slipped en masse onto Galtür, overrunning areas of the village that, until then, were considered completely safe. Avalanche hazard zones—red (high risk), yellow (moderate risk), and green (safe)—which are defined from known events, had to be redefined after this calamity. The village is now better protected by a large snow "dam" and steel fences, built to fragment areas where unstable masses of snow can form.

FEBRUARY 23, 1999	
Location	Galtür, Austria
Type	Powder avalanche
Speed of the falling snow	190mph (306kph)

187,000

ESTIMATED WEIGHT OF THE ICE IN TONS AS IT BROKE AWAY FROM THE MOUNTAIN

> "... THE SNOW HAD SET LIKE CONCRETE... THERE WAS VERY LITTLE CHANCE OF ANYONE SURVIVING..."
>
> **JASON TAIT**, TOURIST WHO VIDEOTAPED THE EVENT FROM HIS HOTEL WINDOW, 1999

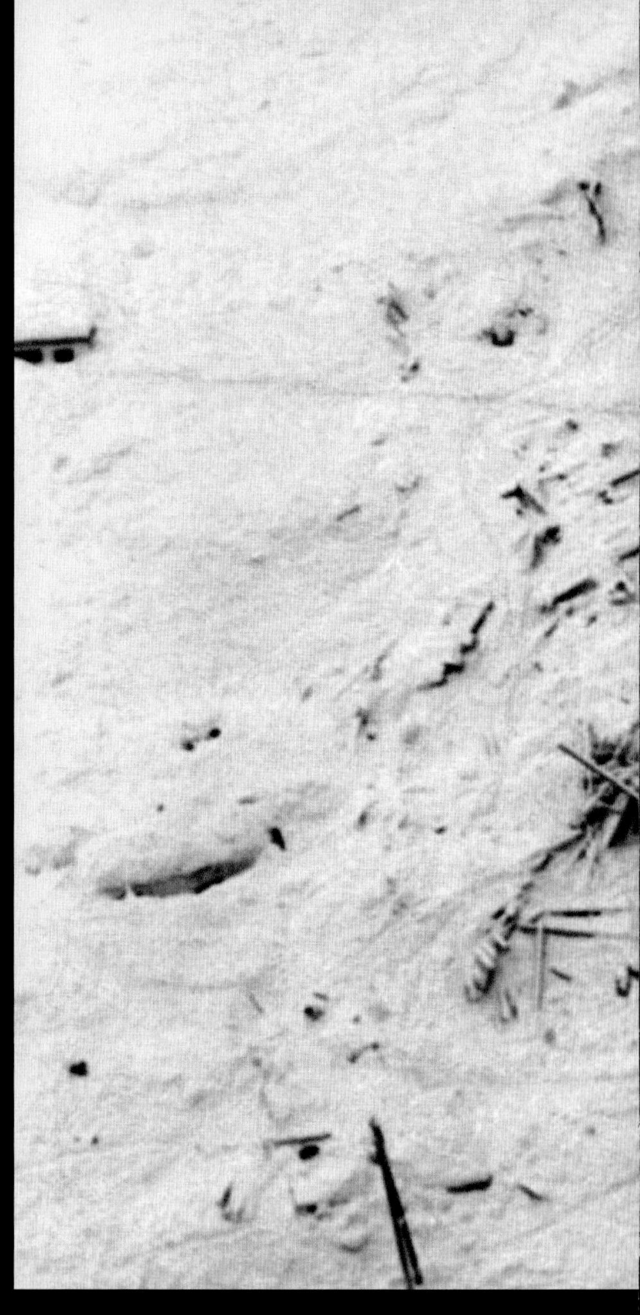

HOW AVALANCHES WORK

Avalanches occur when a large mass of snow cascades swiftly down a mountain slope. Snow gathers in layers throughout the cold season. If these layers are weak, additional snowfall, rain, or warm temperatures can trigger an avalanche by loosening the snow. Substantial rain can moisten packed snow, leading to wet avalanches. These are slow-moving and heavy, and cause massive destruction. Dry avalanches contain cold, powdery snow. They travel at great speeds and produce powder clouds as they gain momentum. These clouds can cause fatalities by suffocation, as witnessed in the 1999 avalanche at Galtür.

Cascading downhill
An avalanche crashes through the Savoia Pass on the northwest side of K2 in the Karakoram Range, in Pakistan. The great speed and mass of an avalanche leaves little chance of survival for those in its path.

VALZUR AVALANCHE
A day after the disaster in Galtür, a second avalanche hit the neighboring community of Valzur. Though not as deadly, it buried homes and killed seven people.

CAUGHT BY SURPRISE

The inhabitants of Galtür were caught unaware when the avalanche struck at 8:01am. With a leading edge estimated to be 330ft (100m) high, it cascaded down the mountainside at an estimated 190mph (306kph). This meant that it took just 50 seconds for the snow to reach and bury the village. There was no effective warning for the villagers that morning. The momentum of the enormous volume of snow forced it into the green zone of the village—the area considered to be the safest in the event of an avalanche. Seven modern buildings in this zone were completely demolished, and many more were severely damaged. A total of 57 people were buried under the snow, of whom 31 died—many suffocating in the fine, powdery snow. The avalanche gathered more snow as it swept down the mountain so final estimates are that up to 330,000 tons of snow had swept across the village.

RESCUE EFFORTS
Rescue workers search for survivors in the partially buried village of Galtür. Research shows that after 45 minutes of being buried under snow, only 20–30 percent of avalanche victims survive.

ICE SHEETS
Massive ice deposits occured at Versoix, Switzerland, in 2005 as the result of very high winds sweeping cold lake water over objects on the shore. In the sub-zero temperatures the ice deposits grew to dramatic effect.

ICE STORMS

Ice storms can bring occasional widespread havoc mainly to mid-latitude continents. When rain falls during freezing temperatures it forms an icy coating over everything it touches, damaging power lines and roads, making travel virtually impossible.

HOW THEY FORM

Ice storms are usually associated with widespread precipitation from traveling low-pressure systems. The freezing rain that is the essential ingredient of ice storms starts high up, perhaps above 1 mile (2km), as snow. The flakes fall through a layer warmer than 32°F (0°C) and melt—for a drop of half a mile or so. This melting layer has a shallow, surface layer beneath it that is at sub-zero temperatures, meaning that the droplets become supercooled and freeze as soon as they impact with anything on the ground. If the precipitation is heavy and prolonged, massive deposits of freezing rain form an icy coating, or "glaze." The process generates dangerous, icy conditions, which can persist for many days until the weather pattern brings milder air.

NOTABLE ICE STORMS

NAME	DATES	REGION	COST
Great Ice Storm of 1998	January 4–10	Atlantic Canada, Quebec, Eastern Ontario, New York, and New England	$5–7 billion
January 2007 North American Ice Storm	January 11–24	Canada and eastern and central US	$380 million
December 2008 Ice Storm	January 11–12	New England and upstate New York	$2.5–4 billion
January 2009 Central Plains and Midwest Ice Storm	January 25–30	Oklahoma, Arkansas, Missouri, Illinois, Indiana, and Kentucky	$125 million

DANGEROUS ICE

Deposits of glaze on sidewalks and roads are known as "black ice" because they are invisible and extremely slippery, removing all traction for vehicles or pedestrians. If the coating of glaze is thick it can severely damage vegetation. The weight of the ice brings down tree branches and deprives crops of carbon dioxide and water so that they do not survive the worst events. Some storms in North America over the last few decades have led to large-scale power outages as the weight of the glaze snaps electricity lines, leaving millions in the dark and cold for days—billions of dollars-worth of damage have been caused by deposits of some 2in (5cm) of glaze. Freezing rain is a real danger to aircraft, meaning that wings and fuselage must be de-iced routinely. Ice deposits change the shape of the wings and affect the aerodynamics of the airplane.

FREEZING WINDS
Icy north easterly winds brought freezing weather to Lake Geneva in January 2005. The freezing blast sculpted an icy coating on trees and buildings.

THUNDERSTORMS

Thunderstorms produce a range of violent weather. They are the source of hail, torrential rain, lightning, severely gusty winds, and tornadoes.

LIFECYCLE OF THUNDERSTORMS

Thunderstorms start life as small cumulus clouds that can grow to monstrous proportions in less than an hour. This rapid formation means that forecasters need to keep a very sharp and continuous eye on information from radar and other data sources to keep up with developments. Thunderstorms are most often created by intense surface heating, so they are common in areas where the weather is hot and humid. The landmasses therefore witness more storms than the oceans, and tropical areas are more at risk than those in higher latitudes. As air heated at the surface is lifted, it cools, and water vapor condenses to produce cumulus cloud. The water vapor in the cloud condenses into minute droplets. As the droplets grow heavy enough to start falling through the cloud, they form a cool downdraft that descends through the entire cloud, and leads to rain falling on Earth's surface. If the cumulus clouds grow in height to become cumulonimbus clouds, the water droplets freeze and the cloud can become charged with electricity. If the difference in charge becomes great enough, giant sparks leap from the cloud as lightning. The intense heating of the air around these explosive flashes generates thunder.

STAGES OF A THUNDERSTORM

A thunderstorm begins its life within a growing cumulus cloud. The total duration of this "single cell" storm is around one hour, during which it will produce precipitation such as small hail, a chilly downdraft with a gust front, and possibly lightning.

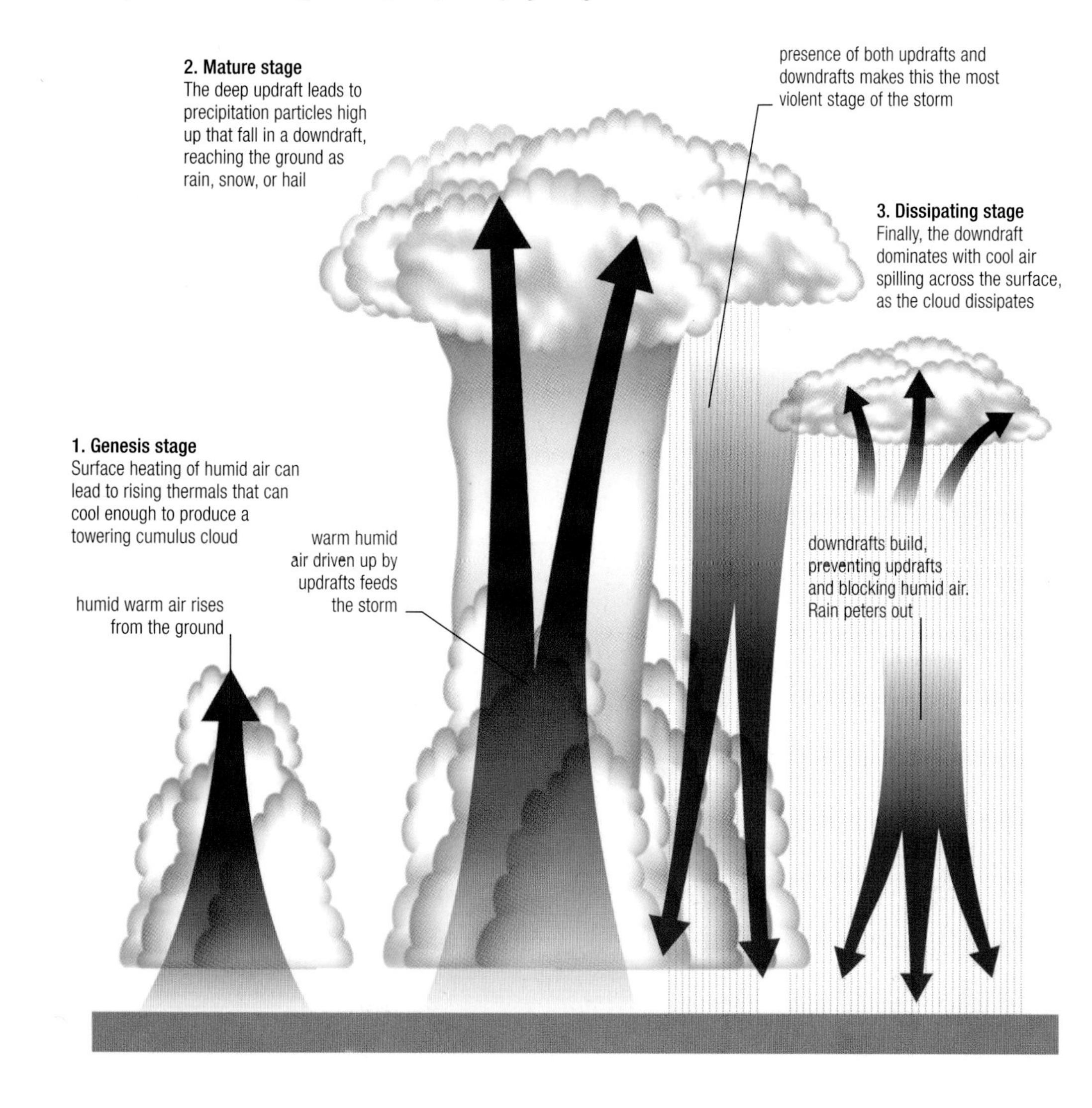

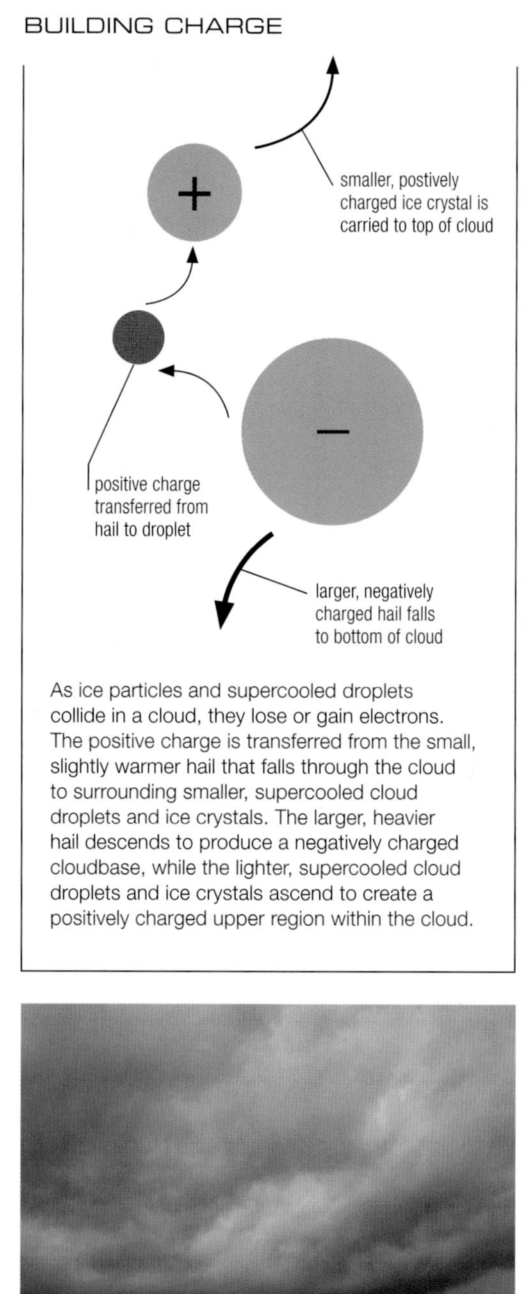

As ice particles and supercooled droplets collide in a cloud, they lose or gain electrons. The positive charge is transferred from the small, slightly warmer hail that falls through the cloud to surrounding smaller, supercooled cloud droplets and ice crystals. The larger, heavier hail descends to produce a negatively charged cloudbase, while the lighter, supercooled cloud droplets and ice crystals ascend to create a positively charged upper region within the cloud.

RAIN SHAFT
The threatening low cloud and distant rain shaft of an active thunderstorm hovers over Miami, Florida. This state sees the most thunderstorms of any in the US, due to its rich supply of humid air and strong surface heating.

SERIOUS FLOODS
A traffic officer directs cars during a thundery downpour in Chongqing municipality, China, on a July day. Thunderstorms create serious urban flooding with disruption to traffic flow. The strong updrafts of humid air in thunderstorms create high-intensity rain.

CLOUD TYPES

Thunderstorms are the products of cumulonimbus clouds that grow right up through the depth of the troposphere until they reach the tropopause. The vigorous upward motion turns to horizontal flow, indicated by the icy anvil that stretches out in the winds at high levels. The head of the updraft within the growing cloud is marked by cumulus congestus clouds with their cauliflower-shaped tops, while at the base a shaft of heavy precipitation will force a plunging, cool downdraft. Showers can fall from cumulus congestus clouds before the cloud flattens out at the tropopause. Cumulonimbus clouds are often randomly scattered, but they sometimes can be organized into a linear pattern, producing a long line of stormy weather.

SINGLE CELL SHOWER

The relatively short-lived, individual shower cloud grows vertically because it develops within a layer where the wind direction and speed hardly change throughout the surrounding atmosphere. Rain is visible, marking the location of the downdraft.

MULTICELL STORM

This is a group of interrelated precipitating cumulus clouds in different stages of their life cycle. The downdraft outflow from a mature cloud can scoop up some surrounding surface humid air to create a new early cumulus stage "partner."

MULTICELL LINE

Multicell storms are often formed into lines that lead to the organization of their downdraft outflows. The gust fronts that mark the leading edge of these storms can be damaging, indicated by the presence of a line of threatening clouds.

SUPERCELL

Inside the longest-lived thunderstorm is a rotating column of air known as a mesocyclone. They last longer since their updraft is vertical, so it is not swamped by the downdraft. Supercells can produce tornadoes.

HAILSTORMS

Hail is a shower of round or irregularly shaped pieces of ice, known as hailstones, which are formed inside cumulonimbus clouds. They originate as small ice particles or frozen raindrops that are caught in the updraft of air inside the cloud. As they ascend, they get bigger by gathering water on their surface. How big they grow depends on how strong and extensive the updraft is and how much water is in the cloud. Eventually, the droplets become too heavy to be supported by the updraft and they start to descend. They either fall as hail, possibly melting a little in the warmer air lower down, or they may become caught up once more in a strong updraft and gain another icy shell. The most intense hailstorms are associated with very deep thunderclouds, which have a very strong, tilted ascending current, this can mean the drops move horizontally within the upper reaches of the cloud, growing within the chilly air. Although it may seem paradoxical, the largest hailstones occur during the warm season. This is because the surface heating is strongest then—producing tall thunderclouds and evaporating more water to provide more layers of ice.

FREAK HAILSTORM Rescue teams had to dig out trapped motorists when a hailstorm hit Bogatá, Colombia, in 2007.

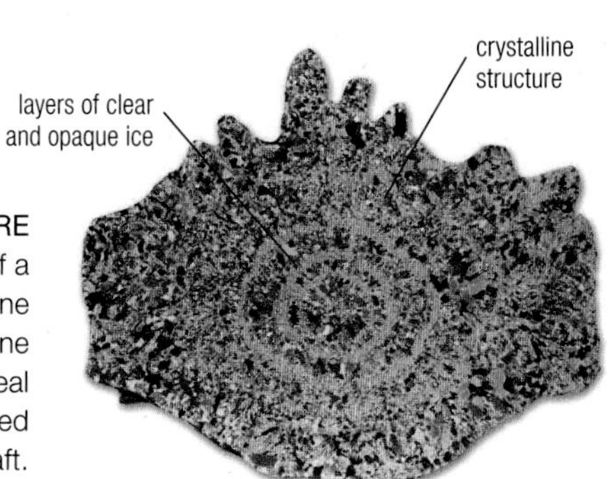

INTERNAL STRUCTURE A polarized light photograph of a thin section of a giant hailstone shows the internal crystalline structure. The ring layers reveal how many times it has traveled around the updraft.

HAILSTONE DAMAGE Large hailstones can cause damage to crops, buildings, and vehicles. Smashed windshields are a risk for drivers in the US, if they are caught in a storm.

The worldwide pattern of the annual average frequency of lightning also reveals where thunderstorms occur. Broadly, the greatest incidence is across the humid tropics, mainly over the strongly heated continents. In Africa, the storms are markedly seasonal, the activity related to the migration of the thundery ITCZ between the two hemispheres. There are other active areas including parts of the US and northern Argentina, for instance, where thunderstorms can be hazardous. Smaller risk is present across higher latitude continents, again with a bias toward greater summer activity. Lightning is least likely across much of the oceans and the Arctic and Antarctica.

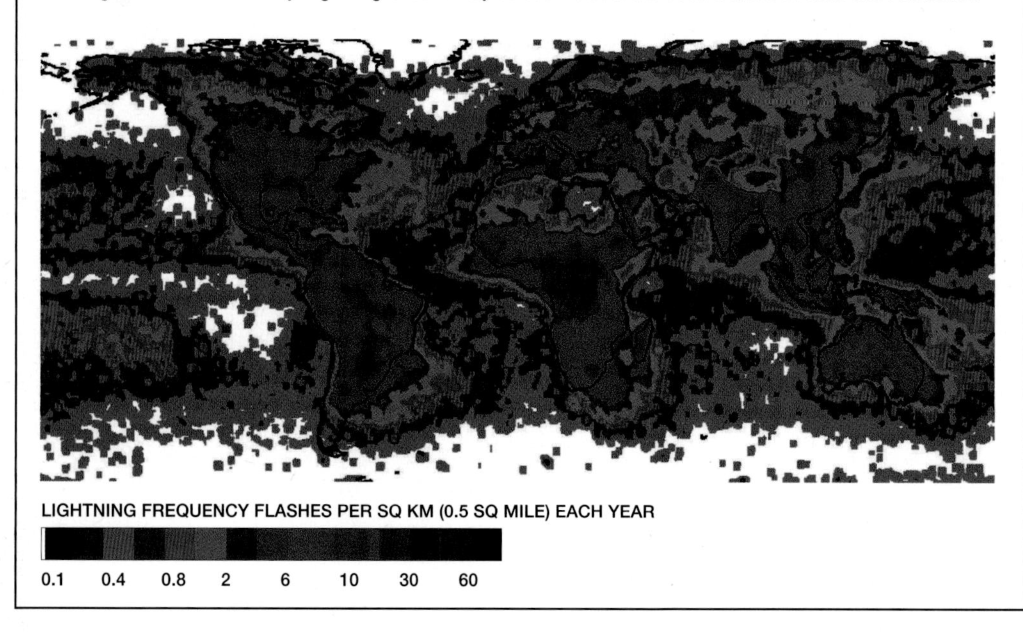

LIGHTNING

Lightning is an electrical discharge either between a cloud and Earth's surface, from cloud to cloud, within a cloud, or from a cloud into the air. The stroke heats the air instantaneously to about 54,000°F (30,000°C) leading to expansion of the air and the tell-tale rumble of thunder. Light reaches our eyes virtually instantly, while sound travels at about 1,000 ft per second (330 m per second), so it is possible to estimate how far away a lightning stroke is by counting the seconds between the stroke and thunder. A three-second difference converts to a distance of about 0.62 miles (1km) for example. A flash of lightning carries a current of up to 100,000 amps that can electrocute humans and animals. Strikes occasionally start fires; it is estimated that some 10,000 such fires are started across the US in an average year. The culprits are often thunderstorms that display "dry lightning"—cumulonimbus clouds that do not produce precipitation. Lightning produces radio waves called "sferics" that are detected by a global network of stations, allowing for study of the frequency and distribution of strokes.

RIBBON LIGHTNING
This type of lightning is formed when a sequence of "return strokes," which is when the current flows back up the path of the first discharge, occur in rapid succession, moving in the wind so that they appear as a ribbon.

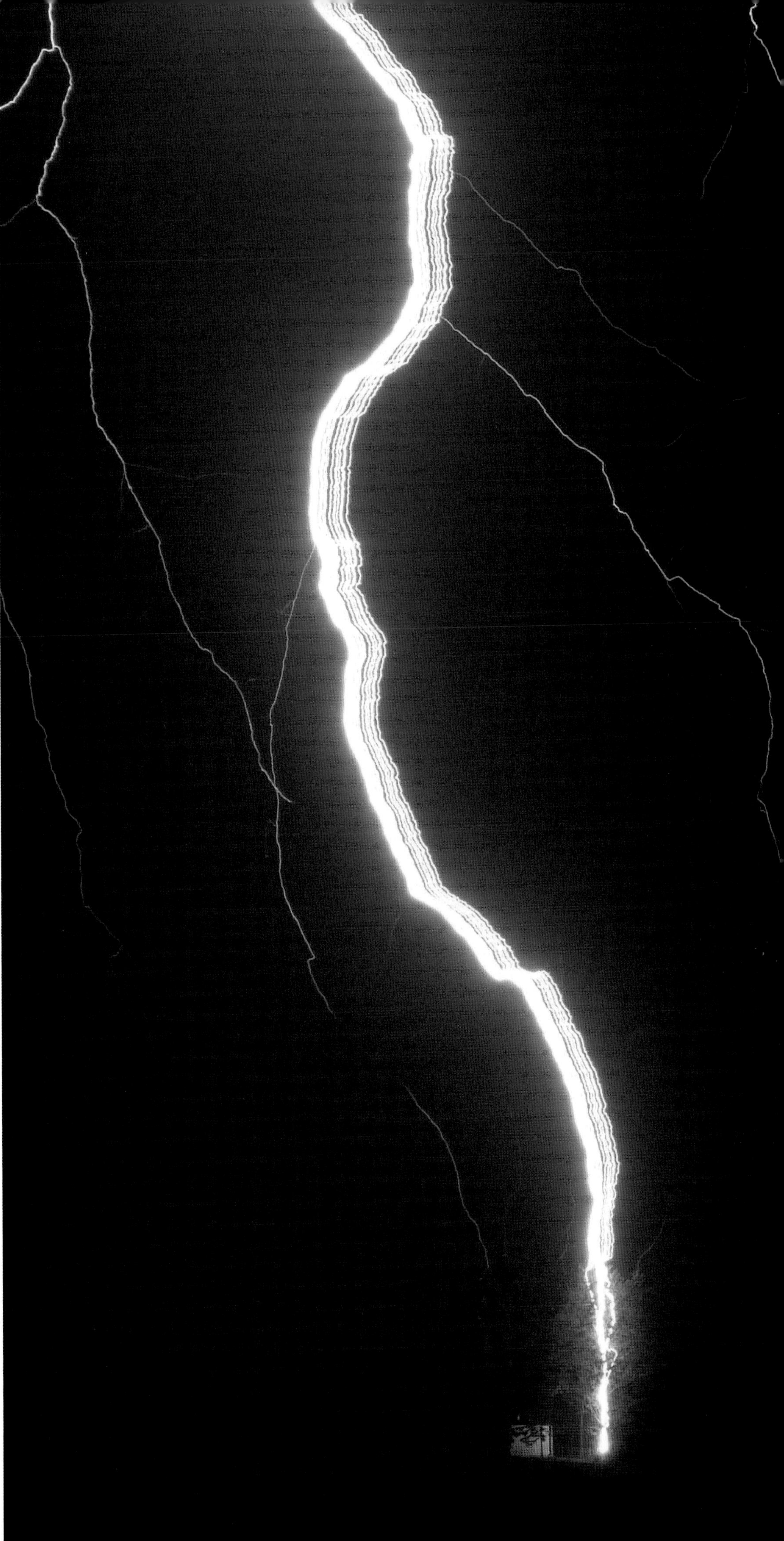

LIGHTNING DAMAGE
Tall buildings without lightning conductors are susceptible to serious damage. The metal conductor projects above the highest point, to channel strikes safely into the ground where they are earthed. Here brickwork has suffered explosive damage.

LIGHTNING RESEARCH
Scientists investigate how the structure of a storm, including its precipitation and updrafts, influences lightning by launching an instrumented weather balloon. It measures pressure, temperature, humidity, wind velocity, and carries an electric-field meter.

SHEET LIGHTNING
When a flash of lightning occurs inside the cloud, it is termed "sheet lightning." Part of the cloud is illuminated from within, looking like a bright white sheet.

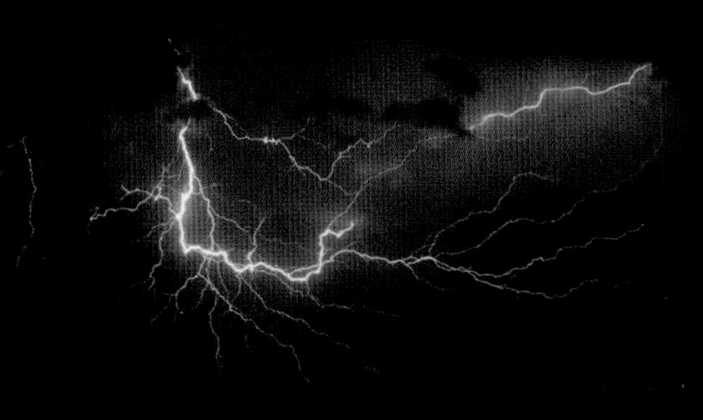

CLOUD-TO-CLOUD LIGHTNING
This fairly common intra-cloud discharge resembles tree branches. The surface of the cloud is unaffected and the lightning dissipates into the atmosphere.

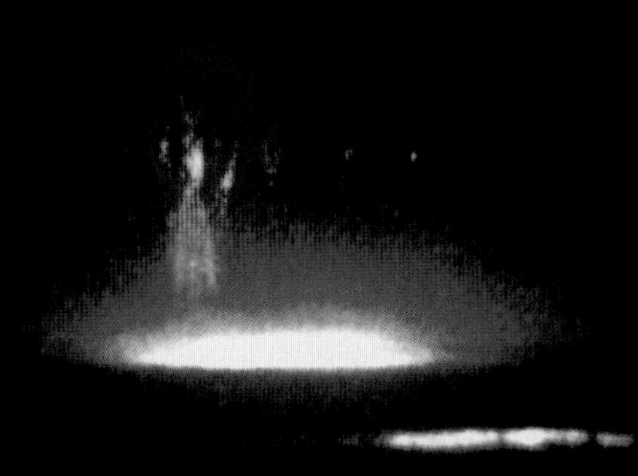

RED SPRITE
These huge but faint illuminations occur at about the same time as lightning in a cloud. They extend from the cloud top to the ionosphere, and last a few milliseconds.

CLOUD TO GROUND
A rush of electrons from a cloud toward the ground are joined by a current rising up from the ground to create cloud-to-ground lighting.

TORNADOES

Tornadoes are rotating columns of air extending from a thundercloud to the ground. Among the most violent of all weather systems, with wind speeds reaching up to 300mph (480kph), they can cause widespread destruction. Typically 165 to 330ft (50 to 100m) wide, they are generally short-lived with average track lengths of 1 to 2 miles (2 to 4km).

FORMATION

Tornadoes form at the base of a parent cumulonimbus thundercloud that has reached supercell status (see pp.312–13). Low air pressure at the base of the cloud causes warm, moist air from the ground to rush up and rub against the cold air in the cloud. This creates a mesocyclone—a wide column of revolving air—which begins to protrude downward, forming a funnel, or vortex. Air drawn into the funnel's center, a low-pressure area, cools and its moisture condenses to form a funnel-shaped cloud. This becomes a tornado when it touches the ground.

1 DOWNWARD EXTENSION
A funnel cloud begins to protrude from the base of a rotating parent cumulonimbus, indicating the possible development of a tornado.

2 NARROWING BASE
The rotating air picks up speed as more warm air is sucked in. The rotation affects the surface, throwing up a cloud of soil.

3 HITTING THE GROUND
The funnel cloud touches the ground, becoming a tornado. The funnel may be obscured by dust picked up by the rising air and swirling winds.

4 DUST AND DEBRIS
The high-speed winds throw dust and debris into the air. The tornado eventually loses energy, and its funnel shrinks back to the parent cloud.

WHIPPING UP DEBRIS
A tornado tears through the seaside village of Lennox Head in New South Wales, Australia. Damage occurs not only due to the high wind speed of the funnel cloud, but also because of the rapidly moving debris whipped up by it.

TORNADO DISTRIBUTION MAP

Tornadoes are usually linked to the Great Plains of the US, particularly in an area known as Tornado Alley, but they also occur elsewhere. Atmospheric conditions that are ripe for the formation of thunderstorms may lead to tornadoes in parts of Europe, Asia, South America, South Africa, and Australia.

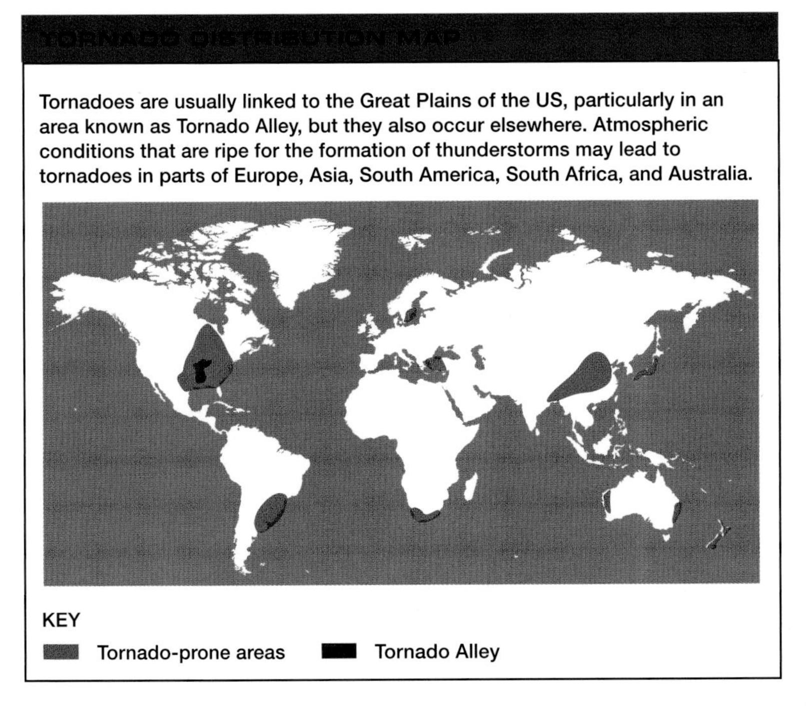

KEY
Tornado-prone areas
Tornado Alley

TRAIL OF DESTRUCTION

Tornadoes are usually about 300ft (100m) wide, so the swathe of damage they cause is relatively thin—but can be extremely severe. The swirling winds pick up and hurl objects, including cars and rooftops, over long distances. The narrowness of the funnel means there may be total destruction in one place next to an area left virtually untouched.

TORNADO OUTBREAK
In April 2011, a devastating outbreak of tornadoes occurred across parts of the southeastern US, killing more than 340 people. This is an aerial view of wreckage caused by the tornadoes in the city of Tuscaloosa in Alabama.

TYPES OF VORTICES

Tornadoes are rapidly rotating air columns associated with cumulonimbus clouds that have made contact with the land surface. They are usually marked by the presence of condensed water droplets in the form of a funnel cloud, but can sometimes be "dry" when they are made visible by solid debris such as dust or soil. Tornadoes over water are called waterspouts. These are weak vortices that consist of a column of cloud droplets that can be amplified above the ocean's surface, spraying large volumes of seawater. They commonly occur in tropical and subtropical seas where there is a ready supply of warm, moist air.

Another weather phenomenon created in the same way as a tornado is a dust devil. Common over intensely heated, dry continental surfaces, they are short-lived, upright vortices that produce no damage. Dust devils are smaller than tornadoes and are not attached to a cloud.

DUST DEVIL
A dust tornado grinds across a farm road in western Kansas. These vortices are made of dust that is drawn into the rotating air.

WATERSPOUT
A typical summertime waterspout is seen off the coast of Florida Keys. These tornadoes occur over the sea, but occasionally make landfall.

ENHANCED FUJITA SCALE

Created by University of Chicago's storm researcher Tetsuya Theodore Fujita, the Fujita scale classifies tornadoes according to their wind speed, based on the amount of damage they cause. It was replaced by the updated, Enhanced Fujita scale in February 2007.

F-SCALE	WIND SPEED	DAMAGE
F0	**65–85mph (105–137kph)**	Small tree branches snapped off; windows and patio doors broken
F1	**86–110mph (138–177kph)**	Trees are uprooted and chimneys collapse; mobile homes overturned
F2	**111–135mph (179–217kph)**	Tree trunks are snapped off and whole houses shift off foundation
F3	**136–165mph (219–265kph)**	Trees are debarked, only stumps are left. Exterior walls of houses collapse
F4	**166–200mph (267–322kph)**	Well-constructed houses levelled; cars blown away
F5	**Over 200mph (Over 322kph)**	Well-constructed houses levelled and swept away

OKLAHOMA 1999

Oklahoma is situated in the disaster-prone Tornado Alley, an area between the Rocky Mountains and Appalachian Mountains that experiences several hundred tornadoes every year. On May 3, 1999, Oklahoma witnessed the worst tornado outbreak in its history. The day started out warm and sunny, but by the afternoon supercell thunderstorms had developed. More than 70 tornadoes were reported across the state that day, with wind speeds reaching more than 200mph (300kph). The deadliest of these struck Bridge Creek, located southwest of Oklahoma City, the state capital. A mobile Doppler radar monitoring the flow of air within its massive funnel cloud measured a world-record wind speed of 302mph (486kph). Measuring F5 on the enhanced Fujita scale (see p.319), the total damage from this tornado alone reached $1.1 billion.

The tornadoes caused widespread destruction across the state, whipping up clouds of debris and leveling thousands of homes. Hailstones larger than golfballs were reported in some areas, and a large number of people were left without power for more than a day. Forty-eight people lost their lives during the tornado outbreak and more than 600 people were injured.

10 DEADLIEST OKLAHOMA TORNADOES

	PLACE	YEAR	F-SCALE	DEATHS
1.	Woodwards	1947	F5	116
2.	Snyder	1905	F5	97
3.	Peggs	1920	F4	71
4.	Antlers	1945	F5	69
5.	Pryor	1942	F4	52
6.	Bridge Creek	1999	F5	36
7.	Oklahoma City	1942	F4	35
8.	Cleveland County	1893	F4	33
9.	Bethany	1930	F4	23
10.	McAlester	1882	F3	21

Two tornadoes in Kansas, Oklahoma, on January 27, 2009

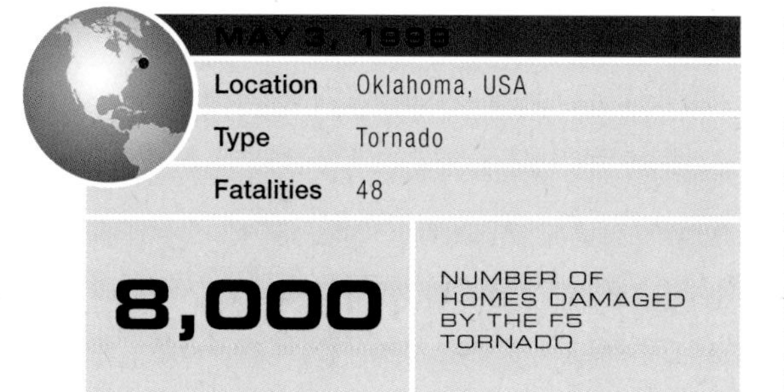

MAY 3, 1999

Location	Oklahoma, USA
Type	Tornado
Fatalities	48

8,000 NUMBER OF HOMES DAMAGED BY THE F5 TORNADO

PREPARING FOR THE TORNADOES

Timely warnings reduced the number of fatalities caused by the tornadoes. Frequent radar monitoring and official storm spotters following the funnel cloud provided constant updates of its intensity and track. Television and radio broadcasts were interrupted by updates of the progress of the tornadoes, linked to site-specific warnings. Residents of Oklahoma are well-trained in storm safety. All sizeable shopping malls and public buildings have tornado shelters and many homes have storm cellars.

THE DESTRUCTIVE PATH OF THE TORNADOES

1 TORNADO WARNINGS
Frequent tornado warnings were sent out by the Oklahoma weather service via TV and radio, giving residents time to seek shelter.

2 TORNADO INTENSIFIES
The killer tornado intensified significantly while approaching Oklahoma City from the southwest. It attained F5 status in Bridge Creek before weakening as it crossed the Canadian River, to reinvigorate to an F5 in the community of Moore.

N
Kingfisher
Payne
Logan
Creek
Lincoln
Oklahoma
Oklahoma City
Canadian
Moore
Bridge Creek
Caddo
Cleveland
Pottawatomie
Grady
McClain
0 miles 20
0 km 20

Key

Tornado path

Affected counties

MAP OF OKLAHOMA TORNADOES
The outbreak of tornadoes followed the typical pattern of funnel clouds tracking from southwest to northeast. Tornadoes move relatively slowly—it is the winds circulating around them that do the damage.

> "IF YOU ARE IN ITS PATH, TAKE COVER IMMEDIATELY... THIS STORM MAY CONTAIN DESTRUCTIVE HAIL THE SIZE OF BASEBALLS... OR LARGER."
>
> **NATIONAL WEATHER SERVICE**, CITY OF NORMAN IN OKLAHOMA

3 TORNADO PATH
The churned-up red earth of Oklahoma traces the narrow but devastating path of destruction on the outskirts of Oklahoma City. Amazingly, devastation was total along its line, but minor or non-existent just a few hundred feet away.

4 AFTERMATH
Thousands of buildings were destroyed, 116,000 homes lost power, and more than 40 people perished due to the storm. The event led to the establishment of the National Weather Center at the University of Oklahoma.

TORNADO LOOMS

The worst tornado outbreak on record for South Dakota hit the state on June 24, 2003. There were a total of 67 tornadoes, including this one that destroyed the community of Manchester. It lasted 20 minutes, whipping up soil debris with winds estimated to have reached 260mph (418kph).

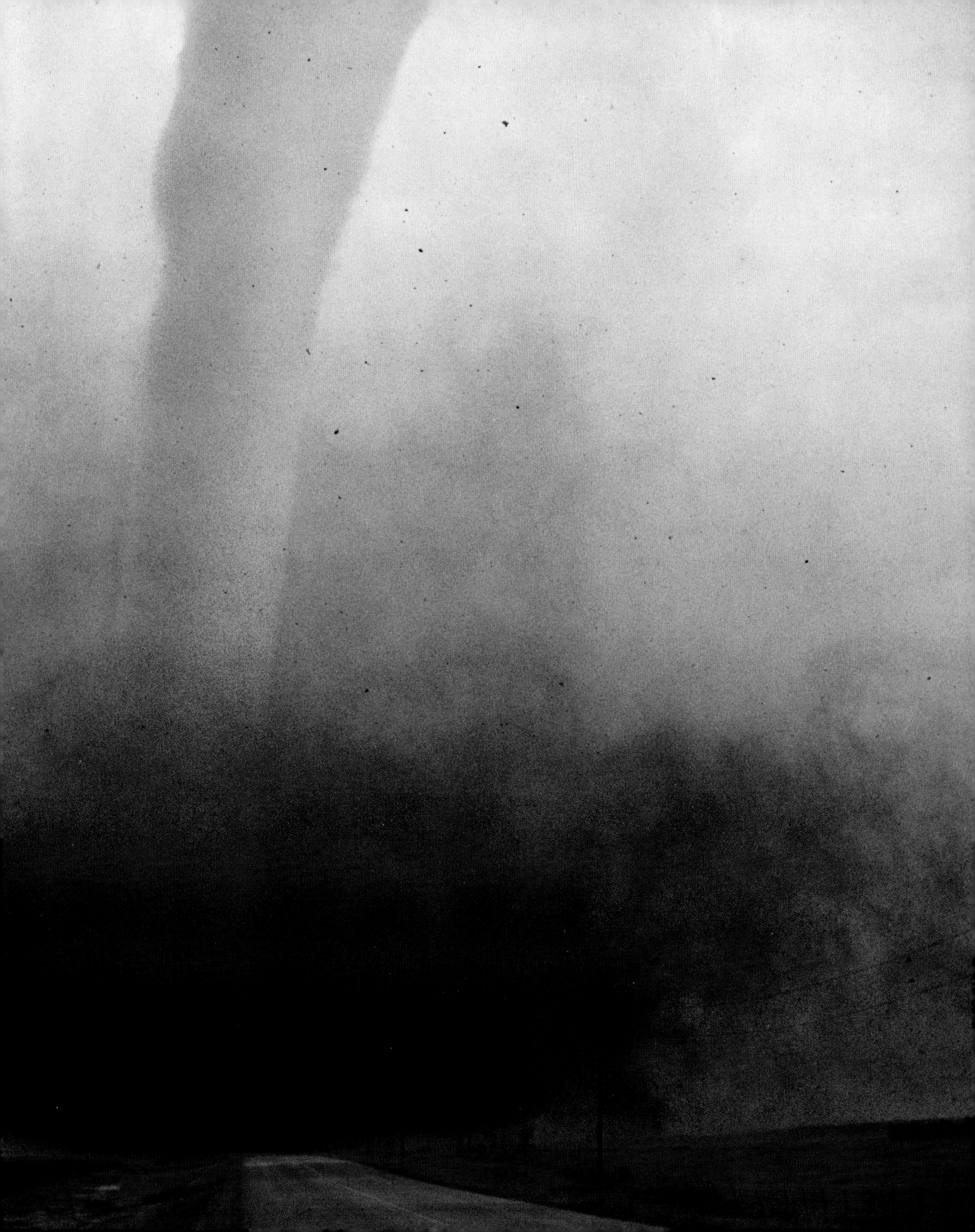

SANDSTORMS AND DUST STORMS

Hot deserts and drylands are the main sources for sand and dust storms. Winds whip up loose sand particles, causing sandstorms. Smaller dust particles can be transported great distances from their source, potentially forming dust storms.

FORMATION AND DISTRIBUTION

Sand and dust storms form in hot, barren deserts and dry semiarid regions. The processes of weathering and occasional fluvial activity, especially in semiarid regions, lead to the deposition of fine particles in dry lakebeds and floodplains. The particles are whipped up by strong winds that can produce sandstorms. These tend to be more localized because of the relatively large size of sand grains. Dust storms are created by the tiniest dust particles that are blown further away from the source. The dust can ascend to 4 miles (6km), or sometimes higher and, if atmospheric conditions are favorable, can be transported a very long way. It is not unusual for dust to cross the North Atlantic, rising from the western Sahara and then settling across the Caribbean, or from China across the Pacific Ocean. There also have been reports of "red rain"—rain mixed with dust—in the UK, and "pink snow"—a combination of snow and dust—in the Alps, showing just how far Saharan dust can travel. Dry cold fronts and downdraft outflows from thunderstorms can help carry the dust along.

TRAVELING DUST STORMS
This massive dust storm in 2000 was generated over northwest Africa. It stretched up to 1,000 miles (1,600km) across the eastern Atlantic.

DESERT CITY SANDSTORM
On May 14, 2010, this huge sandstorm swept across the Chinese city of Golmud, near the edge of the Gobi Desert. Visibility was reduced to around 600ft (183m).

IMPACT AND EFFECTS

A sand or dust storm can spell danger for people using transportation, with a sudden drop in visibility that can last from a few hours to a day or more. The high concentration of small particles of dust is a serious health risk. Dust that is smaller than 0.0004in (0.01mm) can be inhaled to the deepest part of the lung. Health problems such as emphysema and bronchitis can be related to high dust concentrations, as can the incidence of conjunctivitis. Soil erosion is also a problem in dryland regions, as nutrients in the upper layers of the soil are blown away. This dust can flow downwind and cover plants, contaminating supplies of food and water. In strong winds, dust can have abrasive properties that can seriously damage both crops and livestock.

RED DUST STORM

In September 2009, a dust plume measuring more than 620 miles (1,000km) in length spread from Australia's desert regions and dry farmland across to Sydney and the eastern coast of Australia, turning the sky a red-orange color.

AMERICAN DUST BOWL

In the mid-1930s, the North American prairie lands stretching from the Texas Panhandle to southern Nebraska were prone to severe dust storms that caused major agricultural damage. The "Dust Bowl" covered more than 154,440 sq miles (400,000 sq km) at its peak, which led around 250,000 people to flee the drought-ridden region. The disaster was entirely man-made, a consequence of the rush to plow up the High Plains for wheat farming. The lure of quick money led to large-scale dryland farming that was seriously mismanaged. In 1931, a survey by the Agricultural College of Oklahoma discovered that around 20,077 sq miles (52,600 sq km) of the 25,000 sq miles (64,750 sq km) cultivated in that state suffered serious soil erosion. On April 14, 1935—also known as "Black Sunday"—the region suffered the worst day of all, when an estimated 330,700 tons of High Plains topsoil was lifted skyward into high-speed winds traveling at more than 50mph (80kph). Reports of the time suggested that the dust storms created a near blackout in some areas.

CHINESE DUST STORM 2010

In March 2010, a severe drought across much of China was followed by intense dust storms. The dust came from weather systems whipping across the Gobi Desert, the largest in Asia, and blowing over northern China to reach the capital Beijing. By March 20, the orange dust covered 313,000 sq miles (810,000 sq km) of the country, affecting several million people. The huge cloud of dust then progressed to Korea and Japan before being blown on the jet stream across the Pacific to reach the western US.

The period leading up to the widespread dust storm was one of acute drought, matched by higher than average temperatures. The dryness over southwest China was reported to be the worst in a century. However, Beijing is no stranger to dust storms—one massive storm in April 2006 is estimated to have dumped 330,700 tons of dust across the city. According to the Chinese Academy of Science, dust outbreaks in China have increased sixfold in the past 50 years. This is attributed to increasing desertification due to poor agricultural practices and deforestation.

INTENSE DROUGHT
A buffalo-drawn cart on a badly cracked field depicts the severity of drought in the Yunnan province of China. The 2010 drought was the worst to impact the country in 100 years, with rainfall at 60 percent below normal.

“VISIBILITY WAS ABOUT 10 METERS... WE TRIED NOT TO GO OUT. IF WE DID, WE WORE FACE MASKS...”

WANG HAIZHOU, FARMER LIVING IN THE CITY OF KUERLE, NORTHERN XINJIAN

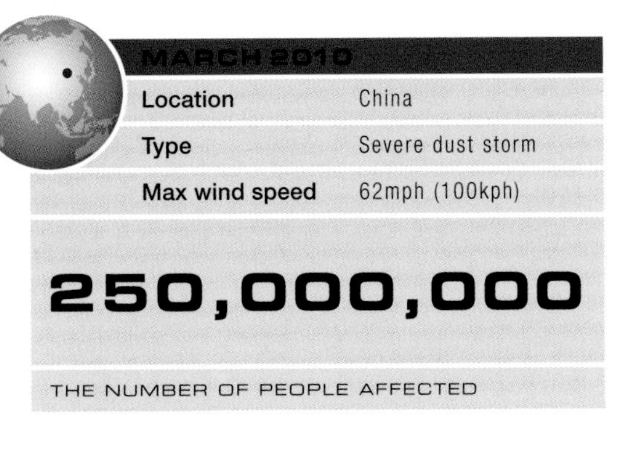

MARCH 2010	
Location	China
Type	Severe dust storm
Max wind speed	62mph (100kph)

250,000,000

THE NUMBER OF PEOPLE AFFECTED

EFFECTS

During the March dust storm, many areas in and around China reported the air quality to be "hazardous"—a rating very rarely used. People were advised to remain indoors, particularly those with coronary, respiratory, or immune system problems. Many flights were delayed or canceled as the dust storm gripped Beijing airport. Concentration of pollutants in Hong Kong peaked, reaching a record 15 times over the maximum level recommended by the World Health Organization (WHO). Similarly, Taiwanese authorities reported unprecedented levels of pollution. The Korean Meteorological Agency served a nationwide warning when concentrations exceeded the serious health hazard level of 8oz per cu ft (800mg per cu m). The storm's severity actually reached a peak measure of 29oz per cu ft (2,847mg per cu m) in the country.

BLANKETED IN DUST
The top image of Beijing shows the clear weather experienced on March 17, which changed to a thick pall of orange dust just five days later on March 22.

DUST STORM IN GOLMUD
Golmud, a new Chinese city near the edge of the Gobi Desert, suffers frequent springtime storms that blow in from the desert. The dust spreads much further, across other parts of China.

PROTECTIVE GEAR
Visitors and locals were advised to wear face masks to protect their lungs from the dust when the storm hit central Beijing.

WILDFIRES

A lightning strike on combustible material, such as trees or grassland following a period of hot, dry weather can cause a wildfire. Strong winds can spread the flames at high speeds across vast areas, causing massive destruction.

CAUSES OF WILDFIRES

The start and spread of wildfires depends on a combination of factors. A vital ingredient is an arid antecedent period during which shrubbery, trees, and vegetation become very dry. Parts of the world that experience a hot, dry summer are most at risk of wildfires. Such dryness is often related to persistently high temperatures and low relative humidity. Strong winds that are hot and dry promote the dessication of vegetation. Once started, the stronger the winds the more rapidly the fire spreads, sometimes at amazingly dangerous rates. Fires can move as fast as 14mph (22kph) across grassland. When the El Niño pressure system develops (see pp.284–85), areas of the world that experience drier weather, such as the eastern half of Australia, are more at risk of wildfires. Even though conditions for fires may be ripe, there has to be an agent to ignite the flames. This is very often lightning. The thunderstorms may also produce rain, but not enough to significantly quell any outbreaks. Otherwise, wildfires can be started simply by human carelessness, or as an act of arson. The latter appears to have been the case in the 2009 Greek wildfire, which engulfed 14 towns on the outskirts of Athens.

INFERNO
Firefighters battle to prevent the wildfires from reaching power lines north of Los Angeles in 2010.

LIGHTNING STRIKE
Lightning is the main ignition source for wildfires. The precipitation that accompanies it is never enough to extinguish flames once they spread.

HUMAN CARELESSNESS
This aerial view shows smoke on the outskirts of Marseille, France, after a wildfire was sparked in 2009 by practice shelling by the military.

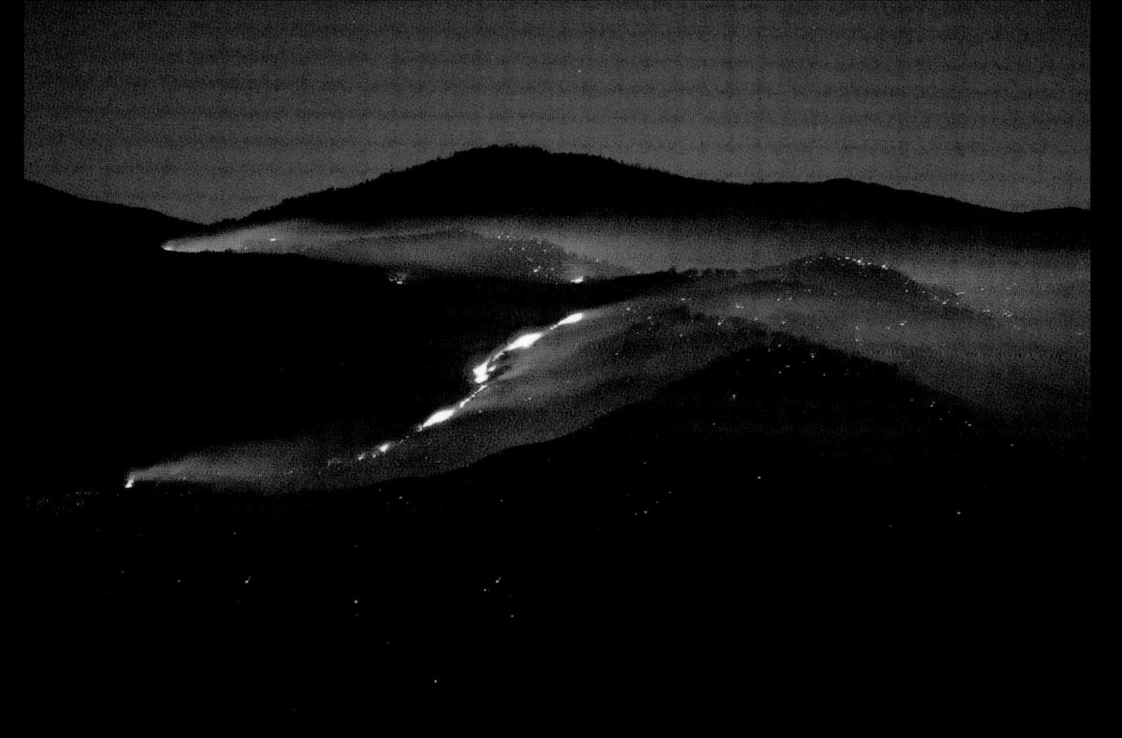

WIND DIRECTION
In 2007 the deadly wildfires in Southern California spread across 770 sq miles (2,000 sq km). The wind carried hot embers that ignited trees far from the original point of ignition. Tragically, nine people died and at least 1,500 homes were destroyed.

AFTER EFFECTS

Wildfires destroy homes and infrastructure, and communities can take years to recover. Intense heat eliminates much vegetation and affects the soil, creating a significant risk for both flash flooding and mudslides, due to changes in the porosity of the land surface. In Hawaii serious soil erosion due to wildfires has led to enhanced deposition of soil material into the surrounding seas, killing some areas of seaweed and reducing fish stocks. Forests do recover, but how quickly depends on the tree species— grassland is the mostly rapidly regenerated vegetation type. Wildfires also release huge amounts of carbon dioxide into the atmosphere.

FOREST REGROWTH
In coniferous forests, intense heat forces pine cones to open, scattering many seeds onto the forest floor, and leading to new growth.

DISTRIBUTION OF WILDFIRES

The areas of the world at highest risk are those that satisfy the climatic and land-cover ingredients for the outbreak and spread of wildfires. They tend not to occur in higher latitudes or across the very hottest regions because of the lack of vegetation. Equatorial rainforests are also relatively low risk because of the normal widespread precipitation.

Wildfires (yellow and red areas) are most common in regions where there is a prolonged dry season accompanied by an occasional thunderstorm. However, some areas of high incidence are reflections of deliberate burning to clear tropical forests, such as throughout parts of South America. In extremely dry summers, fires can occur in southern parts of Europe.

BACK BURNING
This firefighter in Pennsylvania is attempting to halt the bushfire by burning vegetation to clear flammable material ahead of the fire's leading edge.

BLACK SATURDAY BUSHFIRES 2009

FEBRUARY 7, 2009

Location	Victoria, Australia
Type	Wildfires
Fatalities	173

11,800

TOTAL LIVESTOCK KILLED IN THE FIRES

On February 7, 2009 Australian weather forecasters predicted the "worst day" in the state of Victoria's history. They were right. Record high temperatures combined with gale-force winds was a recipe for disaster given the severe and prolonged drought in the area. Vegetation was tinder dry, optimizing the combination of factors for a devastating wildfire. On the day, the maximum temperature peaked at 115.5°F (46.4°C), while the wind reached speeds of up to 56mph (90kph). As the relentless winds coursed through the drought-ridden landscape, nine small-scale fires ignited, most beginning in the middle of the day. A rapid change in wind direction later in the afternoon meant that townships to the northeast were suddenly susceptible, trapping some people in their homes. Emergency servcies were overwhelmed by the scale of the disaster, and the cellular phone networks crashed due to overuse. The fire moved so quickly that dozens of people died while attempting to flee from their homes. The inferno's radiant heat was so intense it was powerful enough to kill anyone within 984ft (300m). A total of 173 people perished with some 500 injured.

THE SMOKE SPREADS
The smoke plume from the persistent fires stretched out a great distance, reaching over the ocean from northeast of Melbourne.

A HELPLESS FEELING
The trees are ablaze in the Bunyip State Forest near the township of Tonimouk. The fire service was overwhelmed by the huge inferno.

FIREFIGHTING

1 AERIAL SUPPORT
Two days after the outbreak, a helicopter releases its water load in support of the firefighters on the ground, tackling flames across the Bunyip Range some 62 miles (100km) east of Melbourne. Aerial observation was critical in mapping the progress and intensity of the wildfires.

AFTERMATH

The scale of the Black Saturday wildfires was unprecedented in Australia. In the immediate aftermath, thousands of volunteers and government workers provided shelter and sustenance for both survivors and families of the victims, but two years after the fire, only about 40 percent of the 1,795 destroyed homes had received permission to rebuild. The state government was advised to ban construction in fire-susceptible areas. Some important lessons were learned after the event, including a change to the "stay or go" policy, which was amended to advise residents caught in future wildfires to leave their homes immediately once warnings are issued. Tragically, 113 people perished in vain attempts to protect their property. The authority in charge of sending warnings to the public did not do so in a timely manner for some communities, meaning that people living there were unaware of the danger until the fire was upon them.

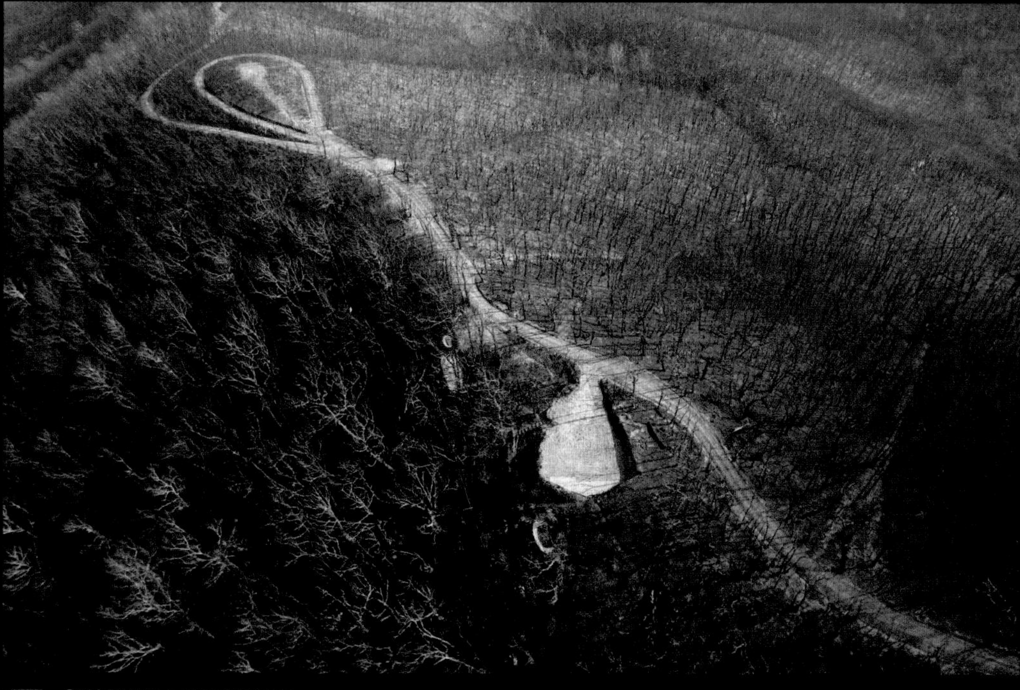

2 CHARRED LANDSCAPE
The landscape outside the community of Kingslake was still completely charred some six months after the February wildfires. The steep slopes and strong winds aided rapid spread of the flames in this area, leading to 38 fatalities.

LAND DAMAGE

Biologists estimate that the number of killed or injured wildlife ran into the millions. Many kangaroos required attention after suffering badly burned feet as they returned to their smouldering habitats. Some 6.6 billion sq ft (620 million sq m) of grazing land and more than 35,300 tons of hay and silage were lost, and insurance claims for the bushfires totalled AUS$1.2 billion. Water engineers were concerned that some of Melbourne's dammed water supply would be polluted with ash from burned catchments.

"THERE WASN'T A FIREFIGHTER IN THE STATE WHO DIDN'T HAVE A BAD FEELING ABOUT SATURDAY."

KEITH BARBOUR, FIREFIGHTER

3 REBUILDING
The framework for a new house is constructed among the ruins. Around 100 towns were affected by the bushfires in Australia, and for many people the process of rebuilding is slow and expensive – it could take many years before the destruction is repaired. Those who choose to rebuild their homes and stay in these areas are at perennial risk of fire outbreaks.

CLIMATE CHANGE

Earth's atmosphere helps keep the planet warm. However, records show that atmospheric temperatures have been slowly increasing over the last century. The mechanisms behind this are still unclear, but scientists believe that human activity may be upsetting the natural balance of atmospheric gases.

GLOBAL WARMING

Earth's climate has swung between extremes of hot and cold throughout its existence. These changes have been influenced by natural events that have affected the composition of the atmosphere and its rate of warming. While natural fluctuations explain some of the recent temperature rise, the burning of fossil fuels by humans is thought to be behind a noticeable increase in concentrations of carbon dioxide (CO_2). Although this gas makes up only a small part of the atmosphere (around 0.038 percent by volume), it is highly efficient at trapping heat. Water vapor and methane also stop heat from the Sun being radiated back into space. This ability to conserve heat has led to them being dubbed "greenhouse gases." Trying to predict what will happen as the planet warms up is not easy. Little is known about the natural processes that remove and store carbon dioxide and how they interact with each other. What is certain is that climate change will affect weather around the world, leading to some areas becoming colder and wetter while others become hotter and drier.

melting sea ice affects ocean circulation

volcanoes add aerosols to atmosphere, reflecting solar energy back into space

bright surfaces such as ice reflect solar energy back into space

clouds release water vapor as rain

solar energy warms land and ocean

vegetation and soil absorb CO_2 from the atmosphere

melting land ice raises sea level and affects ocean circulation

deforestation and biomass burning release CO_2 into the atmosphere

evaporation from land and ocean creates cloud

oceans absorb, store, and release heat and CO_2

CLIMATE IN BALANCE
Interactions between land, ocean, and atmosphere, powered by the Sun, are so finely balanced that an adverse human impact on one system could have a significant effect on the global climate.

IMPACT

Climate change could have significant consequences, not least a higher frequency of extreme weather events, such as intense storms, flooding, and droughts. Tide gauge and satellite measurements show that sea levels are rising around the globe as the water warms and expands in volume, putting many coastal cities and coral islands at risk. Retreating ice sheets at the poles and rapidly melting glaciers are contributing to this rise by adding huge quantities of fresh water. Warming of the oceans could also release methane buried in ocean floor sediments, resulting in sudden climate change. Large amounts of methane are also bound up in tundra soils and would be released if the permafrost melts.

ON THE RETREAT
These two views of Pedersen Glacier, Alaska, show how much it has shrunk in the period between these two photos, taken in 1917 (far left) and 2005 (left). Melting glaciers are making sea levels rise faster than at any time in the last 350 years.

DRIVERS OF CHANGE

Climate change is being driven by a mixture of natural and man-made factors. The seasonal fluctuation of solar radiation caused by the tilt of Earth's axis influences ocean and atmospheric currents, affecting local and global climates. Volcanoes erupt massive quantities of greenhouse gases and sulphate aerosols into the atmosphere. Sulphates and other particulates can block out the Sun, leading to cooling, and provide condensation nuclei for clouds and rain. Water vapor, although a natural greenhouse gas, increases as warmer temperatures evaporate water into the atmosphere. This increases cloud formation, which reflects sunlight back into space.

Although the burning of coal, oil, and natural gas to meet human energy needs is responsible for much of the increase in carbon dioxide, it is not the only factor. Trees and plants remove carbon dioxide during photosynthesis, but deforestation and land clearance have drastically reduced this natural carbon "sink." Burning plant material releases further quantities of carbon dioxide. Exhaust fumes from cars and airplanes contain another greenhouse gas, nitrous oxide, which is also produced during sewage treatment and the application of fertilizers to soils. However, action to reduce airborne pollutants will lessen their cooling effect and warm the atmosphere.

GREENHOUSE GASES

There is strong evidence that human activity is increasing levels of greenhouse gases in the atmosphere. A report for the Intergovernmental Panel on Climate Change in 2007 showed that fossil fuel emissions were responsible for more than 56 percent of the anthropogenic CO_2 added to the atmosphere in 2004. Methane, nitrous oxide, and fluorinated gases used in refrigerants and solvents contributed nearly a quarter of the total.

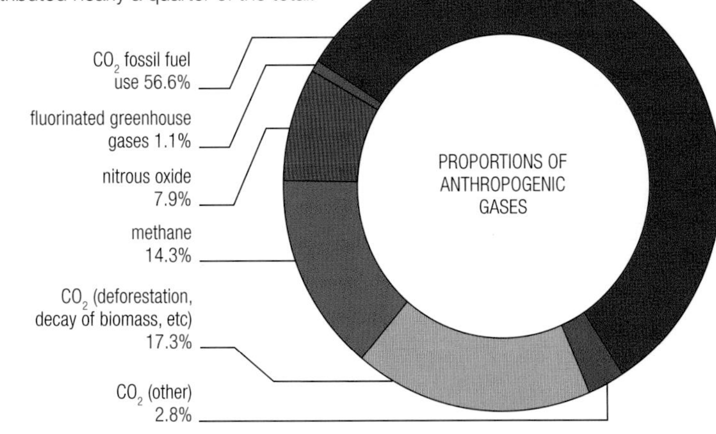

METHANE EMISSIONS
Intensive farming, particularly of rice paddies and cattle herds, is responsible for an increase in methane concentrations. A cow can belch up to 53 gal (200 liters) of methane a day, so scientists are trying to alter their diet as a means to reduce emissions.

7

REFERENCE

<< Earth's atmosphere
The thin line of Earth's atmosphere and the setting Sun were photographed by the crew of the International Space Station.

EARTH

EARTH'S VITAL STATISTICS

Earth is the third planet from the Sun and the only planet known to have an active plate tectonics system, water flowing freely on its surface, and plentiful life forms populating its air, land, and seas.

The planet was born 4.6 billion years ago, along with the Sun and the rest of the Solar System, from a cloud of gas and dust. Earth orbits the Sun at a distance of 93 million miles (149.6 million km).

4.6 BILLION YEARS

Earth's age.

15,760 MILES (40,030KM)

The mean circumference of Earth.

5,022 MILES (12,756KM)

The diameter at the equator.

13 MILES (21KM)

Earth bulges by this amount at the equator, compared with its shape pole to pole. This bulge is caused by Earth's rotation flattening the planet slightly.

10.5 MILES (17KM)

The depth of the troposphere—the part of Earth's atmosphere nearest the surface, where most weather occurs.

23.5

The degree to which Earth's spin axis is tilted from the vertical. This tilt causes the seasons.

66,622MPH (107,218KPH)

The speed at which Earth is orbiting the Sun.

900 MYA

The date the first multicellular animals emerged on Earth.

59°F (15°C)

The average surface temperature.

994MPH (1,600KPH)

The speed Earth is rotating (on the surface at the equator) as it completes one spin every 24 hours.

0.8-8IN (2-20CM)

The average distance that Earth's tectonic plates move in one year.

8.3 MINUTES

The time it takes for light from the Sun to reach Earth.

GEOLOGICAL TIMESCALE

Earth's timescale covers thousands of millions of years. Study of the rocks and fossils of the planet's crust has made it possible to divide this time into segments: eras are divided into smaller segments called periods, which in turn are divided into epochs. The divisions are constantly being refined.

ERA	PERIOD	EPOCH	START (MYA)
Cenozoic	Neogene	Holocene	0.01
		Pleistocene	1.8
		Pliocene	5.3
		Miocene	23
	Paleogene	Oligocene	34
		Eocene	55
		Palaeocene	65
Mesozoic	Cretaceous		145
	Jurassic		202
	Triassic		251
Palaeozoic	Permian		299
	Carboniferous (divided into Lower Mississippian and Upper Pennsylvanian in United States)		359
	Devonian		416
	Silurian		444
	Ordovician		488
	Cambrian		542

EARTH'S CONTINENTS

Earth's land is divided into seven commonly recognized continents. Asia is by far the largest, accounting for nearly one-third of Earth's total land mass.

SHARE OF TOTAL LAND MASS

%	Continent	Area
30%	Asia	17,212,000 sq miles (44,579,000 sq km)
20.5	Africa	11,608,000 sq miles (30,065,000 sq km)
16.5%	North America	9,365,000 sq miles (24,256,000 sq km)
12%	South America	6,880,000 sq miles (7,819,000 sq km)
9%	Antarctica	5,100, 000 sq miles 13,209,000 sq km)
	Europe	3,937,000 sq miles (9,938,000 sq km)
5%	Australia/Oceania	2,968,000 sq miles (7,687,000 sq km)

TECTONIC PLATES

The Earth's outermost rocks are broken into eight or nine major tectonic plates and dozens of smaller ones (see pp.26–27). Over time the plates have moved apart (diverged) and toward each other (converged) to shift continents, open and close oceans, and form mountains. At transform boundaries, plates slide past one another along fault planes, sometimes creating stress that causes earthquakes.

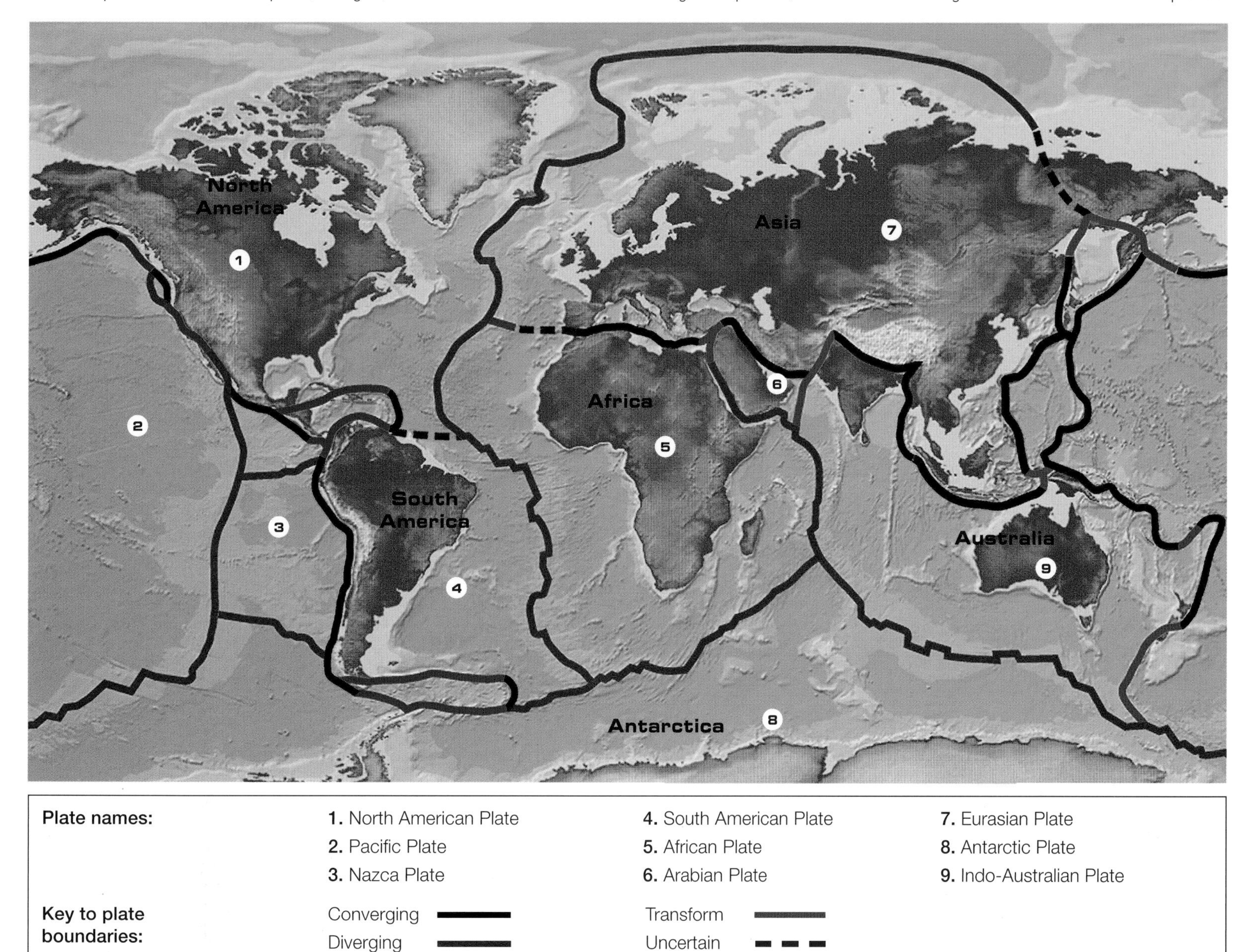

STRUCTURE OF THE EARTH	NAME	AVERAGE DEPTH	DENSITY	TEMPERATURE
	Crust	3–45 miles (5–70km)	0.1lb per cu in (3g per cu cm)	Less than 1,800°F (1,000°C)
	Mantle	1,860 miles (2,990km)	0.1–0.2lb per cu in (3.5–5.5g per cu cm)	1,800–6,300°F (1,000–3,500°C)
	Outer core	3,200 miles (5,150km)	0.36lb per cu in (10g per cu cm)	6,300–7,200°F (3,500–4,000°C)
	Inner core	3,960 miles (6,370km)	0.4lb per cu in (12g per cu cm)	7,200–8,500°F (4,000–4,700°C)

MOUNTAINS

MAJOR MOUNTAIN RANGES

The world's major mountain ranges—such as the Andes, Himalayas, Alps, and Rockies—are comparatively young, having formed within the last few hundred million years. They are situated along the boundaries of tectonic plates that collided in the geological past (see pp.44–45). Mountain ranges were formed as immense pressures caused by the convergence of the plates deformed and uplifted the rocks high above the surrounding land. This process of uplift is still continuing.

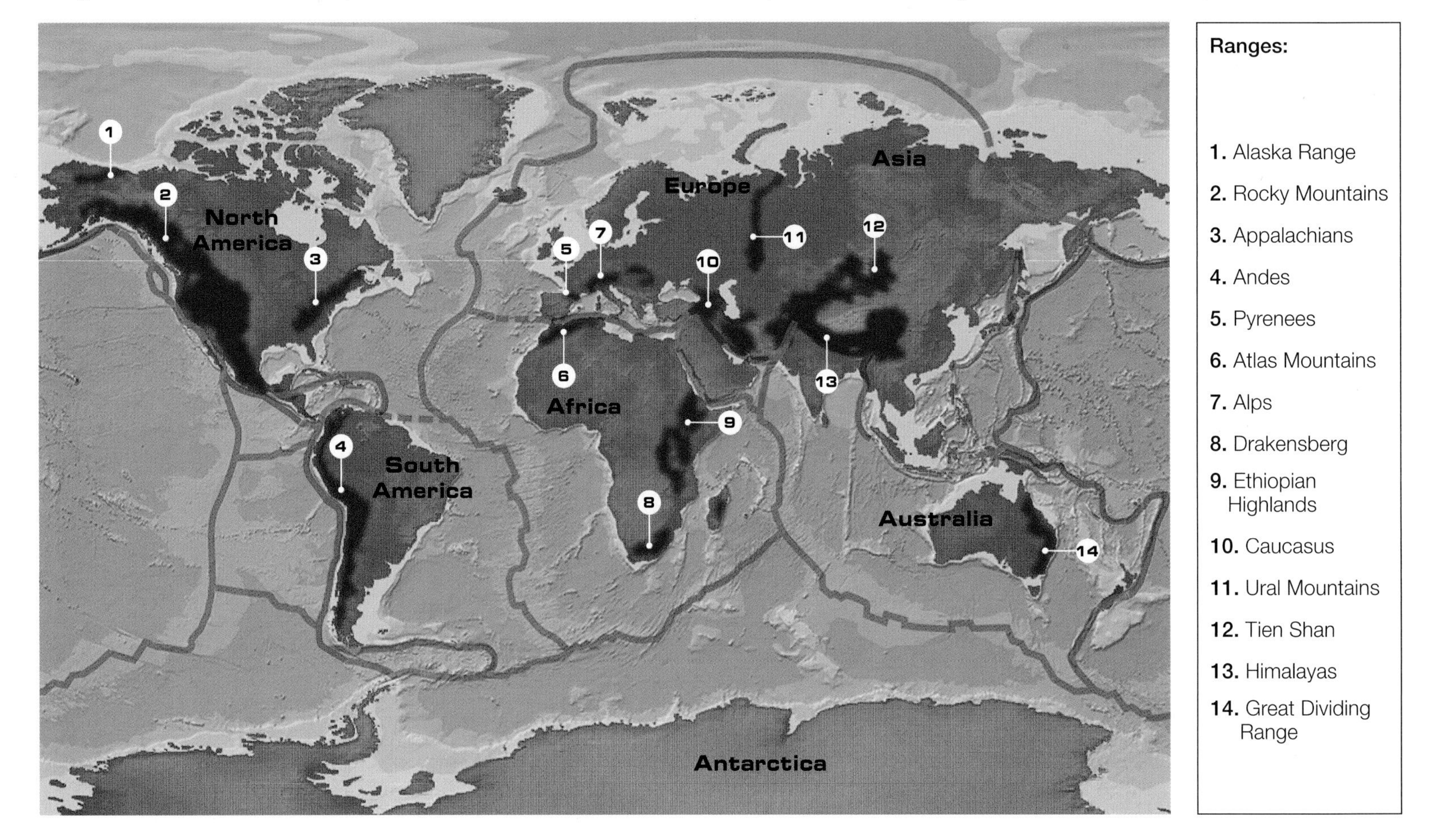

Ranges:

1. Alaska Range
2. Rocky Mountains
3. Appalachians
4. Andes
5. Pyrenees
6. Atlas Mountains
7. Alps
8. Drakensberg
9. Ethiopian Highlands
10. Caucasus
11. Ural Mountains
12. Tien Shan
13. Himalayas
14. Great Dividing Range

THE WORLD'S HIGHEST MOUNTAIN PEAKS

Fourteen peaks rise more than 26,000ft (8,000m) above sea level, and all of which are located in Asia. The vast majority are in the Himalayas, the highest and one of the most recently formed mountain ranges on Earth, often referred to as the "roof of the world." Located in Asia, the Himalayas separate the Indian subcontinent from the Tibetan Plateau.

MOUNTAIN	RANGE	LOCATION	HEIGHT
Everest	Himalayas	Nepal/Tibet	29,035ft (8,850m)
K2	Karakoram	Pakistan/China	28,250ft (8,611m)
Kangchenjunga	Himalayas	Nepal/India	28,169ft (8,586m)
Lhotse	Himalayas	Nepal/India	27,940ft (8,516m)
Makalu	Himalayas	Nepal/India	27,766ft (8,463m)
Cho Oyu	Himalayas	Nepal/India	26,906ft (8,201m)
Dhaulagiri	Himalayas	Nepal	26,795ft (8,167m)
Manaslu	Himalayas	Nepal	26,781ft (8,163m)
Nanga Parbat	Himalayas	Pakistan	26,660ft (8,125m)
Annapurna	Himalayas	Nepal	26,545ft (8,091m)
Gasherbrum I	Karakoram	Pakistan/China	26,470ft (8,068m)
Broad Peak	Karakoram	Pakistan/China	26,400ft (8,047m)
Gasherbrum II	Karakoram	Pakistan/China	26,360ft (8,035m)
Shishna Pangma	Himalayas	Tibet	26,289ft (8,013m)

HIGHEST PEAK BY CONTINENT

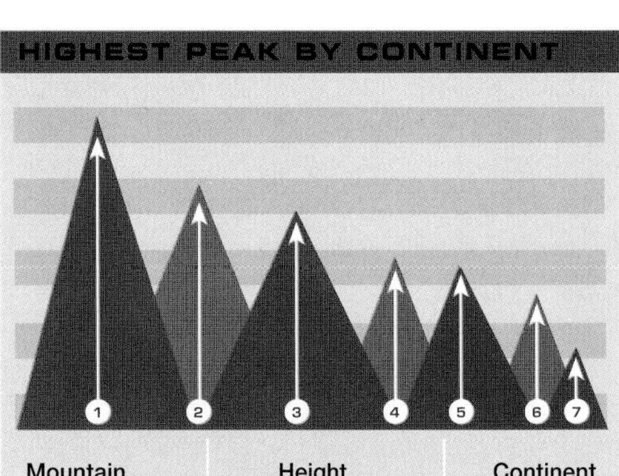

Mountain	Height	Continent
1 Mount Everest	29,035ft (8,850m)	Asia
2 Aconcagua	22,841ft (6,962m)	South America
3 Mount McKinley	20,321ft (6,194m)	North America
4 Kilimanjaro	19,331ft (5,892m)	Africa
5 Elbrus	18,510ft (5,642m)	Europe
6 Vinson Massif	16,050ft (4,892m)	Antarctica
7 Kosciuszko	7,310ft (2,228m)	Australia

OCEANS

OCEAN CURRENTS

The water in the oceans is in constant motion, circulating in persistent currents that flow both near the surface and, more slowly, at great depths. Surface currents are driven by winds but their patterns are modified by the effects of Earth's rotation on its axis, which causes circular water movements called gyres. These gyres rotate clockwise in the northern hemisphere and counterclockwise in the southern hemisphere. Specific components of these gyres are called boundary currents. The boundary currents on the eastern side of oceans are predominantly cold and move toward the equator; those on the western side tend to be warm and move away from the equator. The present global circulation pattern is also partly controlled by the configuration of the ocean basins and surrounding coastlines. In the past, the distribution of oceans and continents, and therefore current circulation, was very different.

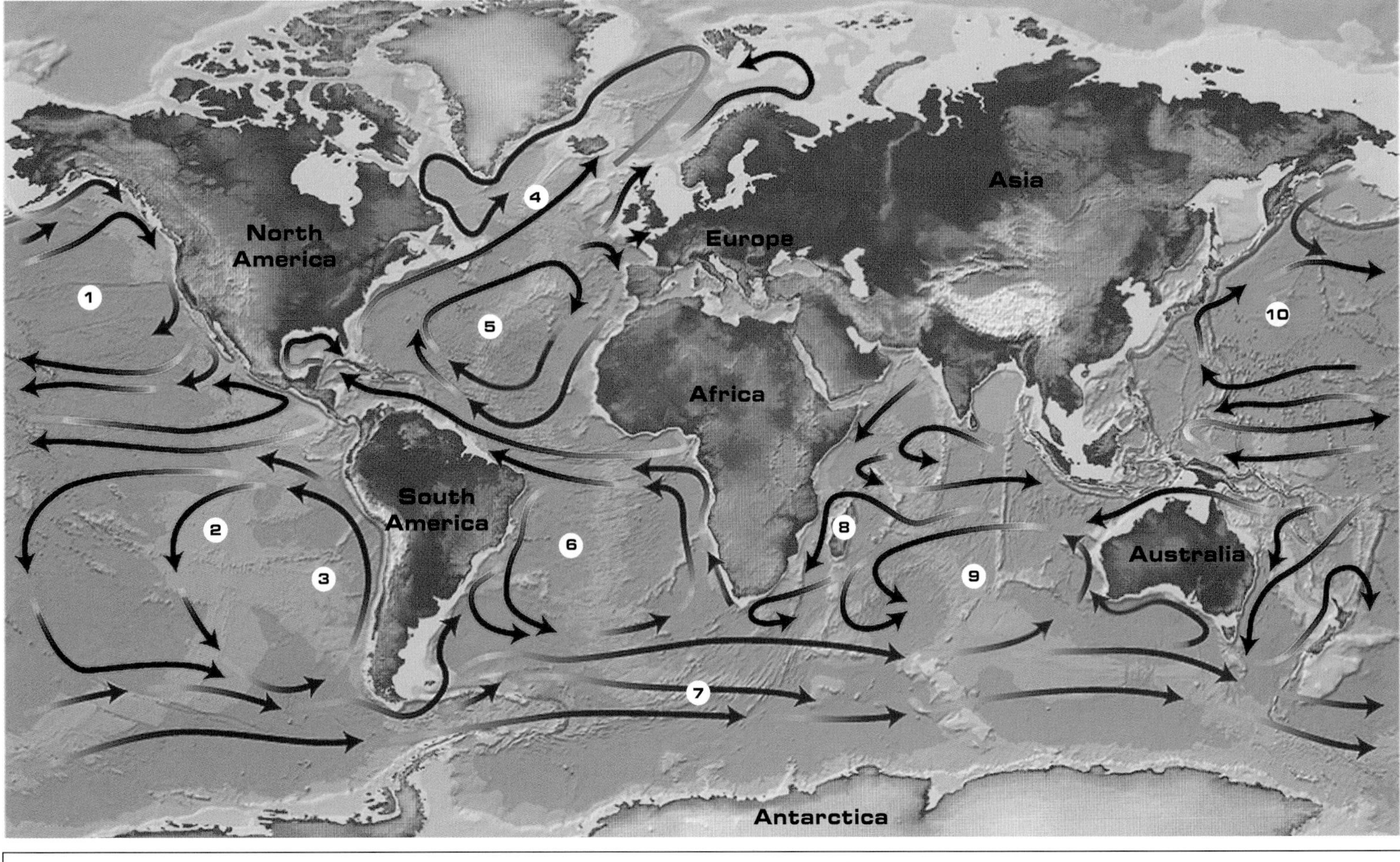

Currents:

1 North Pacific gyre
2 South Pacific gyre
3 Humboldt current
4 Gulf Stream
5 North Atlantic gyre
6 South Atlantic gyre
7 Antarctic circumpolar current
8 Aghulas current
9 South Indian gyre
10 North Pacific gyre

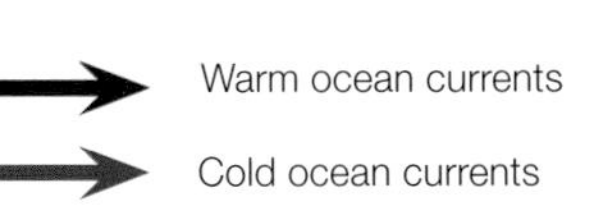

MAJOR TSUNAMIS

Often caused by earthquakes at sea, tsunamis are displacements of water that result in huge waves, which can cause large-scale devastation to coastal areas. This list gives some of the major tsunamis of recent years. Death tolls from tsunamis can be hard to measure because many fatalities may be caused by the earthquake that triggered the tsunami.

YEAR	LOCATION	FATALITIES
1933	Sanriku coast, Japan	More than 3,000
1944	Tonankai earthquake, Japan	More than 1,200
1946	Nankai earthquake, Japan	More than 1,400
1958	Lituya Bay, Alaska	–
1960	Chilean tsunami	6,000
1963	Vajont Dam, Italy	2,000
1976	Moro Gulf, Philippines	93,500 left homeless
1998	Papua New Guinea	2,200
2004	Indian Ocean tsunami	230,000
2006	South of Java Island	800
2011	Tohoku tsunami, Japan	More than 15,000

VOLCANOES

VOLCANO STATISTICS

Volcanic eruptions of molten rock from within Earth are awesome demonstrations of the pent-up heat energy stored within our planet. The world is dotted with volcanoes and their rock products, both ancient and modern—reminders of Earth's long history of volcanism. Most volcanoes occur at plate boundaries. The largest volcanoes are situated above hotspots (see pp. 32–33).

74,000 YEARS
How long it has been since the world's largest historic supervolcanic eruption, known as the Toba catastrophe theory, which plunged the world into a 10-year volcanic winter.

24 MILLION TONS
The amount of sulphur dioxide spewed out by Mount Pinatubo in the Philippines in 1991, which encircled the entire planet and brought down the global temperature by 0.9°F (0.5°C)

2,000°F (1,250°C)
The temperature of pyroclastic flows that roll down the side of a volcano during an eruption. These flows contain ash, rock, and water.

1,550
The approximate number of active volcanoes that span the globe today.

90 PERCENT
The proportion of volcanoes on Earth that exist on the Pacific Ring of Fire.

20
The number of volcanoes that are erupting in the world on any given day.

260,000
The number of people estimated to have died in the past 300 years from volcanic eruptions and their aftermath.

1815
The year of the volcanic eruption in Tambora, Indonesia, which killed a record-breaking 92,000 people.

13,681FT (4,170M)
The height of the tallest volcano on Earth, Hawaii's Mauna Loa. The Hawaiian Islands were created by a hotspot.

300 MILLION
The number of people thought to live within the danger range of an active volcano.

37 MILES (60KM)
The maximum height that an ash column can reach after a Plinian eruption.

10,000 YEARS
A volcano is considered active if it has erupted within this many years.

DECADE VOLCANOES

Sixteen volcanoes have been identified by the International Association of Volcanology and Chemistry of the Earth's Interior (IAVCEI) as being worthy of particular study, due to being very active or very near to heavily populated areas. The group is known as the Decade Volcanoes. It is hoped that study of the Decade Volcanoes will aid understanding of how and when eruptions occur.

Volcano names:

1. Rainier
2. Mauna Loa
3. Colima
4. Santa Maria/Santiaguito
5. Galeras
6. Vesuvius
7. Mount Etna
8. Santorini
9. Teide
10. Nyiragongo
11. Avachinsky-Koryaksky
12. Unzen
13. Sakurajima
14. Taal
15. Ulawun
16. Merapi

VOLCANIC EXPLOSIVITY INDEX

The power of a volcanic eruption can be measured on the Volcanic Explosivity Index (VEI). A VEI value is given to an eruption based on how high the eruption column reaches and how much material is ejected. Certain types of eruption commonly have similar VEI measurements (see classification column). In the whole of human history, no eruption measuring 8 on the VEI has occurred, though there are thought to have been several prehistoric eruptions of that scale. The largest in recent times was the VEI-7 eruption of Mount Tambora, Indonesia, in 1815.

VEI	DESCRIPTION	ERUPTION COLUMN	VOLUME	CLASSIFICATION	HOW OFTEN	EXAMPLE
0	Non-explosive	Up to 330ft (100m)	Up to 350,000 cu ft (10,000 cu m)	Hawaiian	Daily	Kilauea
1	Gentle	330–3,300ft (100–1,000m)	Over 350,000 cu ft (10,000 cu m)	Hawaiian/Strombolian	Daily	Stromboli
2	Explosive	0.5–3 miles (1–5km)	Over 35 million cu ft (1 million cu m)	Strombolian/Vulcanian	Weekly	Galeras, 1992
3	Severe	2–9 miles (3–15km)	Over 350 million cu ft (10 million cu m)	Vulcanian	Yearly	Ruiz, 1985
4	Cataclysmic	6–15 miles (10–25km)	Over 0.02 cu miles (0.1 cu km)	Vulcanian/Plinian	10s of years	Galunggung, 1982
5	Paroxysmal	Over 15 miles (25km)	Over 0.2 cu miles (1 cu km)	Plinian	100s of years	St. Helens, 1981
6	Colossal	Over 15 miles (25km)	Over 2 cu miles (10 cu km)	Plinian/Ultra-plinian	100s of years	Krakatau,1883
7	Super-colossal	Over 15 miles (25km)	Over 25 cu miles (100 cu km)	Ultra-plinian	1,000s of years	Tambora, 1815
8	Mega-colossal	Over 15 miles (25km)	Over 240 cu miles (1,000 cu km)	Ultra-plinian	10,000s of years	Yellowstone, 2 MYA

DEADLIEST ERUPTIONS IN HISTORY

Volcanic eruptions can be deadly if they occur near densely populated areas. Deaths from volcanic eruptions can be due to lava and pyroclastic flows, ash fall, or from related hazards such as mudflows.

VOLCANO	WHEN	DEATHS
Tambora, Indonesia	1815	92,000
Krakatau, Indonesia	1883	36,417
Mount Pelée, Martinique	1902	29,025
Ruiz, Colombia	1985	25,000
Unzen, Japan	1792	14,300
Laki, Iceland	1783	9,350
Santa Maria, Guatemala	1902	c.6,000
Kelud, Indonesia	1919	5,110
Galunggung, Indonesia	1882	4,011
Vesuvius, Italy	1631	3,500
Vesuvius, Italy	79	3,360
Papandayan, Indonesia	1772	2,957
Lamington, Papua New Guinea	1951	2,942
El Chichón, Mexico	1982	2,000
Soufriere, St. Vincent	1902	1,680
Oshima, Japan	1741	1,475
Asama, Japan	1783	1,377
Taal, Philippines	1911	1,335
Mayon, Philippines	1814	1,200
Agung, Indonesia	1963	1,184
Cotopaxi, Ecuador	1877	1,000
Pinatubo, Philippines	1991	800
Komatagtake, Japan	1640	700
Ruiz, Colombia	1845	700

BIGGEST ERUPTIONS SINCE 1000

Listed here are the largest eruptions, by the estimated volume of material ejected, that have occurred in the past 1,000 years. They would be dwarfed by the supervolcanic eruptions of prehistory.

VOLCANO	WHEN	VOLUME
Tambora, Indonesia	1815	36 cu miles (150 cu km)
Kolumbo, Greece	1650	14.4 cu miles (60 cu km)
Kuwar, Vanuatu	1452–53	8 cu miles (33 cu km)
Long Island, Papua New Guinea	1660	7.2 cu miles (30 cu km)
Huaynaputina, Peru	1600	7.2 cu miles (30 cu km)
Krakatau, Indonesia	1883	5 cu miles (21 cu km)
Quilotoa, Ecuador	1280	5 cu miles (21 cu km)
Santa Maria, Guatemala	1902	4.8 cu miles (20 cu km)
Grimsvötn and Laki, Iceland	1783–84	3.4 cu miles (14 cu km)
Novarupta, Alaska	1912	3.4 cu miles (14 cu km)
Billy Mitchell, Papua New Guinea	1580	3.4 cu miles (14 cu km)
Pinatubo, Philippines	1991	2.6 cu miles (11 cu km)

EARTHQUAKES

EARTHQUAKE ZONES

Movement of Earth's crust and tectonic plates produces stress and fracture of its rocks along fault planes. The build-up of stress in these brittle crustal rocks results in sudden failure and the release of massive amounts of energy as earthquakes (see pp.206–07). The shock waves are transmitted through the rocks, often over great distances—those of major quakes reach the other side of the world. Earth's main earthquake zones follow the boundaries where tectonic plates move relative to one another.

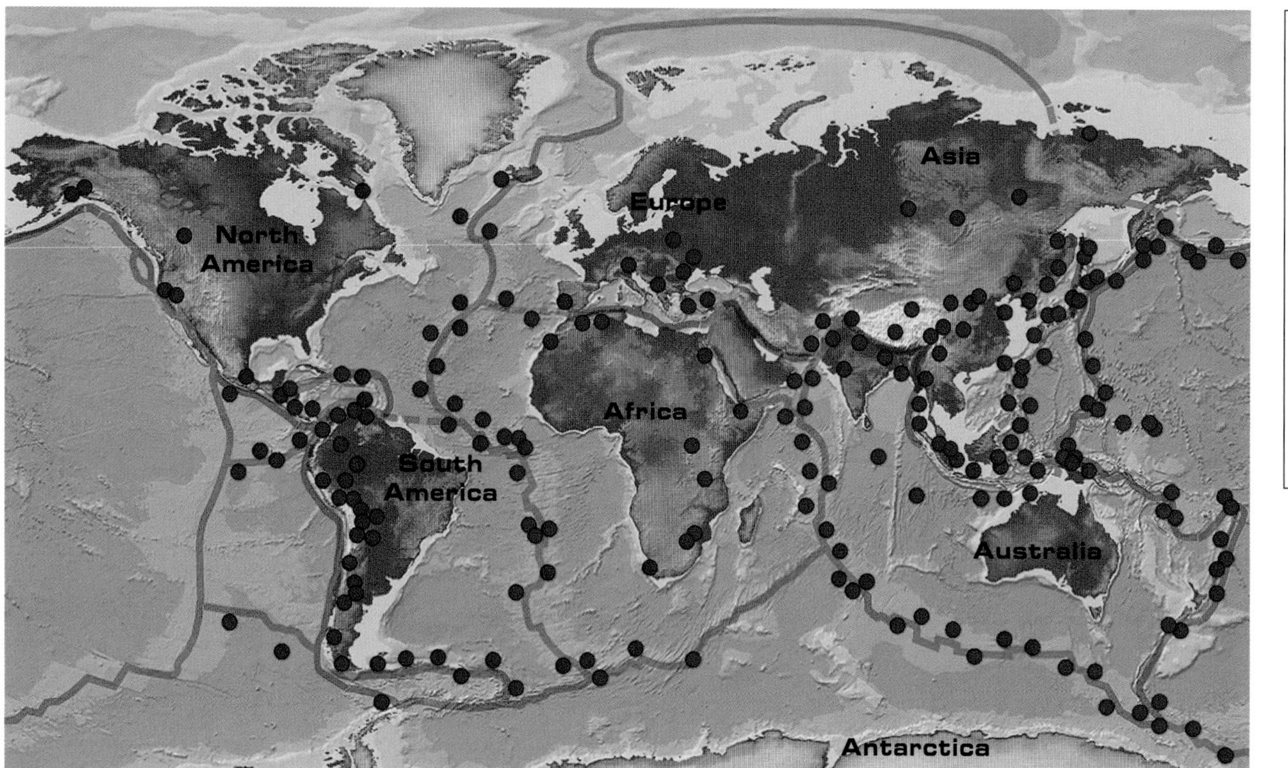

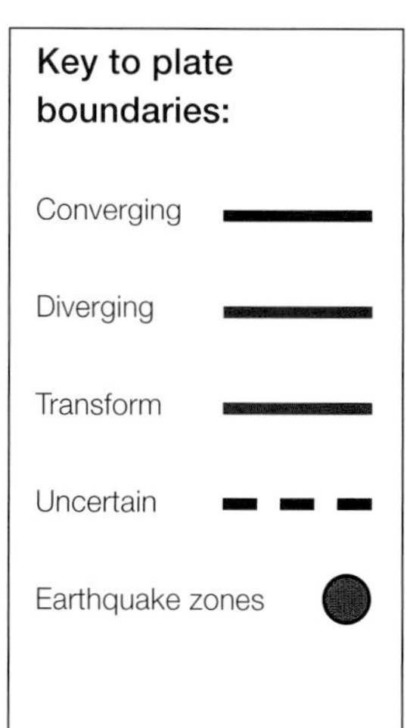

EARTHQUAKE SEVERITY SCALES

The size of an earthquake can be measured by magnitude and by intensity (see pp.212–13). Two traditionally used scales are the Mercalli Intensity Scale (which measures the effects of an earthquake on the surface, at 12 levels) and the Richter Scale (which measures the amount of energy released). However, in scientific use these have largely been superseded by the Moment Magnitude scale. Almost every day, there are hundreds of minor earthquakes around the world. Very large disturbances causing widespread damage are rare.

MOMENT MAGNITUDE	AVERAGE NUMBER PER YEAR	EXAMPLE	DATE
9	0.1	M-9.5 southern Chile M-9.1 Sunda Trench, Sumatra M-9.0 offshore Honshu (Tohoku), Japan	May 22, 1960 December 26, 2004 March 11, 2011
8	1	M-8.8 offshore Bio Bio, Chile M-8.5 Lisbon, Portugal M-8.1 Samoa	February 27, 2010 November 1, 1755 September 29, 2009
7	18	M-7.9 Chengdu, Sichuan, China M-7.8 San Francisco, California, US M-7.5 Muzzafarabad, Kashmir M-7.0 Port au Prince, Haiti	May 12, 2008 April 18, 1906 October 8, 2005 January 12, 2010
6	134	M-6.9 Loma Prieta, California, US M-6.3 L'Aquila, Italy M-6.1 Christchurch, New Zealand	October 18, 1989 April 6, 2009 February 21, 2011
5	1,300	M-5.5 Newcastle, Australia M-5.4 Illinois, US M-5.3 San Giuliano di Puglia, Italy	December 27, 1989 April 18, 2008 October 31, 2002
4	13,200	M-4.5 Seattle-Tacoma, Washington, US M-4.2 Kent, England M-4.1 Melton Mowbray, England	January 30, 2009 April 28, 2007 October 28, 2001
3	130,000	M-3.9 Virginia, US M-3.6 Dumfries, Scotland M-3.0 Bargoed, Wales	May 5, 2003 December 26, 2006 October 10, 2001

THE MERCALLI INTENSITY SCALE		
I	Instrumental	No Earth movement felt; detectable only by seismograph
II	Feeble	Movement felt only by people at rest or on upper floors of buildings
III	Slight	Noticeable indoors, especially on upper floors; hanging objects swing
IV	Moderate	Felt indoors by most people and outdoors by a few; hanging objects swing; crockery, doors, and windows rattle; sensation of a heavy truck hitting the walls; parked cars rock
V	Rather strong	Felt by nearly all; sleeping people waken; doors swing; small objects move; crockery and glassware break; liquids may spill
VI	Strong	Felt by all; difficult to walk; furniture moves; objects fall from shelves and pictures from walls; plaster may crack; little structural damage
VII	Very strong	Difficult to stand; drivers feel cars shaking; furniture may break; bricks or chimneys may fall; some structural damage depending on construction of building
VIII	Destructive	Car drivers cannot steer; considerable damage in poorly constructed buildings or tall structures such as factory chimneys; branches broken from trees; wet or sloping ground may crack; water levels in wells may be affected
IX	Ruinous	Considerable structural damage to all building types; houses may shift off their foundations; the ground cracks and reservoirs are seriously damaged
X	Disastrous	Most buildings destroyed; bridges and dams seriously damaged; large cracks in ground; major landslides, rail tracks bend
XI	Very disastrous	Few structures remain standing; pipelines and rail tracks destroyed
XII	Catastrophic	Total destruction; upheavals of ground and rock

RICHTER SCALE: TYPICAL RANGES	
<3.5	Recordable, but not generally felt
3.5–5.4	Felt by many, but causes little or no damage
5.5–6.0	May be slight damage to well-constructed buildings; more severe damage to poorly constructed buildings
6.1–6.9	Can cause severe and widespread damage up to a range of 62 miles (100km)
7.0–7.9	Major effects, causing severe damage over larger areas
>8.0	Can cause devastation over several hundred miles

INTENSITY (RICHTER)	EARTHQUAKES PER YEAR (ESTIMATED)
2.5 or less	900,000
2.5 to 5.4	30,000
5.5 to 6.0	500
6.1 to 6.9	100
7.0 to 7.9	20
8.0	One every 5 to 10 years

LARGEST EARTHQUAKES SINCE 1900

The largest magnitude earthquakes occur on average once every five to 10 years. Since 1900 there have been five Earth movements with a magnitude of 9 or more, and several more measuring between 8 and 9.

LOCATION	DATE	MAGNITUDE
Chile (South America)	May 22, 1960	9.5
Prince William Sound (Alaska)	March 28, 1964	9.2
West coast of Northern Sumatra (Indonesia)	December 26, 2004	9.1
Near east coast of Honshu (Japan)	March 11, 2011	9.0
Kamchatka (Russia)	November 4, 1952	9.0
Offshore Maule (Chile)	February 27, 2010	8.8
Off the coast of Ecuador	January 31, 1906	8.8
Rat Islands (Alaska)	February 4, 1965	8.7
Northern Sumatra (Indonesia)	March 28, 2005	8.6
Assam (Tibet)	August 15, 1950	8.6
Andreanof Islands (Alaska)	March 9, 1957	8.6
Southern Sumatra (Indonesia)	September 12, 2007	8.5
Banda Sea (Indonesia)	February 1, 1938	8.5
Kamchatka (Russia)	February 3, 1923	8.5
Chile–Argentina border (South America)	November 11, 1922	8.5
Kuril Islands (Russia)	October 13, 1963	8.5

MOST DESTRUCTIVE EARTHQUAKES

Earthquakes near urban populations cause more fatalities regardless of their magnitude, due to the number of people nearby and, in recent times, the hazards of falling buildings in heavily built up areas.

YEAR	LOCATION	DEATHS	MAGNITUDE
1976	Tangshan, China	255,000	7.5
2004	Sumatra	227,898	9.1
2010	Haiti region	222,570	7.0
1920	Haiyuan, Ningxia, China	200,000	7.8
1923	Kanto, Japan	142,800	7.9
1948	Ashgabat, Turkmenista	110,000	7.3
2008	Eastern Sichuan, China	87,587	7.9
2005	Pakistan	86,000	7.6
1908	Messina, Italy	72,000	7.2
1970	Chimbote, Peru	70,000	7.9
1990	Western Iran	40,000 to 50,000	7.4
1783	Calabria, Italy	50,000	–
1755	Lisbon, Portugal	70,000	8.7
1727	Tabriz, Iran	77,000	-
1693	Sicily, Italy	60,000	7.5
1667	Shemakha, Caucasia	80,000	-

WEATHER

CLIMATE REGIONS

Earth's landmasses can be divided into climate regions, based on average temperatures and rainfall, and type of vegetation. Tropical areas are hot all year-round and often very wet, while polar regions and the tops of high mountains are characterized by extreme cold. In the temperate regions between the poles and the tropics, climates are more moderate. Areas classified as desert have an annual rainfall of less than 10in (25cm). Climate is influenced by latitude, but also by factors such as height above sea level and distance from the ocean.

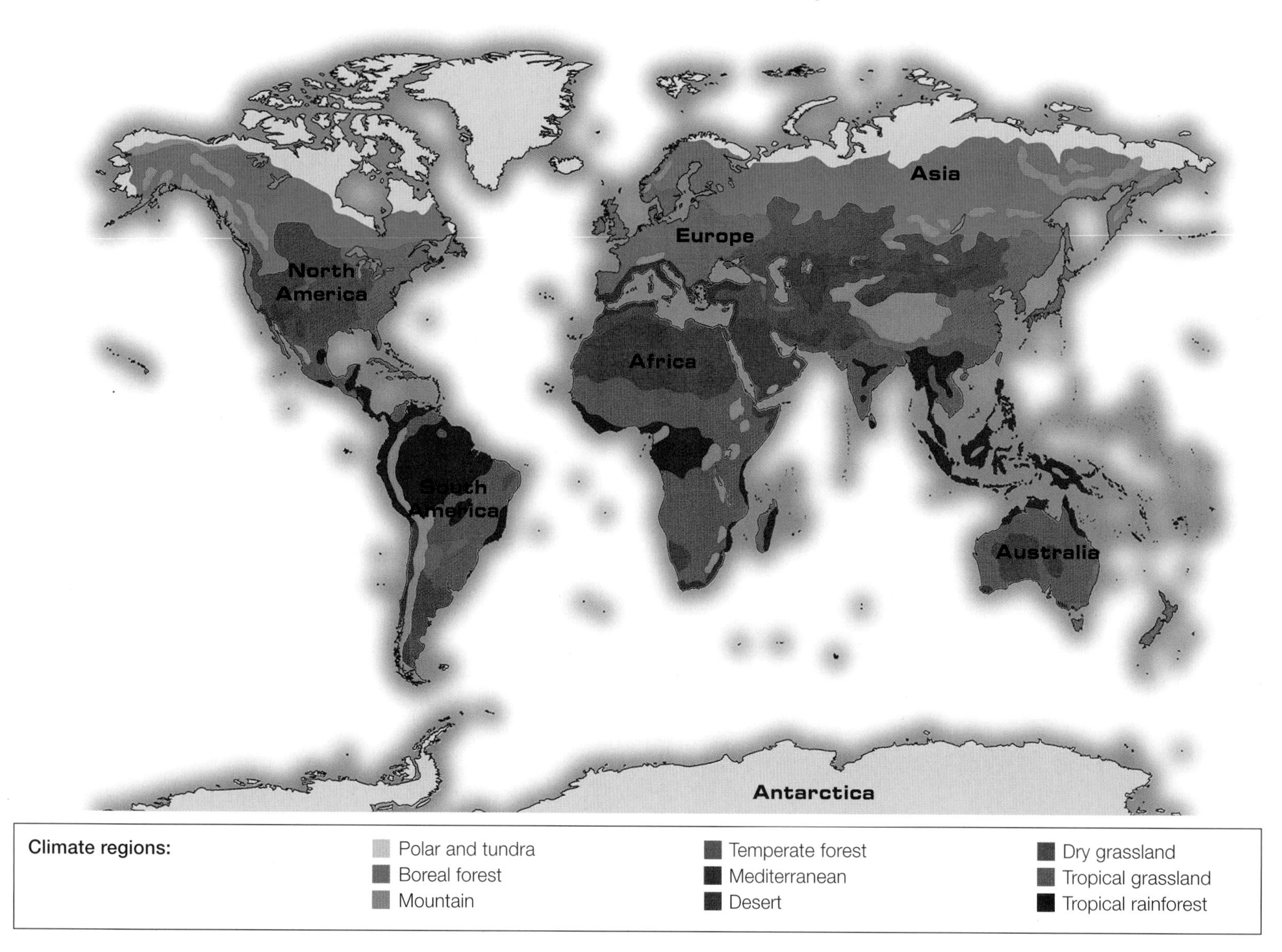

WEATHER SYMBOLS

The symbols shown here are an internationally recognized system used for classifying types of weather. They were devised by the World Meteorological Organization and represent varying intensities of precipitation and other conditions. They are used on maps to provide an instantly understandable snapshot of weather conditions at a particular time.

WEATHER MAP SYMBOLS			
Intermittent light rain	•	Freezing rain	⁀∼
Continuous light rain	••	Freezing drizzle	⁀∼
Intermittent moderate rain	⁚	Thunderstorm	ꓘ
Continuous moderate rain	∴	Tornado	)(
Continuous heavy rain	⁘	Fog	≡
Light rain shower	▽	Haze	∞
Moderate rain shower	▽	Stationary front	▲▼
Heavy rain shower	▽	Cold front	▼▼▼
Snow	*	Occluded front	▲▲▲
Drizzle	,	Warm front	▲▲▲

EXTREME WEATHER STATISTICS

134.6°F (57°C)
The hottest recorded temperature, measured in El Aziz, Libya, on September 13, 1922.

-128°F (-89°C)
The coldest temperature on record, recorded at Vostok, Antarctica, on July 21, 1983.

536IN (1,361CM)
The highest recorded annual precipitation, measured in Lloro, Colombia.

0.03IN (7MM)
The lowest recorded average annual precipitation, measured in Arica, Chile.

102FT (31M)
The most snowfall in a year recorded at Mount Rainier, Washington, US, between February 19, 1971 and February 18, 1972.

UP TO 350 DAYS
Most rainy days in a year, recorded at Mount Waialeale, Kauai, Hawaii, US. The total rainfall is about 460in (11,684mm).

280MPH (450KPH)
The fastest tornado wind speed, recorded at Wichita Falls, Texas, US, on April 2, 1958.

231MPH (372KPH)
The fastest non-tornado gust of wind, recorded at Mount Washington, New Hampshire, US, on April 10, 1934.

182
The number of days per year without sunshine at the South Pole.

THE BEAUFORT WIND FORCE SCALE

Beaufort's scale was originally devised for sailors as a means of judging wind strength by observations of sea disturbance and the way sails, and therefore steerage, were affected. The scale was later modified for use on land, substituting features such as trees and cars.

BEAUFORT NUMBER	WIND SPEED	DESCRIPTION
0	0–1mph (0–2kph)	Calm—smoke rises vertically, air feels still
1	1–3mph (2–6kph)	Light air—smoke drifts
2	4–7mph (7–11kph)	Slight breeze—wind detectable on face, some leaf movement
3	8–12mph (12–19kph)	Gentle breeze—leaves and twigs move gently
4	13–18mph (20–29kph)	Moderate breeze—loose paper blows around
5	19–24mph (30–39kph)	Fresh breeze—small trees sway
6	25–31mph (40–50kph)	Strong breeze—difficult to use an umbrella
7	32–38mph (51–61kph)	High wind—whole trees bend
8	39–46mph (62–74kph)	Gale—twigs break off trees, walking into wind is difficult
9	47–54mph (75–87kph)	Severe gale—roof tiles blow away
10	55–63mph (88–101kph)	Whole gale—trees break and are uprooted
11	64–74mph (102–119kph)	Storm—damage is extensive, cars overturn
12	75+mph (120+kph)	Hurricane—widespread devastation

RETIRED HURRICANE NAMES

There is a list of hurricane names that is recycled every seventh year. However, if a storm causes significant damage its name is retired for the sake of clarity.

YEAR	NAMES
1990	Diana Klaus
1991	Bob
1992	Andrew
1995	Luis Marilyn Opal Roxanne
1996	Cesar Fran Hortense
1998	Georges Mitch
1999	Floyd Lenny
2000	Keith
2001	Allison Iris Michelle
2002	Isidore Lili
2003	Fabian Isabel Juan
2004	Charley Frances Ivan Jeanne
2005	Dennis Katrina Rita Stan Wilma
2007	Dean Felix Noel
2008	Gustav Ike Paloma
2010	Igor Tomas

GLOSSARY

A

abyssal Relating to oceanic depths greater than about 6,500ft (2,000m). The abyssal plain is the flattish plain at 13,000–20,000ft (4,000–6,000m) that forms the bed of most ocean basins. The abyssal zone is the region of both seabed and open water between 6,500ft (2,000m) and the abyssal plain.

air mass A mass of air of uniform characteristics that may extend over thousands of miles.

anemometer An instrument for measuring wind speed.

anticline An archlike, upward fold of originally flat strata often caused by horizontal compression.

anticyclone A weather system in which winds circulate around an area of high pressure.

archipelago A group of islands forming a chain or cluster.

arête A narrow mountain ridge separating two adjacent cirques.

asthenosphere The layer of the mantle immediately below the rigid lithosphere. It is sufficiently nonrigid to flow slowly in a solid state and plays a key part in allowing the movement of tectonic plates.

atmosphere The layer of gases that surrounds a planet.

aureole The area around an igneous intrusion where thermal alteration (metamorphism) of country rock has taken place.

B

barometer An instrument for measuring air pressure.

basalt Earth's most common volcanic rock, usually in the form of solidified lava. Basaltic lava can erupt on continents, but basalt also forms the top of the oceanic crust. Basalt is glassy to fine-grained (composed of very fine crystals).

batholith A very large igneous intrusion, 60 miles (100km) or more across, which originates deep underground. Earth's great batholiths can be seen only if exposed on the surface by erosion.

Beaufort scale A means of estimating wind speed by observing criteria such as sea state and swaying trees.

bergschrund A deep crack formed at the back of a cirque glacier caused by the glacier moving away from the cirque wall.

biome A biological community existing on a large scale and defined mainly by vegetation type. The extent of any particular biome is determined by climatic conditions.

black smoker A hydrothermal vent in which the emerging hot water is made black by dark mineral particles.

blizzard Seriously reduced visibility due to heavy snow and high winds.

blockfield An area of broken rocks, usually in a mountainous region.

boreal Relating to or coming from the colder parts of the northern hemisphere, between the Arctic and temperate zones.

boss An igneous intrusion that is roughly circular in horizontal cross-section.

breeze A wind of moderate strength.

C

caldera A bowl-shaped volcanic depression larger than a crater, typically greater than 0.6 miles (1km) in diameter caused by subsidence into a partially emptied magma chamber after it has been evacuated by a volcanic eruption.

canyon A deep valley, usually not as narrow as a gorge; also a deep, irregular passage in a cave system.

cirque A steep-sided, rounded hollow carved out by a glacier. Many glaciers have their origin in a mountain cirque, from which they flow to lower ground.

clast A fragment of rock, especially when incorporated into another newer rock.

climate The prevailing weather conditions in an area over a long period of time.

cold front The leading, elongated edge of a cold dry mass of air.

condensation The conversion of a substance from the gaseous to the liquid state—for example, water vapor condensing to liquid water.

continental margin The continental shelf and continental slope taken together.

continental rise The slightly sloping area around the edge of the deep seafloor where it abuts the continental slope.

continental shelf The gently sloping, submerged portion of continental crust seaward of most continental coasts.

continental slope The sloping ocean bottom between the edge of the continental shelf and the continental rise.

convection The movement and circulation of fluids (gases, liquids, or hot and ductile rocks) in response to differences in temperatures, which in turn cause density variations between different parts of the fluid.

core The innermost part of Earth. It consists of a liquid outer core (mostly iron with probably some sulphur) and a solid inner core of nickel-iron.

Coriolios effect An apparent force that explains the deflection of the air or ocean currents on the rotating Earth.

corrasion Erosion of rocks by scraping—for example by rock-laden glacial ice.

country rock Existing rock into which an igneous intrusion is emplaced.

crater A bowl-shaped depression through which an erupting volcano discharges gases, pyroclasts, and lava. The crater walls form by accumulation of ejected material. This term also refers to a circular depression in the landscape caused by a large meteorite impact.

craton A stable area of Earth's continental crust, made of old rocks largely unaffected by mountain-building activity since Precambrian times. Also called a shield.

crust The rocky outermost layer of Earth. The continents and their margins are made of thicker but less dense continental crust, while thinner, denser oceanic crust underlies the deep ocean floors.

current In oceanography, a flow of ocean water. Both surface and deep-sea currents exists. Some are driven by the wind; others (thermohaline currents) by differences in temperatures and salinity that create differences in density between water masses.

cyclone A pressure system in which air circulates around an area of low pressure. A tropical cyclone is another name for a hurricane or typhoon.

D

deep-sea trench A canyon-like, linear depression in the ocean floor, the site of subduction of one tectonic plate below another. Trenches are the deepest regions of the oceans.

deformation The ductile flow or brittle fracture of existing rocks due to geological movements.

deposition The laying down of material such as sand and gravel in new locations, usually by wind or moving water and ice.

desertification The transformation of a formerly more fertile region into desert.

dike A sheet of intrusive igneous rock at a steep angle to the surface. A large number of dikes in a region is termed a dike swarm.

drift A broad, slow-moving flow of surface water; for example, the North Atlantic Drift.

drumlin A hill-sized, usually streamlined mound of debris left behind by a retreating glacier.

dust-devil A whirlwind that carries dust or sand aloft by convection from a hot surface.

E

El Niño The phenomenon in which the normally strong currents of the equatorial Pacific weaken and warm water that has piled up in the western Pacific floats eastward, making water in the eastern Pacific warmer than usual. It is part of a larger phenomenon that can trigger changes to weather patterns worldwide.

eon A large span of time in geological history. Earth's history is divided into four eons, which are subdivided in turn into eras and periods.

epeirogenesis The raising or lowering of parts of Earth's crust due to vertical movements only, rather than the tectonic movements involved in orogenesis.

erosion The process by which rocks or soil are loosened and worn or scraped away from a land surface. The main agents of erosion are wind, water, and moving ice, and the sand grains and other rock particles that they carry.

erratic A rock that has been transported from its original location, usually carried by ice.

eruption The discharge of lava, pyroclasts, gases, and other material from a volcano. Hawaiian eruptions involve large flows of relatively fluid lava but little explosive activity. Strombolian eruptions involve frequent ejections of rocks and gases, but without sustained explosions. Surtseyan eruptions take place in shallow water and are partly powered by the conversion of water to steam. Vulcanian eruptions involve a series or explosive events that feed significant eruption columns. Plinian eruptions involve massive sustained explosions to form towering eruption columns, in which much of the volcanic cone may be destroyed. Fissure eruptions take place through long cracks in the ground rather than via a crater.

esker A long line of raised debris left behind by a glacier. Eskers are thought to mark the position of former meltwater channels.

exfoliation A weathering process that involves the splitting off of outer layers of rocks.

F

fault A fracture where the rocks masses on either side have moved relative to one another. If the fault is at an angle to the vertical, and the overhanging rocks have slid downward, it is called a normal fault. If the overhanging rock has slid upward (in relative terms) it is called a reverse fault. A strike-slip fault is one where movement is horizontal. A transform fault is a large-scale strike-slip fault associated with plate movements.

flash-flood A sudden flood occurring after heavy rain.

flood basalt An extensive area of basalt resulting from massive volumes of lava erupting and flowing widely over the landscape in relatively short intervals of geological time. It is resistant to erosion and commonly forms extensive plateaus after its formation.

floodplain A flat plain next to a river that is liable to be covered with water when the river floods.

fold A geological structure in which originally flat-lying rocks appear to have been flexed and bent. In reality, the rocks behaved in a ductile manner and flowed into these shapes, like a warm caramel bar. They may bend upward in the middle to form a ridge (anticline) or downward to form a ridge (syncline). A symmetrical fold has slopes that are the same steepness on either side of the vertical; an asymmetrical fold does not. A recumbent fold is a fold that is lying on its side. In an overturned fold, a syncline is partly tucked underneath a neighboring anticline.

foliation The arrangement of minerals in parallel bands in some deformed metamorphic rocks.

forced wave A water wave created by storm winds at sea. Forced waves are taller and have a shorter wavelength than swell waves.

front In meteorology, the forward-moving edge of an air mass.

frost Icy deposits that occur when the temperature of a surface is below 32°F (0°C).

Fujita scale A scale that expresses the severity of tornadoes based on the damage they inflict.

fumarole In volcanic regions, a small opening in the ground through which hot volcanic gases can escape.

funnel A tube of whirling air descending from a cloud. It becomes a tornado when its end touches the ground.

G

gabbro A dark, coarse-grained igneous rock that forms underground when basalt magma (molten rock) slowly cools.

gale A wind speed of between 34 and 40 knots, or force 8 on the Beaufort scale.

geothermal energy Energy obtained by tapping the heat generated in Earth's interior.

geyser A jet of boiling water and steam that rises at intervals from the ground. It is powered by hot rocks heating groundwater.

glacier A mass of semipermanent ice capable of flowing downhill. Glaciers come in many varieties. The largest are ice sheets, such as the Antarctic ice sheet. Similar, but slightly smaller, are ice caps. A valley glacier is smaller and flows down a valley, eroding rock as it does so. An outlet glacier is one that flows down from an ice cap or ice sheet. A piedmont glacier spreads out from higher ground onto a broad plain. A surge-type glacier is one that periodically increases its speed much above normal. A cirque glacier does not spread behind the mountain hollow (cirque) that it has carved out. In a polythermal glacier, a cold surface of ice overlies much warmer ice.

greenhouse effect The tendency of the atmosphere to contribute to warming of Earth, by letting through the Sun's radiation but absorbing some that is re-radiated by Earth. A greenhouse gas (such as water vapor, carbon dioxide, or methane) is any gas that promotes this process, whether naturally or via human action.

groundmass The fine-grained material surrounding larger crystals or clasts in a rock. The term is most commonly applied to igneous rocks and some heterogeneous sedimentary rocks.

gust front An area of strong winds that travels ahead of a line of storm clouds.

guyot A flat-topped submarine mountain, also called a tablemount.

gypsum A common mineral found in sedimentary rock. Gypsum consists mostly of calcium sulphate dihydrate and is used to make plaster and cement.

gyre A large-scale circular movement of ocean currents.

H

hanging valley A valley, usually carved by a glacier, that enters high up the side of a larger valley.

hotspot A long-lived zone of volcanic activity thought to originate deep in Earth's mantle. Tectonic plates passing over hotspots are marked by linear chains of volcanoes that become progressively older with increased distance from the hotspot.

hot spring A bubbling up of hot water and steam from the ground, caused by hot rocks heating the groundwater from beneath.

hurricane A large-scale low-pressure system of tropical regions involving powerful circulating winds and torrential rain. It is also termed a tropical cyclone and (especially in east Asia) a typhoon. Its energy comes from the latent heat of water that has evaporated from warm oceans and later condenses.

hydrothermal system Any natural system involving the heating and circulation of underground water and steam, powered by nearby hot or molten rocks. A hydrothermal vein is a mineral vein laid down by past hydrothermal action. A hydrothermal vent is an outlet for heated water from a hydrothermal system, especially one at the bottom of an ocean. When the heated water is colored with dark material, sulphides, and other minerals precipitated when the hot fluid comes into contact with cold seawater, the vent is known as a black smoker.

I

iceberg A floating mass of ice derived from a glacier.

igneous rock Rock that originates from the solidification of molten magma. Extrusive (volcanic) igneous rocks have solidified on Earth's surface after volcanic activity and commonly have small, hardly visible crystals. Intrusive (plutonic) igneous rocks have solidified below the surface, cooling slowly enough to allow larger crystals to form. A body of intrusive igneous rock is called an igneous intrusion.

instability An atmospheric state in which a small disruption, such as heating, will lead to a displacement of the warmed air.

intertropical convergence zone The region where the main air masses north and south of the equator come into contact.

island arc, volcanic A line of volcanic islands associated with an ocean–ocean subduction zone.

isostasy The concept that all columns of rock on Earth have the same weight at some depth in the mantle. Continents rise high above the sea floor because continental crust is less dense than oceanic crust.

J

jet stream Any of several winds that blow for long distances high in the troposphere.

joint A crack running through a rock. Unlike a fault, the rocks on either side of a joint remain in the same position with respect to each other. A cooling joint results when an igneous rock shrinks as it cools.

K

katabatic wind A flow of air down slopes on nights when the air is chilled by radiation.

L

La Niña A situation in which the waters of the eastern Pacific become unusually cold; the opposite of the El Niño phenomenon.

lahar A mudflow of water mixed volcanic ash and other debris.

lava Molten rock that has reached Earth's surface. Basaltic a'a lava (45–52 percent by weight of silica) forms a rough-textured rock when it cools. Pahoehoe lava, of similar basaltic composition, flows slowly and creates a smooth or rope-textured surface. Basaltic pillow lava is lava ejected underwater, forming pillow-shaped mounds of rock. More silica-rich and andesitic-dactic lavas (more than 57 percent by weight of silica) typically form block lava, jostling masses of angular blocks, some of them many feet across.

lightning The visible flash that accompanies electrical discharges from storm clouds. Discharges within or between clouds are sheet lightning; those between clouds and the ground are forked lightning.

lithosphere Earth's crust together with the rigid uppermost layer of the underlying mantle. Each of Earth's tectonic plates is made of a section of lithosphere.

M

maar A wide volcano crater formed by explosive eruptions that have excavated into underlying bedrock. They are commonly filled with water.

magma Molten rock rising from the interior of Earth.

magma chamber A region just below Earth's surface where magma (molten rock) has collected.

mantle The rocks lying between Earth's crust and its core. It makes up 84 percent of Earth's volume.

mantle plume A pipelike upwelling of hot rocks rising through the mantle to the base of the lithosphere, hypothesized as an explanation for a hotspot at Earth's surface.

massif An area of mountains or a large section of continental crust bounded by faults.

mass movement The movement of rocks, soil, or mud down a slope in response to gravity.

mesosphere The layer of Earth's atmosphere between the stratosphere and thermosphere, at an altitude of about 30–50 miles (50–80km).

metamorphic rock A rock that has been transformed underground by heat or pressure to a new texture or new set of minerals. For example, marble is metamorphosed limestone.

metamorphism The process by which rocks are transformed by heat, pressure, or chemical reactions underground.

microclimate The distinctive climate of a particular place, such as a narrow valley or hillside.

mid-ocean ridge Any of the submerged mountain ranges running across the floors of the major oceans (although not all occur exactly in mid-ocean). They are sites where two tectonic plates are spreading apart and material is rising up from the deeper mantle to form new oceanic lithosphere. Mid-ocean ridges vary widely in their spreading rates, from fast- to slow-spreading ridges.

monsoon A pattern of winds, especially in southern Asia, that blow from one direction for about half the year, and from the other direction for the other half. The term is also used to refer to the heavy rains carried by these winds at certain times of the year.

N

nuée ardente A highly destructive, incandescent pyroclastic flow, typical of Peléan volcanoes.

O

ocean basin A region of low-lying oceanic crust within which a deep ocean (or part of one) is contained, and usually surrounded by land or shallower seas.

oceanic crust The type of Earth's crust that forms the deep ocean bed. Made mainly of basalt and grabbo, it is thinner and denser than continental crust.

ocean trench Elongated low-lying region of the ocean floor, marking a subduction zone. Trenches are the deepest parts of the ocean.

occlusion In meteorology, the situation where a moving mass of colder air catches up with a warmer air mass, pushing the latter up and away from the surface of Earth.

orography The study of large-scale mountainous regions.

orogensis The process of mountain building, resulting from pressures generated by the horizontal convergence of tectonic plates. An orogeny is a particular episode of mountain building.

P

phreatic eruption A type of volcanic eruption or geyser activity where groundwater is turned to steam by contact with hot rocks.

plate One of the major fragments into which Earth's surface is divided.

plate boundary A boundary between two tectonic plates. The plates concerned can be diverging (constructive boundary), converging (destructive boundary), or sliding past one another (conservative boundary).

plateau A large area of relatively flat land that stands topographically above its surroundings.

precipitate Any material that has come out of a solution to form a solid deposit.

precipitation Water that reaches Earth's surface from the atmosphere, including rain, snow, hail, and dew.

pressure system Any pattern of weather in which air circulates around an area of high or low pressure. The pressure gradient is the pressure difference between two given points.

pumice A light-colored, low-density volcanic rock containing innumerable bubbles formed by expanding gases, ejected during some volcanic eruptions.

pyroclastic Consisting of, or containing, volcanic rock fragments. Pyroclastic flows are fast-moving, sometimes deadly, clouds of hot gases and debris.

Q

quartzite A hard metamorphic rock derived from sandstone. During the formation of quartzite, quartz grains recrystallize to form a more rigid, interlocking mass.

R

resurgence The emergence of water from an underground aquifer at the surface as a spring.

Richter scale A scale used to express the amplitude of seismic waves at an earthquake's origin.

ridge of high pressure A long, narrow area of high pressure extending out from an anticyclone.

rift valley A large block of land that has dropped vertically downward compared with the surrounding regions, as a result of horizontal extension and normal faulting. A rift valley is also called a graben.

rock Any solid material made up of one or more minerals and occurring naturally on Earth or other planetary bodies.

S

schist A type of metamorphic rock that consists of thin layers of minerals such as mica that can flake apart. Schist is derived from fine-grained sedimentary rocks.

scoria A dark-colored volcanic rock containing innumerable bubbles formed by expanding gases ejected during some basaltic volcanic eruptions. Scoria is also known as cinder.

scree Broken rock fragments found at the base of cliffs or on mountainsides, caused by physical or chemical weathering.

seafloor spreading The creation of new oceanic lithosphere by the upwelling of magma at mid-ocean ridges and consequent spreading of the seafloor on either side.

seamount An undersea mountain, usually volcanic in origin.

sediment Solid particles that have been transported by water, wind, volcanic processes, or mass movement, and later deposited.

sedimentary rock Rock formed when small particles are deposited—by wind, water, volcanic processes, or mass movement—and later harden.

seismic wave A vibrational wave generated by an earthquake. P waves can travel through both solid and liquid portions of Earth's interior, S waves only through solid parts.

silicate Any rock or mineral composed of groups of silicon and oxygen atoms in chemical combinations with atoms of various metals. Silicate rocks make up most of Earth's crust and mantle.

stratosphere A layer of Earth's atmosphere extending from the top of the troposphere, from 5 to 10 miles (8 to 16km) up to about 30 miles (50km).

subduction The descent of an oceanic tectonic plate under another plate when two plates converge. Subduction zones can be classified as either ocean–ocean or ocean–continent depending upon the nature of the two converging plates.

syncline A downward fold of originally flat strata, often caused by horizontal compression.

T

tectonic plate Any of the large rigid sections into which Earth's lithosphere is divided. The relative motions of different plates leads to earthquakes, volcanic activity, continental drift, and mountain building.

temperate Relating to the regions of Earth between the tropics and the polar regions.

terrace A flat region in a river valley that is higher than the present floodplain. It represents a former floodplain created before the river had eroded down to its present level.

thermosphere The layer of Earth's atmosphere above the mesosphere. It extends above altitudes of about 50–400 miles (80–640km).

tide The rise and fall (usually twice each day) of water on the shore caused by the gravitational pull of the Moon and the Sun. The difference in level between high and low tide is called the tidal range, and tends to vary on a monthly basis: spring tides (when the effects of the Moon and Sun reinforce each other) produce the highest high tide and the lowest low tide; neap tides (when the effects of the Moon and Sun oppose each other) have the smallest tidal range.

tornado A narrow, rapidly whirling tube-shaped column of air, which is often highly destructive.

tributary Any river or stream that flows into a larger river.

troposphere The lowest, densest layer of the atmosphere, where most weather phenomena occur. The height of the upper limit of the troposphere (known as the tropopause) varies from the equator to the poles.

trough In meteorology, a long, relatively narrow area of low pressure.

tsunami A fast-moving, often destructive sea wave generated most often by earthquake activity; popularly but incorrectly known as a tidal wave. It rises in height rapidly as it reaches shallow water.

tuff A rock made of fine-grained pyroclastic material produced by an explosive volcanic eruption.

typhoon, see hurricane

V

viscosity Resistance to flow in fluids. The higher the viscosity of a fluid, the more sluggishly it flows.

volatile In volcanology, a term used to describe water, carbon dioxide, and other potentially gaseous compounds that are dissolved in molten rocks.

volcano An opening in Earth's crust where magma reaches the surface; also, a mountain that contains such an opening. A shield volcano has shallow slopes and is built from lava that flowed easily. A stratovolcano is steeper and built from alternate layers of ash and lava. A dome volcano is rounded, steep, and built from viscous lava. A cinder cone is built from scoria that has fallen from the explosion clouds of an eruption.

W

wave A regular motion or disturbance that transfers energy. For a wave crossing the open ocean, the water itself does not move significantly except up and down as the wave passes. The high point of a wave is its crest and the low point its trough. For waves breaking on shores (breakers) water motion becomes more complex and chaotic (turbulent).

weathering The alteration of rocks caused by their being exposed at or close to Earth's surface. Usually the original rock eventually crumbles or becomes weakened.

INDEX

Page numbers in **bold** indicate feature profiles or extended treatments of a topic. Page numbers in *italics* indicate pages on which the topic is illustrated.

T

ACKNOWLEDGMENTS

Dorling Kindersley would like to thank the following people for their assistance in the preparation of this book: Alka Ranjan, Dharini, Priyaneet Singh, and Rupa Rao for editorial assistance; Govind Mittal and Shriya Parameswaran for design assistance; Chris Bernstein for compiling the index; Caitlin Doyle for proofreading.

Picture credits
Dorling Kindersley would also like to thank the following for their kind permission to reproduce their photographs:

(Key: a-above; b-below/bottom; c-center; f-far; l-left; r-right; t-top)

1 Corbis: Douglas Peebles (c). **2-3 Science Photo Library:** Bernhard Edmaier. **4-5 Reuters:** Kyodo (t). **6-7 NASA:** JSC. **8-9 Science Photo Library:** Mark Garlick. **9 NASA:** (bc). **10 Alamy Images:** David Hutt (cl). **10-11 Science Photo Library:** Planetobserver. **13 akg-images:** IAM (br). **Alamy Images:** Interfoto (crb). **14 Geoffrey F. Davies, Australian National University:** (tr). **16 GeoScience Features Picture Library:** D Bayliss (bc); Dr.B.Booth (bl). **17 GeoScience Features Picture Library:** Dr.B.Booth (cra). **18-19 National Geographic Stock:** Carsten Peter / Speleoresearch & Films. **20-21 SuperStock:** Robert Harding Picture Library. **21 Corbis:** Frans Lanting (c). **Getty Images:** James P. Blair / National Geographic (bc). **Photolibrary:** Yann Arthus Bertrand (cr). **Science Photo Library:** Fred Mcconnaughey (br). **24 Corbis:** Song Weiwei / Xinhua Press (br). **James Kirkikis Photography:** (cr). **25 Corbis:** Frank Krahmer (cr). **Photolibrary:** (cla). **Science Photo Library:** British Antarctic Survey (tr). **28 NASA:** Visible Earth (cr). **Photolibrary:** John Warburton-Lee Photography (cl). **Science Photo Library:** Daniel Sambraus (bl). **29 NASA:** Courtesy of Earth Sciences and Image Analysis Laboratory, NASA Johnson Space Center (t). **30 Corbis:** NASA (bcl). **NASA:** (tr); Visible Earth (bl). **Photolibrary:** (crb); Robert Harding Travel (br). **31 Corbis:** Michael S. Yamashita (cl). **Getty Images:** Philippe Bourseiller (tc). **naturepl.com:** Doug Allan (cra). **Photolibrary:** Peter Arnold Images (cr). **32-33 Corbis:** Fred Hirschmann / Science Faction. **33 NASA:** Visible Earth (crb). **34-35 Getty Images:** Steve Allen (t). **35 4Corners Images:** Guido Cozzi (br). **36 Photolibrary:** (br, l). **37 UNAVCO:** (t). **38-39 Corbis:** Victor Fraile. **40 Corbis:** David Muench (tr). **40-41 Getty Images:** Jorg Greuel (b). **41 Corbis:** Carolina Biological / Visuals Unlimited (bc); moodboard (crb). **42-43 Corbis:** DLILLC. **46-47 Alamy Images:** Robert Harding Picture Library Ltd . **46 © NERC. All rights reserved. IPR/136-15CT:** (bl). **47 Martin Redfern:** (cb). **48 Corbis:** Roger Ressmeyer (cl). **Getty Images:** De Agostini (br). **49 Corbis:** George Steinmetz (br). **NASA:** (tr). **50 Corbis:** Visuals Unlimited (b). **UNAVCO:** Dr. Reilinger (cra). **51 Corbis:** Roger Ressmeyer (cr). **ESA:** (bl). **Science Photo Library:** Friedrich Saurer (tl). **52-53 Photolibrary:** (t). **52 NASA:** Space Imaging / Earth Observatory (cl). **Science Photo Library:** Zephyr (bl). **53 Corbis:** Hanan Isachar (cr). **Photolibrary:** Robert Harding Travel (br). **54-55 Corbis:** Han Chuanhao / XinHua / Xinhua Press. **56-57 ESA. 56 Photolibrary:** (bl). **58 Corbis:** Paul Souders (cra). **59 NASA:** Visible Earth. **60-61 Corbis:** Stefen Chow (c). **61 Alamy Images:** James Brunker (tr). **Corbis:** Geoff Renner / Robert Harding World Imagery (bc). **Science Photo Library:** Worldsat International (br). **62-63 Photolibrary:** Thomas Hallstein (c). **63 NASA:** (cra). **Science Photo Library:** Ron Sanford (crb). **64-65 Science Photo Library:** Bernhard Edmaier. **66 NASA:** (bl). **66-67 Getty Images:** Tim Fitzharris (b). **67 Corbis:** Micha Pawlitzki (ca). **NASA:** (cra). **68 NASA:** JPL (c). **Photolibrary:** Radius Images (bc). **68-69 Photolibrary:** Rafael Macia (t). **69 Corbis:** Kevin Schafer (cr). **70 Science Photo Library:** European Space Agency (br). **70-71 Getty Images:** Philippe Bourseiller (t). **71 FLPA:** David Hosking (br). **NASA:** Earth Observatory (bc). **72-73 NASA:** USGS. **74-75 Photolibrary:** (b). **74 Getty Images:** Stocktrek Images (bc). **75 GeoScience Features Picture Library:** Dr.B.Booth (tc, cla). **Photolibrary:** Michael Andrews (cra). **Science Photo Library:** Michael Szoenyi (tr). **76 NASA:** (bl). **76-77 Corbis:** Serguei Fomine / Global Look. **77 Alamy Images:** David Gowans (c). **Jim Taylor:** (crb). **Vladimir Kholostykh:** (br). **78-79 NASA:** GeoEye / Space Imaging. **79 Corbis:** Momatiuk - Eastcott (tr, br); William James Warren / Science Faction (cr). **80 Science Photo Library:** US GEOLOGICAL SURVEY (bl). **80-81 Getty Images:** Gordon Wiltsie (c). **81 Corbis:** Galen Rowell (cr, tc, cra). **82-83 Hervé Douris. 88-89 Getty Images:** Toshi Sasaki (t). **90 Corbis:** Pablo Corral Vega (cl). **Getty Images:** Barcroft Media (bc); Arlan Naeg / AFP (tr). **91 Alamy Images:** Greg Vaughn (bl). **Corbis:** Arctic-Images (tl). **Martin Rietze:** (tr). **92-93 Corbis:** Roger Ressmeyer. **94 Alamy Images:** CuboImages srl (cra). **Corbis:** Alaska Volcano Observatory - dig / Science Faction (clb); John and Lisa Merrill (br). **95 Corbis:** Ed Darack / Science Faction (clb); Jim Wark / Visuals Unlimited, Inc. (tr); Tony Roberts (cr). **96 Corbis:** G. Brad Lewis / Science Faction. **97 Alamy Images:** Tom Pfeiffer (br). **Corbis:** (tr); G. Brad Lewis / Science Faction (fbl, bl). **Getty Images:** G. R. 'Dick' Roberts / NSIL (bc). **Science Photo Library:** Herve Conge, Ism (cr); Pasieka (cl). **98-99 Corbis:** Frans Lanting. **99 Science Photo Library:** Dr. Richard Roscoe, Visuals Unlimited (bl); Bernhard Edmaier (cr); G. Brad Lewis (tr). **100-101 Martin Rietze. 101 Corbis:** (br); Atli Mar Hafsteinsson / Nordicphotos (cr). **Martin Rietze:** (tr, c, tc). **U.S. Geological Survey:** Tim Orr (bc). **102-103 Getty Images:** Barcroft Media (c). **103 Getty Images:** AFP (tr). **104-105 Corbis:** Alberto Garcia. **106-107 Corbis:** Jacques Langevin / Sygma. **106 Corbis:** Ocean (c); STR / epa (bl). **108-109 Photolibrary:** (b). **108 Corbis:** Pablo Corral Vega (cr). **NASA:** Visible Earth (cl). **110-111 NASA:** Visible Earth. **110 Getty Images:** NASA-JSC / Science Faction (c). **111 Getty Images:** Astromujoff (tc). **Photolibrary:** (br). **112 Corbis:** Frans Lanting (cr). **113 Alamy Images:** Hemis (bl). **114 Masterfile:** (bc). **Science Photo Library:** NASA (clb). **114-115 Martin Rietze:** (c). **115 Hervé Douris:** (cr). **116 Corbis:** Gary Fiegehen / All Canada Photos (clb); G. Brad Lewis / Science Faction (cb). **U.S. Geological Survey:** John Pallister (bc). **116-117 SuperStock:** Robert Harding Picture Library. **118-119 Corbis:** Yann Arthus-Bertrand. **120-121 Photolibrary:** Adalberto Rios (b). **121 NHPA / Photoshot:** Ross Nolly (br); Kevin Schafer (cr). **122 Science Photo Library:** Bernhard Edmaier (bl). **122-123 Getty Images:** Tom Pfeiffer / VolcanoDiscovery (t). **123 Corbis:** Olivier Coret / In Visu (bc). **Getty Images:** Giuseppe Finocchiaro (br). **Science Photo Library:** Dr Juerg Alean (bl); Miriam And Ira D. Wallach Division Of Art, Prints And Photographs / New York Public Library (cr); Royal Astronomical Society (c). **124-125 Corbis:** Adi Weda / epa. **124 Corbis:** Adi Weda / epa (bc). **Getty Images:** AFP (br, bl). **125 Corbis:** Adi Weda / epa (br, bl). **Getty Images:** AFP (tr, bc). **126 Alaska Volcano Observatory / USGS:** Game McGimsey (t). **127 NASA:** Earth Observatory (br). **Science Photo Library:** USGS (cla). **SuperStock:** Radius (bl). **128-129 Science Photo Library:** NASA (t). **129 Corbis:** (cr); Roger Ressmeyer (br). **Photolibrary:** Paul Nevin (cra). **130-131 Alamy Images:** blickwinkel (t). **130 Alamy Images:** Emmanuel Lattes (br). **131 Corbis:** Atlantide Phototravel (br); Ashley Cooper (bl). **132 Getty Images:** Eric Bouvet / Gamma-Rapho (b). **133 Corbis:** Louise Gubb (cla, bc). **NASA:** Visible Earth (br). **134 Alamy Images:** Greg Vaughn (crb). **Getty Images:** Pete Oxford (clb). **naturepl.com:** Jack Dykinga (bl). **135 Photolibrary:** Robert Harding Travel. **136-137 Martin Rietze:** (c). **136 Tom Pfeiffer / VolcanoDiscovery:** (br). **137 Getty Images:** Raphael Van Butsele (bl). **Ulrich Kueppers:** (br). **Hugh Tuffen:** (cra). **U.S. Geological Survey:** Jim Vallance (tc). **138-139 Science Photo Library:** Bernhard Edmaier (b). **138 NASA:** Visible Earth (tr). **139 Corbis:** Jim Wark / Visuals Unlimited, Inc. (bc). **NASA:** JSC (ca). **SuperStock:** Photononstop (cr). **140 NASA:** (bl). **Photolibrary:** Robert Harding Travel (br). **141 Masterfile:** Frank Krahmer. **142 Corbis:** Arctic-Images (br). **Photolibrary:** (bl); Robert Harding Travel (clb). **143 Corbis:** David Jon Ogmundsson / Nordicphotos (l). **144 ESA:** Envisat (bl). **144-145 Photolibrary:** Pacific Stock (t). **145 Corbis:** Roger Ressmeyer (bc). **Getty Images:** Adastra (br); Greg Vaughn (bl). **Photolibrary:** Pacific Stock (fbl). **146-147 Corbis:** G. Brad Lewis / Science Faction. **148 Photolibrary. 149 Dorling Kindersley:** Colin Keates / Courtesy of the Natural History Museum, London (bl). **Getty Images:** Richard Roscoe / Visuals Unlimited, Inc. (bc); Stocktrek Images (crb). **150-151 Photolibrary. 151 NASA:** JPL (br). **Rex Features:** (c). **U.S. Geological Survey:** David Wieprecht (crb). **152-153 Corbis:** Michael S. Yamashita (b). **153 Alamy Images:** (cl). **Corbis:** (bc). **Mary Evans Picture Library:** Rue des Archives / Tallandier (cr). **Photolibrary:** (cla, ca). **154-155 NASA:** Earth Observatory (b). **155 NASA:** Earth Observatory (br). **156-157 NASA:** GSFC / MITI / ERSDAC / JAROS, and U.S. / Japan ASTER Science Team. **156 Dorling Kindersley:** James Stevenson / Courtesy of the Museo Archeologico Nazionale di Napoli (b, crb); James Stevenson (c). **157 Corbis:** Stapleton Collection (tr). **Getty Images:** Hulton Collection (cr); (br). **158-159 Photolibrary:** (b). **158 Corbis:** Frank I. Jones / National Geographic Society (tr). **Science Photo Library:** Library Of Congress (tc). **159 Alamy Images:** Loetscher Chlaus (cr). **Corbis:** Frank I. Jones / National Geographic Society (tc). **Photolibrary:** Robert Harding Travel (tr). **160 Science Photo Library:** US Geological Survey (clb, bl). **160-161 Science Photo Library:** David Weintraub (c). **161 Corbis:** Gary Braasch (tr); Douglas Kirkland (bl). **162-163 Corbis:** Gary Braasch. **164 Corbis:** Mitchell Kanashkevich (br). **164-165 Corbis:** Pablo Corral Vega (t). **165 Getty Images:** Design Pics / Corey Hochachka (cb); Keisuke Iwamoto (br). **166 Corbis:** Arctic-Images (cr). **NASA:** Earth Observatory (cb). **SuperStock:** Nordic Photos (l). **167 Bryan & Cherry Alexander / ArcticPhoto:** (cr). **Corbis:** HO / Reuters (b). **168 Corbis:** Arctic-Images. **169 Copyright 2011, EUMETSAT / the Met Office:** (b). **Getty Images:** Arctic-Images (cra); Mehdi Fedouach / AFP (cl). **170-171 Corbis:** Paul Souders. **172-173 SuperStock:** Wolfgang Kaehler (b). **172 Corbis:** George Steinmetz (tr). **Photolibrary:** (ca). **173 www.photo.antarctica.ac.uk:** (cla). **174 Corbis:** Nigel Pavitt / JAI (cr). **Masterfile:** Westend61 (bl). **175 Satellite image courtesy of GeoEye. Copyright 2008. All rights**

reserved.: (tr). **NASA:** The ASTER Volcano Archive (br). **Photolibrary:** (bl). **176-177 Olivier Grunewald**. **177 Getty Images:** Marco Longari / AFP (bc). **Olivier Grunewald:** (c). **Press Association Images:** Karel Prinsloo / AP (cr). **178 Photolibrary:** (cl). **Science Photo Library:** Simon Fraser (br). **179 Getty Images:** Michele Falzone (tr). **SuperStock:** Science Faction (br). **180 Getty Images:** Patrice Coppee (tr); Image Makers (c); SSPL (br). **Olivier Grunewald:** (clb). **181 Science Photo Library:** Jeremy Bishop (t). **182 Corbis:** Justin Guariglia (bc). **Getty Images:** Pedro Ugarte / AFP (tr). **182-183 Photolibrary:** (cla). **183 Corbis:** Arctic-Images (bl); George Steinmetz (tr). **Getty Images:** Frank Krahmer (br). **184-185 Panos Pictures:** Georg Gerster. **186 Corbis:** Bo Zaunders (cr). **Photolibrary:** Tips Italia (br); Xavier Font (bl). **Science Photo Library:** Bernhard Edmaier (bc). **187 Alamy Images:** LOOK Die Bildagentur der Fotografen GmbH. **188-189 Corbis:** Christophe Boisvieux (b). **Science Photo Library:** George Steinmetz (t). **189 Getty Images:** Kelly Cheng Travel Photography (tr). **Photolibrary:** (bc). **190-191 Getty Images:** Heath Korvola. **192-193 Corbis:** Arctic-Images. **192 Photolibrary:** Robert Harding Travel (c). **194-195 Photolibrary**. **196-197 Photolibrary:** (tc). **196 Photolibrary:** (b). **197 Alamy Images:** Michele Falzone (br); Peter Arnold, Inc. (tc). **Photolibrary:** Robert Harding Travel (ca). **Rex Features:** KeystoneUSA-ZUMA (clb). **198 Getty Images:** Dimas Ardian (br). **198-199 Getty Images:** Eka Dharma / AFP (t). **199 Corbis:** Sigit Pamungkas / Reuters (bl). **Getty Images:** Dimas Ardian (cra); Ulet Ifansasti (tc). **NASA:** Earth Observatory (br). **200-201 Getty Images:** Dario Mitidieri / Contributor. **202 Corbis:** Anthony Asael / Art in All of Us (bl); STR / epa (bc). **203 Corbis:** Imaginechina (bl); Arif Sumbar / epa (br). **Reuters:** KYODO Kyodo (crb). **206 NASA:** MODIS (cl). **207 Courtesy of KiwiRail (New Zealand Railways Corporation):** (l). **208-209 Corbis:** Katie Orlinsky (c). **208 Corbis:** Yuan Man / Xinhua Press (c). **Getty Images:** Logan Abassi / AFP (bl). **210 Corbis:** Roger Ressmeyer (clb). **U.S. Geological Survey:** (ca). **210-211 Alamy Images:** Roy Garner. **211 Reuters:** Fatih Saribas (br). **212 Getty Images:** Dimas Ardian (cr). **Science Photo Library:** James King-Holmes (br). **213 Corbis:** Arctic-Images (cr). **IRIS - Incorporated Research Institutions for Seismology / www.iris.edu:** (t). **214-215 Corbis:** Sergio Dorantes / Sygma. **216-217 Corbis:** Bettmann (c). **216 Science Photo Library:** US Geological Survey (ca). **218 Getty Images:** Martin Bernetti / AFP (br). **Photolibrary:** EPA / Claudio Reyes (bl). **218-219 Photolibrary:** EPA / Ian Salas (t). **219 Photolibrary:** EPA / Leo La Valle (bl). **Tokyo Institute of Technology:** (crb). **U.S. Geological Survey:** (tr). **220-227 Getty Images:** Peter Parks / AFP (c). **220 U.S. Geological Survey:** (bl). **221 Corbis:** David Gray / Reuters (tr); Chen Xie / Xinhua Press (cr); Li Ziheng / Xinhua Press (crb). **222-223 GNS Science:** Richard Jongens. **224 Corbis:** Bettmann (bl). **224-225 Corbis:** Masaharu Hatano / Reuters. **226 Corbis:** Kai Pfaffenbach / Reuters (tr). **226-227 Corbis:** Yannis Kontos / Sygma (b). **227 Corbis:** Louisa Gouliamaki / epa (tl). **Getty Images:** Pierre Verdy / AFP (tr, tc). **228-229 Corbis:** David Wethey / epa (b). **228 Getty Images:** Kurt Langer (tr). **229 Getty Images:** Marty Melville / AFP (tr). **230-231 Press Association Images:** Shuzo Shikano / Kyodo News / AP. **232 Corbis:** Dianne Manson / epa (bl). **Reuters:** Crack Palinggi (cl). **233 Getty Images:** AFP (br). **Reuters:** Kyodo (t). **234 Corbis:** Raheb Homavandi / Reuters (c). **Getty Images:** AFP (bl). **234-235 NASA:** Courtesy of Space Imaging (c). **235 Getty Images:** Behrouz Mehri / AFP (br). **236-237 Corbis:** HO / Reuters (t). **236 Corbis:** Issei Kato / Reuters (br). **237 Getty Images:** China Photos (tr). **NASA:** Earth Observatory / DigitalGlobe (cl). **238-239 Corbis:** Imagemore Co., Ltd. (c). **239 Alamy Images:** Peter Tsai Photography (c); Stock Connection Blue (br). **240-241 Bluegreen Pictures:** David Fleetham. **242 Corbis:** Kevin Schafer (b). **243 Alamy Images:** Mike P Shepherd (br). **Dr. Asfawossen Asrat / Addis Ababa University:** (tr). **246 Science Photo Library:** Dr Ken Macdonald (b). **247 Science Photo Library:** US Geological Survey (tr). **248-249 naturepl.com:** Wild Wonders of Europe / Lundgre. **250 Press Association Images:** NOAA Ocean Exploration Program (cra). **251 National Oceanography Centre, Southampton:** (br). **Courtesy of the Natural Environment Research Council:** (bc). **NOAA:** PMEL Vents Program (bl). **Science Photo Library:** B. Murton / Southampton Oceanography Centre (t). **252-253 Corbis:** HO / Reuters (t). **252 National Oceanography Centre, Southampton:** (br). **253 National Oceanography Centre, Southampton:** OAR / National Undersea Research Program (NURP); Univ. of Hawaii (bl). **NOAA:** NSF (br). **Photolibrary:** (bc). **Reuters:** STR New (fbl). **254 Getty Images:** Gamma-Rapho (b). **255 Getty Images:** Royal New Zealand Navy / AFP (cl); STR / AFP (br). **NASA:** Earth Observatory (tr); GSFC / METI / ERSDAC / JAROS, and U.S. / Japan ASTER Science Team (tc). **Science Photo Library:** Pasquale Sorrentino (bl). **256 Alamy Images:** FLPA (br). **Corbis:** (cb); Sygma (ca). **Getty Images:** Popperfoto (tr). **257 Corbis:** Arctic-Images (br). **Science Photo Library:** Omikron (t). **258 Corbis:** Norbert Wu / Science Faction (tr). **SeaPics.com:** Jez Tryner (bl). **259 Corbis:** George Steinmetz (t). **Science Photo Library:** Bernhard Edmaier (br). **260 Getty Images:** Matt Cardy. **261 Corbis:** Huang Zongzhi / XinHua / Xinhua Press (cl). **Rex Features:** Bruce Adams (ca). **Victoria Hillman:** (b). **262 Getty Images:** Nicholas Kamm / AFP (bl). **264-265 Getty Images:** AFP (t). **264 NASA:** Space Imaging (br, bc). **265 Corbis:** Reuters (cra); Xinhua Photo / Enwaer (tr). **Getty Images:** AFP (br, crb). **266-267 Reuters:** Ho New. **268 Getty Images:** Toshifumi Kitamura / AFP (bl). **Press Association Images:** NOAA Pacific Tsunami Warning Center / ABACA USA / Empics Entertainment (c). **Reuters:** KYODO Kyodo (bc). **268-269 Reuters:** KYODO Kyodo (c). **269 Kevin Jaako:** (t). **Reuters:** Yuriko Nakao (cr); Ho New (br). **270-271 Corbis:** Douglas Keister. **272 Getty Images:** Barcroft Media (b). **273 Bryan & Cherry Alexander / ArcticPhoto:** B&C Alexander (cr). **Corbis:** Mike Grandmaison (cl); Gerd Ludwig (fcl); Keenpress / National Geographic Society (bc). **Getty Images:** Sylvester Adams (cl); Paul Mansfield Photography (br). **274 Courtesy of ECMWF:** (cl, cr). **NASA:** Earth Observatory (crb). **278 Corbis:** (b). **NASA:** Earth Observatory (ca); Ronald Vogel, SAIC for NASA / GSFC (cra). **279 Masterfile:** J. A. Kraulis (cr). **281 Corbis:** Larry W. Smith / epa (bc). **Getty Images:** Anthony Bradshaw (fbl). **Rex Features:** Canadian Press (br); Darren Greenwood / Design Pics Inc. (bl). **282-283 Alamy Images:** Gene Rhoden. **284 Getty Images:** Time & Life Pictures (cb). **NASA:** TOPEX / Poseidon, NASA JPL (cl). **284-885 Masterfile:** Bill Brooks (t). **285 Photolibrary:** Alan Majchrowicz (cb). **286-287 Corbis:** Stringer / epa (b). **286 NASA:** GSFC / METI / ERSDAC / JAROS, and U.S. / Japan ASTER Science Team (bl). **Reuters:** Stringer Australia (tr). **287 Corbis:** Keira Lappin / epa (tr). **Getty Images:** Mark Ralston / AFP (cr); Jonathan Wood (tl). **288 NASA:** Earth Observatory (cr, b). **289 Getty Images:** Khin Maung Win / AFP (tr). **Press Association Images:** Channi Anand / AP (b). **290-291 Corbis:** Matiullah Achakzai / epa (c). **290 Corbis:** Rashid Iqbal / epa (bl). **291 NASA:** Earth Observatory (r). **292 NASA:** Earth Observatory (b). **293 NASA:** Earth Observatory (r). **Wikipedia:** (bl). **294 NASA:** (b). **295 Corbis:** Mick Roessler (cb). **NOAA:** (cla). **Photolibrary:** Peter Arnold Images (bc). **Science Photo Library:** Chris Sattlberger (tc). **296 Getty Images:** Stocktrek Images. **297 Corbis:** STR / Reuters (c); Stringer / Thailand / Reuters (cr). **Getty Images:** Khin Maung Win / AFP (br). **NASA:** Earth Observatory (cl). **Wikipedia:** (bl). **298-299 Corbis:** Mike Theiss / Ultimate Chase (c). **298 NOAA:** (bl). **299 Corbis:** John O'Boyle / Star Ledger (br); Rick Wilking / Reuters (bc). **Getty Images:** Jerry Grayson / Helifilms Australia PTY Ltd (cr). **NOAA:** (tc). **300-301 Getty Images:** Michael Appleton / NY Daily News Archive. **302-303 Getty Images:** Hulton Archive (b). **302 NASA:** Earth Observatory (cra). **303 Courtesy of The Weather Channel:** (br). **Press Association Images:** Mike Howarth, National Trust (cra). **Rex Features:** (tc). **304 NOAA:** (b). **304-305 NOAA:** (t). **305 Alamy Images:** AF archive (bl). **306-307 Corbis:** Colin McPherson (b). **306 Science Photo Library:** University Corporation For Atmospheric Research (tl). **307 Corbis:** Larry W. Smith / epa (cr). **NOAA:** (br). **308 Corbis:** Galen Rowell (br). **308-309 Corbis:** Jacques Langevin / Sygma (t). **309 Corbis:** epa (br). **310-311 Photolibrary**. **311 Rex Features:** Action Press (bc). **312 Corbis:** Seth Resnick / Science Faction (br). **313 Corbis:** Mike Hollingshead / Science Faction (br); Gordon Wiltsie / National Geographic Society (fbl); Imaginechina (t); Tom Jenz / Graphistock (bl). **Science Photo Library:** Jim Reed Photography (bc). **314 Getty Images:** Mauricio Duenas / AFP (t). **Science Photo Library:** NASA (bl); Roger Hill (c); National Center For Atmospheric Research (cl). **315 Przemyslaw Wielicki, http://1x.com/artist/wielicki:** (r). **Science Photo Library:** James H. Robinson (cl); Peter Menzel (bl). **316-317 Chris Kotsiopoulos**. **317 Science Photo Library:** Daniel L. Osborne, University Of Alaska / Detlev Van Ravenswaay (br); Thomas Wiewandt / Visuals Unlimited, Inc. (cr); Pekka Parviainen (tr). **318 Getty Images:** Ross Tuckerman / AFP (b). **Gene E. Moore:** (ca). **319 Corbis:** Christine Prichard / epa (bl); Jim Reed (c). **National Geographic Stock:** Mike Theiss (cr). **320 Corbis:** Visuals Unlimited (c). **Press Association Images:** J. Pat Carter / AP (bl); Paul Hellstern / AP (bc). **320-321 Reuters:** Sue Ogrocki (c). **321 Corbis:** Jim Reed (br). **322-323 Gene Rhoden / weatherpix.com**. **324 NASA:** Observatory / Image courtesy Norman Kuring, SeaWiFS Project (cra). **324-325 Press Association Images:** ChinaFotoPress / Photocome (b). **325 Reuters:** Tim Wimborne (cl). **Rex Features:** Photo by Courtesy Everett Collection (tr). **326 Corbis:** Zhang Jian Tao / Redlink (cl). **Rex Features:** Sipa Press (r). **327 Getty Images:** Goh Chai Hin / AFP (tr). **Press Association Images:** AP / Imagine China (br). **328 Corbis:** Andrew Gombert / epa (br). **Getty Images:** AFP (bl). **Science Photo Library:** David R. Frazier Photolibrary, Inc. (cl). **328-329 Corbis:** Gene Blevins / LA DailyNews (t). **329 Corbis:** Pittsburgh Post-Gazette / Zumapress.com (br). **NASA:** GSFCV / Jacques Descloitres, MODIS (bl). **Science Photo Library:** Andrea Balogh (cr). **330 Corbis:** Andrew Brownbill / epa (b). **NASA:** MODIS (cra). **331 Getty Images:** AFP (tl); Lucas Dawson (bl). **Newspix Archive/ Nationwide News:** Mark Smith (cl). **332-333 Glacier Photograph Collection / National Snow and Ice Data Center:** Bruce F. Molnia, 2005 (r); Louis H. Pedersen, 1917 (l). **333 Corbis:** Steve Woit / AgStock Images (br). **334-335 Corbis:** NASA / Bryan Allen
Front Endpapers: **Corbis:** NASA; *Back Endpapers:* **Corbis:** NASA